Springer Texts in
Electrical Engineering

Consulting Editor: John B. Thomas

Springer Texts in Electrical Engineering

John B. Thomas

An Introduction to
Communication Theory and Systems

With 61 Illustrations

Springer-Verlag
New York Berlin Heidelberg
London Paris Tokyo

A Dowden &
Culver Book

John B. Thomas
Department of Electrical Engineering
Princeton University
Princeton, NJ 08544
USA

Library of Congress Cataloging-in-Publication Data
Thomas, John Bowman, 1925–
  An introduction to communication theory and systems.
  (Springer texts in electrical engineering)
  1. Statistical communication theory. I. Title.
II. Series.
TK5102.5.T516   1988   621.38   87-32064

Camera-ready copy provided by the author.

9 8 7 6 5 4 3 2 1

ISBN-13: 978-0-387-96672-4   e-ISBN-13: 978-1-4612-3826-3
DOI: 10.1007/978-1-4612-3826-3

PREFACE

This book was written as a first treatment of statistical communication theory and communication systems at a senior-graduate level. The only formal prerequisite is a knowledge of elementary calculus; however, some familiarity with linear systems and transform theory will be helpful.

Chapter 1 is introductory and contains no substantial technical material.

Chapter 2 is an elementary introduction to probability theory at a nonrigorous and nonabstract level. It is essential to the remainder of the book but may be skipped (or reviewed hastily) by any student who has taken a one-semester undergraduate course in probability.

Chapter 3 is a brief treatment of random processes and spectral analysis. It includes an introduction to shot noise (Sections 3.14-3.17) which is not subsequently used explicitly.

Chapter 4 considers linear systems with random inputs. It includes a considerable amount of material on narrow-band systems and on the representation of random processes.

Chapter 5 treats the matched filter and the linear least mean-squared-error filter at an elementary level but in some detail.

Numerous examples are provided throughout the book. Many of these are of an elementary nature and are intended merely to illustrate textual material. A reasonable number of problems of varying difficulty are provided. Instructors who adopt the text for classroom use may obtain a *Solutions Manual* for most of the problems by writing to the author.

The book contains more than enough material for a one-term course, depending on how much time is spent on Chapter 2 and Chapter 5. I have always covered Chapter 2 rather quickly, assuming either that the material is a review or that the students are taking a probability course concurrently. If Chapter 2 is used as the students only exposure to probability, then this material should be covered slowly enough to allow the concepts to be absorbed. Also, Chapter 5 may contain more detail then some instructors would consider necessary.

John B. Thomas
B-330 Engineering Quadrangle
Princeton University
Princeton, NJ 08544

TABLE OF CONTENTS

Chapter 1

INTRODUCTION

1.1 *Communication Systems*- A communication system can be defined as a system designed to transmit information in the form of messages or data. This definition is vague since, among other shortcomings, it does not define precisely the meaning of the term *information*. The usual problem in a communication system is to eliminate or minimize the effect of unwanted disturbances or distortions and to maximize the amount of information transmitted for a given system complexity, cost, amount of power used, etc.

A typical communication system could be represented by the block diagram of Fig. 1.1. An information source produces messages which are processed, and possibly corrupted, before transmission. The transmission system itself introduces

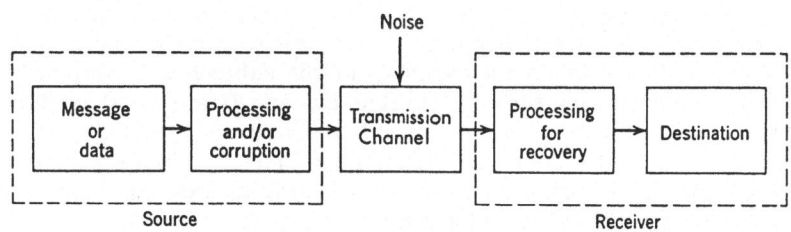

Fig. 1.1 - A typical communication system

additional disturbances. These are often modeled as an additive noise process. The received message then is processed in an attempt to recover it in its original form before passing it on to its destination. Thus the overall problem is to study the effect of the communication system on the messages. To do this the characteristics of the system must be known. In turn, these characteristics are determined, or at least constrained, by (1) the physical environment; (2) the nature of the communication process; (3) practical factors of cost, size, and reliability.

1.2 *Statistical Communication Theory* - A working definition of *statistical communication theory* would be that it is the application of probability and statistics to the analysis and synthesis of communication systems. For our purposes we might subdivide this field as follows:

1. *A mathematical background* in linear analysis, elementary probability and statistics, and random processes.

2. *System analysis* - analysis of linear and nonlinear systems with random inputs, including modulation and demodulation.

3. *System synthesis* - constructive procedures to obtain systems satisfying some statistical criterion of performance. Two major areas here are signal detection and signal extraction (waveform estimation).

4. *Information theory and coding* - development of the concept of information measure and its application to communication systems and to signal encoding.

5. *Noise physics* - a study of the characteristics of physically realizable random processes based on a knowledge of the mechanisms generating them.

Despite a plethora of new advances, many of the underlying notions of (statistical) communication theory remain unchanged. Fundamental to any understanding of the subject is a familiarity with the elementary theory of probability and (stationary) random processes. This material is covered in Chapters 2 and 3 of this book in sufficient detail to make the remaining chapters understandable. However, students with a lasting interest in communication theory should consider formal courses in these topics.

Aside from Chapters 2 and 3, this text is devoted mainly to those aspects of communication theory dealing with linear systems. Thus Chapter 4 treats linear operations on random processes including considerable material on the representation of random processes by orthogonal series whose coefficients are obtained by *linear* operations on the process. Not only are linear operations important in their own right as representing a major class of systems, but as we will see, linear operations on collections of normal (or Gaussian) random variables yield other collections of normal random variables. Thus sets of normal random variables (including normal random processes) are invariant in their overall statistical properties to linear transformations. The almost universal assumption in elementary communication theory that unwanted random variables (noises) can be modelled as normal random variables lends increased importance to *linear* systems with random inputs.

Chapter 5 is concerned with two of the most important optimal linear systems, the matched filter and the least-mean-squared-error linear filter (the Wiener filter). Not only are these optimal systems, but they occur as subsystems in optimal signal detection and estimation.

1.3 *Additional Reading* - We provide at the end of each chapter a list of references to textbooks and, in some cases, periodical literature pertaining to various aspects of the topics discussed. The lists are not intended to be exhaustive. Also, no claim is made for the merit of works included or for the lack of merit of works not included. Each student should become reasonably familiar with the existing literature and, ultimately, make his own decisions as to which references he finds valuable.

Chapter 2

PROBABILITY AND RANDOM VARIABLES

2.1 *Introductory Remarks*- During World War II the increased need for automatic fire control, radar, and sonar lead to the formal development of statistical communication theory. This new step forward has associated with it the names of Wiener and Shannon, among many others. The basic idea developed was that signals, if they are to convey information, must have associated with them certain characteristics which must be indeterminable by the receiver before reception. As a consequence the signals to be considered in the analysis and synthesis of communication systems should be statistical in nature rather than the deterministic standard signals used heretofore. Thus signals, as well as noises, are to be treated as random processes. The behavior of these random processes in a system can be described in terms of their average properties which frequently are known or can be estimated accurately. The determination of such average properties comprises the fields of probability and statistics, which are essential mathematical background to the study of communication systems. Although a rigorous treatment of probability requires considerable mathematical sophistication and some knowledge of measure and integration theory, it is possible to develop an elementary theory using very simple concepts. Such an approach will be given in this chapter.

It will be convenient to use some of the language and elementary ideas from set theory; therefore, a brief introduction to this subject will be our first topic.

2.2 *Elements of Set Theory* - We begin with the undefined notion of a *set,* which will be taken to mean *a collection, an aggregate, a class, or a family of any objects whatsoever.* Each of the objects which make up the set is called an *element,* or *member,* or *point* of the set.

It will be convenient to denote sets by upper case letters and elements of sets by lower case letters. For example, the letter A will denote a given set and

$$a \in A$$

will indicate that a is an element of the set A. It will be said that a "belongs" to A. If the element a does not belong to the set A, we write

$$a \notin A$$

It is conventional to specify a set in one of two ways:

    (1) by listing the elements of the set between braces, or

    (2) by listing some property common to all the elements of the set.

For example, let I be the set of all integers and consider the set B with elements 1, 2, and 3. We write:

$$B = \{1, 2, 3\}$$

or

$$B = \{b \in I \mid 0 < b < 4\}$$

The last expression is read "B is the set of all integers b which are greater than zero and less than four."

It should be noted that the order of the elements is immaterial and that no element can appear more than once. Thus the set $\{1,2,3\}$ and the set $\{3,1,2\}$ are the same, and the collection $\{1,1,2,3,1\}$ is a set only if the repeated elements are deleted.

We shall use equality in the sense of identity. Two sets A and B are equal if, and only if, they contain exactly the same elements. In this case we write

$$A = B$$

If the elements in the two sets A and B are not the same, we write

$$A \neq B$$

and say that A and B are *distinct*.

If every element of B is also an element of A, we say that B *is contained in* A or B is a *subset* of A and write

$$B \subseteq A$$

We might also write

$$A \supseteq B$$

and say that A *contains* B. If $B \subseteq A$ and at least one element of A is not in B, we call B a *proper subset* of A and say that A *properly contains* B

$$A \supset B$$

or that B is *properly contained* in A

$$B \subset A$$

Since it may be necessary to consider a particular set which will turn out to have no elements, it is convenient to define the *void,* or *empty,* or *null* set ϕ as the set which contains no elements.

In most discussions of sets a particular class of objects will exist which comprise an all-embracing set. The sets with which we concern ourselves will then be sets of elements from this fixed set. This fixed set is known as the *universal set* Ω or the *sample space* S. The two terms will be used interchangeably. Then, for every set A, it follows that

$$\phi \subseteq A \subseteq S,$$

and, for every element a, that

$$a \in S \text{ and } a \notin \phi$$

It is sometimes convenient to attach a geometric significance to the subsets of a given sample space. By this convention areas are associated with sets and points with elements. The sample space S is often indicated by a rectangle and subsets of the space by areas within the rectangle. Fig. 2.1 illustrates the technique.

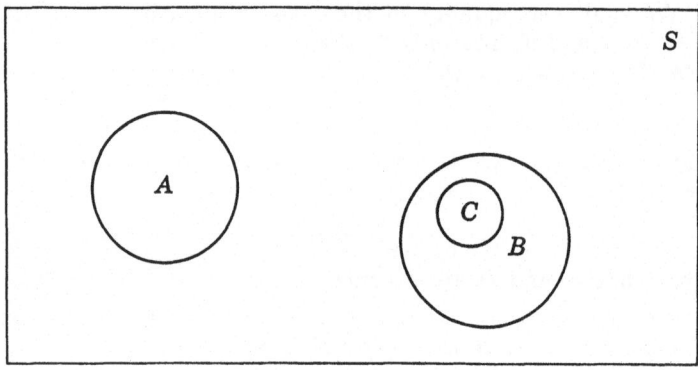

Fig. 2.1 - A Venn diagram

Here A, B, and C are subsets of the sample space S. In addition, C is a proper subset of B and we could write

$$C \subset B$$

These diagrams are called *Venn diagrams*.

At this point it is convenient to consider some of the ways in which sets can be combined with one another and with some of the properties of such combinations.

The *intersection* of two sets A and B is denoted by $A \cap B$ *or* AB and is the set of all elements common to both A and B; that is,

$$A \cap B = AB = \{x \mid x \in A \text{ and } x \in B\}.$$

The intersection AB is sometimes called the *event A and B*. The *union* of two sets A and B is denoted by $A \cup B$ and is the set of all elements in A or B or both; that is,

$$A \cup B = \{x \mid x \in A \quad or \quad x \in B\}.$$

The union $A \cup B$ is sometimes called* the *event A or B*. The Venn diagram of Figure 2.2 shows the union and intersection of A and B for a typical situation.

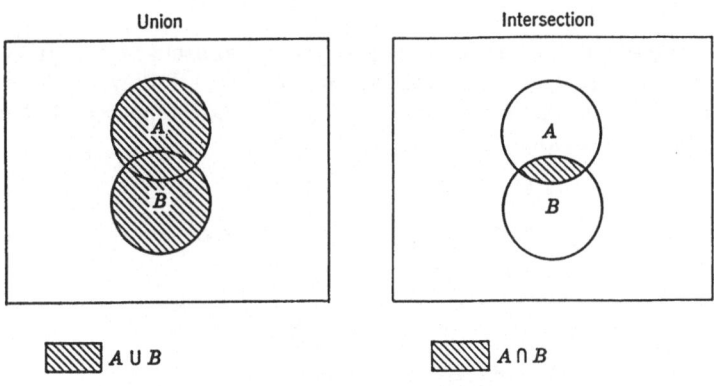

Fig. 2.2 - Union and intersection of sets

The *complement* of the set A is denoted by A^c (*or* A') and is the set of all elements not in A ; that is,

$$A^c = \{x \mid x \notin A\}$$

Some useful relationships involving the complement are

$$(A^c)^c = A$$

$$\phi^c = S \qquad S^c = \phi$$

$$A \cup A^c = S \qquad A\,A^c = \phi$$

$$(A \cup B)^c = A^c\,B^c \qquad (A\,B)^c = A^c \cup B^c$$

---

*Note that "or" is used here in the sense of "either..or ..or both".

If two sets A and B have no elements in common they are said to be *disjoint* or *exclusive* and

$$A \ B = \phi$$

It has already been emphasized that the elements of a set can be anything whatsoever; in particular, they can be sets themselves. To avoid confusion, it is conventional to call

$$a \ set \ of \ sets \qquad a \ class \ of \ sets$$

and

$$a \ set \ of \ classes \qquad a \ family \ (of \ classes).$$

A set will be called a *finite set* if it is empty or contains exactly n elements, for n a positive integer; otherwise, the set will be called an *infinite set*.

Based on the previous definitions, all subsets of the universal set S form an algebraic system, for which the following theorems hold. For completeness, some relationships are repeated that have been given previously.

$$A \cup \phi = A \qquad\qquad A \cup A^{c} = S$$
$$A \ \phi = \phi \qquad\qquad A \ S = A$$
$$A \cup A = A \qquad\qquad S \cup \phi = S$$
$$A \ A = A \qquad\qquad S \ \phi = \phi$$

If $A \subset B$, then $A \cup B = B$ and $AB = A$.

For all A, B; $A \subseteq A \cup B$ and $AB \subseteq A$.

If $B \subset C$, then $AB \subset AC$.

If $A \subset C$, and $B \subset C$, then $A \cup B \subset C$.

If $A \subset C$ and $B \subset D$, then $A \cup B \subset C \cup D$.

The commutative law

$$AB = BA$$
$$A \cup B = B \cup A$$

The distributive law

$$A(B \cup C) = (AB) \cup (AC)$$
$$A \cup (BC) = (A \cup B)(A \cup C)$$

The associative law

$$(A \cup B) \cup C = A \cup (B \cup C) = A \cup B \cup C$$
$$(AB)C = A(BC) = ABC$$

The ideas of union and intersection may be extended in a rather obvious fashion to n sets. If $A_1, A_2,...,A_n$ are subsets of S, the union of $A_1, A_2,...,A_n$ is the set of all elements which belong to at least one of the sets $A_1, A_2,...,A_n$ and may be denoted by

$$A_1 \cup A_2 \cup \cdots \cup A_n \text{ or } \bigcup_{i=1}^{n} A_i$$

The intersection of $A_1, A_2,...,A_n$ is the set of all elements common to $A_1, A_2,...,A_n$ and is written as

$$A_1 A_2...A_n \text{ or } A_1 \cap A_2 \cap \cdots \cap A_n \text{ or } \bigcap_{i=1}^{n} A_i$$

An infinite set is said to be *countable* or *denumerable* if its elements can be indexed by the natural numbers (positive integers); that is, placed in one-to-one correspondence with them. Otherwise an infinite set is said to be *non-countable* or *non-denumerable*. As an example, the set of natural numbers is a countable set; the set of all real numbers is not.

We proceed now to develop briefly a concept of probability based on the idea of *equally likely* outcomes. Later, it will be shown that the relationships developed can be obtained more rigorously from an axiomatic approach.

2.3 *A Classical Concept of Probability-* Much of the historical foundation of probability arose from a consideration of games of chance where the possible outcomes could be reduced to elementary events which were *equally likely*. The term "equally likely" is taken to have the following meaning. If the possible outcomes of a random experiment are the M (elementary) events $x_1, x_2,...,x_M$, then the probability of each event is

$$P(x_1) = P(x_2) = \cdots = P(x_M) = \frac{1}{M} \qquad (2.3\text{-}1)$$

A set of these elementary events is said to be *favorable* to an event E if the event E occurs when, and only when, one of the members of the set occurs. Then the probability $P(E)$ of the event E becomes

$$P(E) = \frac{m}{M}$$

where m is the total number of elements in the set favorable to E. Thus the calculation of the probability of any event reduces to finding two numbers; 1) the total number M of possible outcomes in the experiment and 2) the number m of outcomes favorable to the particular event.

Example 2.1

An unbiased coin is tossed three times: Find the probability that *at least* two of the tosses yield "heads". Find the probability that *exactly* one head is thrown.

The set of all possible outcomes consists of the following eight members:

$$HHH \quad HHT \quad HTT \quad TTT$$
$$HTH \quad THT$$
$$THH \quad TTH$$

Of these, four satisfy the condition of containing at least two heads. The required probability is

$$P(at\ least\ two\ heads) = \frac{4}{8} = \frac{1}{2}$$

There are three outcomes favorable to the event "exactly one head"; hence

$$P(exactly\ one\ head) = \frac{3}{8}$$

This classical concept of probability suffers from a rather obvious shortcoming. It depends completely on *a priori* analysis of the random experiment. All possible outcomes must be found and a clear relation must be established between these outcomes and a set of equally likely events comprising the elementary events (points) of the sample space.

Nevertheless, some of the tools necessary to apply the classical concept will be treated in the following section. These ideas will be useful, also, for the remainder of the text.

2.4 *Elements of Combinatorial Analysis*- Whenever equal probabilities are assigned to the elements of a finite sample space, the computation of probabilities of events reduces to the counting of the elements comprising the events. As mentioned previously, most of the problems involving games of chance fall into this category. For such purposes, some of the introductory concepts of combinatorial analysis will prove immensely useful [1].

(a) Pairs and Multiplets

A *pair* $\{a_j, b_k\}$ is an *unordered set of two elements* a_j and b_k. Thus $\{a_j, b_k\}$ and $\{b_k, a_j\}$ will be considered the same pair. With m elements $a_1, a_2, ..., a_m$ and n elements $b_1, b_2, ..., b_n$ it is possible to form exactly mn pairs $\{a_j, b_k\}$ containing one element from each group. A *multiplet* of size r is an *unordered set of* r *elements*. Given n_1 elements $a_1, a_2, ..., a_{n_1}$; n_2 elements $b_1, b_2, ..., b_{n_2}$; etc.; up to

n_r elements $x_1, x_2, ..., x_{n_r}$, it is possible to form exactly $n_1 n_2 ... n_r$ multiplets $\{a_j, b_k, ..., x_l\}$ containing one element from each group.

(b) Sampling

A *sample* of size r is an *ordered set of r elements* obtained in some way from a universal set of n elements. Two possible policies to follow in obtaining samples are 1) to *sample with replacement* of the elements drawn and 2) to *sample without replacement*.

Consider first the case of sampling with replacement. There are n possible elements any one of which can be chosen on the first drawing to form the first element in the sample. Since the chosen element is replaced prior to the second drawing, there are also n possible elements available to be chosen as the second element in the sample. Thus, after r drawings, the sample (of length r) could be any one of

$$n \cdot n \cdot n \ ... \ n = n^r \qquad (2.4\text{-}1)$$

distinct possible samples.

In sampling without replacement it is apparent that $r \leq n$. After the first drawing, there are only $n-1$ elements available from which to choose and, after r drawings, only $n-r$. Thus the number of possible samples of length r in this case is only

$$n(n-1)(n-2) \ ...... \ (n-r+1) = \frac{n!}{(n-r)!} \qquad (2.4\text{-}2)$$

We note in passing that

$$\frac{n!}{(n-r)!} \leq n! \leq n^n \qquad (2.4\text{-}3)$$

where r and n are non-negative integers and $r \leq n$. An (ordered) sample taken without replacement (so that the same element does not occur twice) is often called a *permutation*. The symbol p_r^n is sometimes used to indicate the number of permutations of n things taken r at a time:

$$p_r^n = \frac{n!}{(n-r)!} \qquad (2.4\text{-}4)$$

Suppose we sample without replacement to obtain a sample of size n from a set of n elements. Then the number of such samples is just $n!$ since

$$\frac{n!}{(n-n)!} = n! \qquad (2.4\text{-}5)$$

Furthermore this is the number of ways of *ordering* the n elements since each sample is a unique ordering and the $n!$ possible samples exhaust the possibilities.

(c) Combinations

A *combination* is a set of elements without repetition and without regard to order. Thus the set {1,2} and the set {2,1} are different permutations but only one combination. Out of n elements it has already been shown that the possible permutations of size r are

$$\frac{n!}{(n-r)!}$$

We have seen that r elements can be ordered in $r!$ ways so that there are $r!$ permutations of any particular set of r elements. Therefore the number of possible combinations of n elements taken r at a time must be

$$\frac{n!}{(n-r)!} \div r! = \frac{n!}{(n-r)!r!} \qquad (2.4\text{-}6)$$

This quantity is frequently denoted by C_r^n or by $\begin{pmatrix} n \\ r \end{pmatrix}$ and is often called the *binomial coefficient* since it occurs in the binomial expansion

$$(a+b)^n = \sum_{r=0}^{n} \begin{pmatrix} n \\ r \end{pmatrix} a^{n-r} b^r \qquad (2.4\text{-}7)$$

It follows directly from Eq. (2.4-6) that

$$\begin{pmatrix} n \\ r \end{pmatrix} = \begin{pmatrix} n \\ n-r \end{pmatrix} \qquad (2.4\text{-}8)$$

and that

$$\begin{pmatrix} n \\ n \end{pmatrix} = 1, \quad \begin{pmatrix} n \\ 0 \end{pmatrix} = 1 \qquad (2.4\text{-}9)$$

Example 2.2

What is the probability that a bridge hand will contain all thirteen card values: that is, $A, K, Q, \ldots, 3, 2$?

The number of possible bridge hands is given by

$$\begin{pmatrix} 52 \\ 13 \end{pmatrix}$$

The number of possible bridge hands with all thirteen card values is

$$4^{13}$$

Thus the required probability is

$$4^{13} \div \begin{pmatrix} 52 \\ 13 \end{pmatrix} \approx 0.0001057$$

(d) Multinomial Coefficients

Let $r_1, r_2, ..., r_k$ be non-negative integers such that

$$r_1 + r_2 + \cdots + r_k = n$$

It will now be shown that the number of ways in which n objects can be divided into k groups, the first containing r_1 objects, the second r_2, etc. is given by the *multinomial coefficient*

$$\begin{pmatrix} n \\ r_1, ..., r_k \end{pmatrix} = \frac{n!}{r_1! ... r_k!} \tag{2.4-10}$$

For the case where $k = 2$, the multinomial coefficient reduces to the binomial coefficient $\begin{pmatrix} n \\ r_1 \end{pmatrix}$ since $r_2 = n - r_1$. In the general case, we begin by selecting the first group of size r_1 from the set of n objects. There will be $\begin{pmatrix} n \\ r_1 \end{pmatrix}$ combinations possible. Next we select the second group of size r_2 from the remaining $n - r_1$ objects. There will be $\begin{pmatrix} n - r_1 \\ r_2 \end{pmatrix}$ combinations possible. We continue in this manner until the $(k-1)$ group has been chosen. After this choice, there are only r_k objects left and they must comprise the k-*th* group. Thus the total number of ways in which the n objects can be divided into the k groups is

$$\begin{pmatrix} n \\ r_1 \end{pmatrix} \begin{pmatrix} n - r_1 \\ r_2 \end{pmatrix} \cdots \begin{pmatrix} n - r_1 - r_2 - ... - r_{k-2} \\ r_{k-1} \end{pmatrix}$$

This product may be written explicitly as

$$\frac{n!}{r_1!(n - r_1)!} \frac{(n - r_1)!}{r_2!(n - r_1 - r_2)!} \cdots \frac{(n - r_1 - r_2 - ... - r_{k-2})!}{(r_{k-1})! r_k!}$$

On canceling common terms it is clear that this last expression is the multinomial coefficient originally defined, by Eq. (2.4-10).

Example 2.3

Twelve dice are thrown. What is the probability that each face occurs twice; that is, that there are two 1's, two 2's, etc?

The total number of possible outcomes is 6^{12} since each die has six distinct faces. Each face value can occur twice in the twelve throws in as many ways as twelve dice can be arranged in six groups of two each. In terms of the multinomial coefficient, $n = 12$ and $r_1 = r_2 = \cdots = r_6 = 2$. Thus the total number of favorable outcomes is

$$\frac{12!}{2 \times 2 \times 2 \times 2 \times 2 \times 2} = \frac{12!}{2^6}$$

and the probability desired is

$$p = \frac{12!}{2^6 6^{12}} \approx 0.003438$$

(e) The Hypergeometric Distribution

Consider a set of n elements. Suppose n_1 of these elements have a particular attribute while the remainder $n_2 = n - n_1$ have some other attribute. A group of r elements is chosen at random without replacement and regard to order. It is desired to find the probability $q(k)$ that, of the group chosen, exactly k elements have the first attribute. We proceed as follows:

The chosen group has k elements with the first attribute and $r-k$ elements with the second. The subgroup with the first attribute can be chosen in $\begin{pmatrix} n_1 \\ k \end{pmatrix}$ ways and the subgroup with the second attribute in $\begin{pmatrix} n-n_1 \\ r-k \end{pmatrix}$ ways. Therefore the total number of favorable outcomes is

$$\begin{pmatrix} n_1 \\ k \end{pmatrix} \begin{pmatrix} n-n_1 \\ r-k \end{pmatrix}$$

The total number of possible outcomes is just the number of combinations of n things taken r at a time and is given by $\begin{pmatrix} n \\ r \end{pmatrix}$. Thus the probability $q(k)$ is

$$q(k) = \frac{\begin{pmatrix} n_1 \\ k \end{pmatrix} \begin{pmatrix} n-n_1 \\ r-k \end{pmatrix}}{\begin{pmatrix} n \\ r \end{pmatrix}} \qquad (2.4\text{-}11)$$

This $q(k)$ is called the *hypergeometric distribution function*.

Example 2.4:

Find the probability that, among r cards drawn at random from a bridge deck, there are exactly four aces.

In this case $n = 52$, $n_1 = 4$, and $k = 4$. We have

$$q(4) = \frac{\begin{pmatrix} 4 \\ 4 \end{pmatrix} \begin{pmatrix} 52-4 \\ r-4 \end{pmatrix}}{\begin{pmatrix} 52 \\ r \end{pmatrix}} = \frac{\begin{pmatrix} 48 \\ r-4 \end{pmatrix}}{\begin{pmatrix} 52 \\ r \end{pmatrix}} \quad , \quad r = 4,5,...,52$$

(f) Stirling's Formulas

In the previous sections the factorial expression

$$n! = n (n-1)(n-2)....(3)(2)(1)$$

has occurred frequently. For small n, this expression is easily evaluated, but the computation may become prohibitively difficult as n increases. Furthermore a number of limiting expressions will be encountered later where it will be convenient to have available approximations to $n!$. An approximation for $n!$ which converges to $n!$ in the limit is*

$$n! \approx (2\pi)^{1/2} n^{n+1/2} e^{-n} \qquad (2.4\text{-}12)$$

This expression is frequently called *Stirling's first approximation* and underestimates $n!$. A *second approximation* which overestimates $n!$ is

$$n! \approx (2\pi)^{1/2} n^{n+1/2} e^{-n+1/12n} \qquad (2.4\text{-}13)$$

Even for small n, the *first approximation* is remarkably accurate.

2.5 *The Axiomatic Foundation of Probability Theory* - The theory of probability can be developed formally from a set of axioms. The approach to be used here was due originally to Kolmogorov** and follows his work closely. The language used will be that of the set theory discussed in Section 2.2. The *elements* of the *sample space* S will be called *outcomes* or *elementary events* or *points* of a random experiment and will be denoted as before by lower case letters $a, b, ...$ or sometimes by ω. Certain subsets of the sample space will be called *events* and will be denoted by upper case letters $A, B, ...$; included will be the *certain event* S and the *impossible event* ϕ. The sample space may be *finite, denumerably infinite,* or *non-denumerably infinite*. Of these the finite sample spaces offer the least conceptual difficulty and will be discussed first.

2.6 *Finite Sample Spaces*- A finite sample space has been defined previously as a set which is either empty or contains n elements where n is a positive integer. A finite sample space S will be called a *probability space* if, for every event $A \in S$, there is defined a real number $P(A)$ with the following properties:

Axiom 1.

$$P(A) \geq 0. \qquad (2.6\text{-}1)$$

---

* W. Feller, *An Introduction to Probability Theory and its Applications, Vol. I,* John Wiley & Sons, Inc., New York, N.Y., 1968; pp. 52-54.
** A.N. Kolmogorov, *Foundations of the Theory of Probability,* 2nd English edition, Chelsea Publishing Company, New York, N.Y.; 1956; pp. 2-20.

Axiom 2.

$$P(S) = 1 \qquad (2.6\text{-}2)$$

Axiom 3.

If A_1 and A_2 are two disjoint events in S (that is, if $A_1 A_2 = \phi$), then

$$P(A_1 \cup A_2) = P(A_1) + P(A_2) \qquad (2.6\text{-}3)$$

The function $P(A)$ is called the *probability of the event A*. Note that an immediate consequence of Axiom 3 is

Axiom 3(a).

Consider the mutually disjoint events $A_1, A_2, ..., A_m$; then

$$P(A_1 \cup A_2 \cup \cdots \cup A_m) = P(A_1) + P(A_2) + \cdots + P(A_m) \quad (2.6\text{-}4)$$

For infinite sample spaces some modifications of the axioms are necessary and some additional concepts of set theory are required.

2.7 *Fields, σ-Fields, and Infinite Sample Spaces* - As mentioned in Section 2.2, a set, whose elements are sets, will be called a *class* of sets. These classes will usually be denoted by bold-faced letters such as **A** or **B**. When the set operations performed on the sets in a class **A** give as a result sets which also belong to this class, then the class **A** is said to be *closed* under those set operations. A *field* (or *algebra*) **F** is a class of sets which is closed under complementation and finite intersections and unions. Thus a field **F** is a nonempty class of sets of S such that

(1) If $A \in \mathbf{F}$ and $B \in \mathbf{F}$, then $A \cup B \in \mathbf{F}$.

(2) If $A \in \mathbf{F}$, then $A^c \in \mathbf{F}$.

These two properties are sufficient to define a field. A number of other properties follow immediately. For example:

(3) If $A \in \mathbf{F}$ and $B \in \mathbf{F}$, then $AB \in \mathbf{F}$

Proof: From (1) and (2) it follows that $A^c \in \mathbf{F}, B^c \in \mathbf{F}$, and $A^c \cup B^c \in \mathbf{F}$; therefore $(A^c \cup B^c)^c = AB \in \mathbf{F}$.

(4) $S \in \mathbf{F}$

Proof: Since the class is nonempty, at least one set $A \in \mathbf{F}$. Hence $A^c \in \mathbf{F}$ and $A \cup A^c = S \in \mathbf{F}$.

(5) $\phi \in \mathbf{F}$.

Proof: Since $S \in \mathbf{F}$, its complement $S^c = \phi \in \mathbf{F}$.

(6) If $A \in \mathbf{F}$ and $B \in \mathbf{F}$, then $A - B \in \mathbf{F}$.

Proof: It has been shown that $B^c \in \mathbf{F}$, and, hence, $AB^c = A - B \in \mathbf{F}$.

A field is sometimes called an *additive* class of sets. It is clear from the properties just developed that all of the set operations can be performed any (finite) number of times on the members of the field **F** without obtaining a set not in **F**.

Example 2.5

Consider the set $S = 1,2$. Two possible fields F_1 and F_2 are

$$F_1 = \{\phi, \{1,2\}\}$$

$$F_2 = \{\phi, \{1\}, \{2\}, \{1,2\}\}$$

Any field **B** on S is called a *completely additive class* (of sets) or σ-*field* (sigma field)* on S if it is closed under *denumerable* intersections and unions. Thus, a σ-field **B** is a *field* on S such that

(1) If $A_1, A_2, \ldots, A_n, \ldots, \in \mathbf{B}$, then $\bigcup_{i=1}^{\infty} A_i \in \mathbf{B}$

It follows that

(2) If $A_1, A_2, \ldots, A_n, \ldots, \in \mathbf{B}$, then $\bigcap_{i=1}^{\infty} A_i \in \mathbf{B}$

Proof: If $A_n \in \mathbf{B}$, then $A_n^c \in \mathbf{B}$ and $\bigcup_{n=1}^{\infty} A_n^c \in \mathbf{B}$. In the same way the complement $(\bigcup_{n=1}^{\infty} A_n^c)^c \in \mathbf{B}$. It follows from one of DeMorgan's rules [2] that

$$(\bigcup_{n=1}^{\infty} A_n^c)^c = \bigcap_{n=1}^{\infty} A_n \in \mathbf{B}$$

The elements of the σ-field will be the *events* of the random experiments. In a given sample space S it is generally possible to define many σ-fields. To distinguish between members of several σ-fields, the members of a given σ-field **B** are usually called **B** -*measurable* sets.

A *probability space* or *field of probability* can now be defined as the triple** $(S, \mathbf{B}, P)$ where S is a sample space (finite or

---

* The completely additive class of sets on the real line containing all intervals is usually called the *Borel class* or *Borel field*. A more complete discussion of the construction of this field is given in [6] and [9].

** Many treatments of probability use the notation Ω instead of S for the sample space or *fundamental* probability set. The points or elementary events of Ω are then usually denoted by ω.

infinite), **B** is a given σ-field defined on S, and P is a real-valued *set function** such that, for every event $A \in \mathbf{B}$,

Axiom 1.

$$P(A) \geq 0 \tag{2.7-1}$$

Axiom 2.

$$P(S) = 1 \tag{2.7-2}$$

Axiom 3.

If $A_1, A_2, ..., A_n, ...$ is any countable sequence of mutually disjoint events in **B**, then

$$P(A_1 \cup A_2 \cup ... \cup A_n \cup ...) = \sum_{i=1}^{\infty} P(A_i) \tag{2.7-3}$$

The function $P(A)$ is called the *probability* of the event A. If S is a finite space and if **B** is chosen to be all of the events in S, then these axioms are equivalent to those used in Section 2.6 to define a finite probability space. In all future discussions of events and their associated probabilities, it should be kept in mind that there is an underlying probability space $(S, \mathbf{B}, P)$ consisting of a sample space, a σ-field, and a probability measure whether these are mentioned specifically or not.

Example 2.6

Let the sample space S consist of the single point b so that

$$S = \{b\}$$

A σ-field **B** on S is given by

$$\mathbf{B} = \{S, \phi\}$$

and a probability set function is

$$P(S) = 1 \quad , \quad P(\phi) = 0$$

The probability $P(\cdot)$ possesses certain properties that are an immediate consequence of its definition.

Corollary 1.

$$P(\phi) = 0 \tag{2.7-4}$$

---

* A *set function* is a function whose domain of definition is a class of sets; that is, a rule exists which relates some object or group of objects to each member of the class. In this case P is a real-valued function which assigns to every $A \in \mathbf{B}$ a number $P(A)$.

Proof: Since $S = S \cup \phi$ and $\phi = S \phi$, it follows from Axiom 3 that $P(S) = 1 = P(S) + P(\phi) = 1 + P(\phi)$; thus $P(\phi) = 0$.

Corollary 2.

$$P(A^c) = 1 - P(A) \qquad (2.7\text{-}5)$$

Proof: The sample space S can be written as $S = A \cup A^c$ and $AA^c = \phi$; consequently, we see that $P(A \cup A^c) = P(A) + P(A^c) = 1$ and Corollary 2 follows. Note that Corollary 1 results if $A = S$.

Corollary 3. For any finite set $E_1, E_2, ..., E_n$ of disjoint events, then

$$P(E_1 \cup E_2 \cup \cdots \cup E_n) = \sum_{i=1}^{n} P(E_i) \qquad (2.7\text{-}6)$$

Proof: Let the null set $\phi = E_{n+1} = E_{n+2} = \cdots$. It follows from Axiom 3 and Corollary 1 that

$$P(E_1 \cup \cdots \cup E_n) = P(\bigcup_{i=1}^{\infty} E_i) = \sum_{i=1}^{\infty} P(E_i) = \sum_{i=1}^{n} P(E_i)$$

Corollary 4. Let E and F be events and let $E \subseteq F$; then

$$P(E) \leq P(F) \qquad (2.7\text{-}7)$$

Proof: The event F can be written as $E \cup E^c F$ and E and $E^c F$ are disjoint; that is, $E \cap E^c F = \phi$. Then, by Axiom 1 and Corollary 3, we have

$$P(F) = P(E) + P(E^c F) \geq P(E)$$

Corollary 5. For every event $E \in \mathbf{B}$,

$$P(E) \leq 1 \qquad (2.7\text{-}8)$$

Proof: The relation $E \in \mathbf{B}$ implies $E \subseteq S$. Then, by Axiom 2 and Corollary 4, it follows that $P(E) \leq P(S) = 1$.

Corollary 6. (Boole's Inequality). Consider any sequence of events $A_1, A_2, ..., A_n, ...$ not necessarily disjoint; then

$$P(\bigcup_{i=1}^{\infty} A_i) \leq \sum_{i=1}^{\infty} P(A_i) \qquad (2.7\text{-}9)$$

Proof: The proof follows from the equality

$$\bigcup_{i=1}^{\infty} A_i = A_1 \cup (A_2 - A_1) \cup [A_3 - (A_1 \cup A_2)] \cup \cdots \qquad (2.7\text{-}10)$$

Note that the right side of this expression consists of a union of disjoint events; hence

$$P(\bigcup_{i=1}^{\infty} A_i) = P(A_1) + P(A_2 - A_1) + P[A_3 - (A_1 \cup A_2)] + \cdots \qquad (2.7\text{-}11)$$

Since $(A_2-A_1) \subseteq A_2$, $[A_3-(A_1 \cup A_2)] \subseteq A_3$, etc., it follows that $P(A_2-A_1) \leq P(A_2)$, $P(A_3-(A_1 \cup A_2)] \leq P(A_3)$, etc., and Eq. (2.7-9) follows immediately.

Consider two events A and B such that $A \in \mathbf{B}$ and $B \in \mathbf{B}$. Assume that $P(B) > 0$. The *conditional probability* of A given B will be denoted by $P(A/B)$ and will be defined as

$$P(A/B) = \frac{P(AB)}{P(B)} \tag{2.7-12}$$

If $B \in \mathbf{B}$ and $P(B) > 0$, then the function $P(\cdot/B)$ on $\mathbf{B}$ is a probability. In other words it satisfies the three axioms used to define a probability; that is, for every event $A \in \mathbf{B}$,

Axiom 1(a).

$$P(A/B) \geq 0 \tag{2.7-13a}$$

Axiom 2(a).

$$P(S/B) = 1 \tag{2.7-13b}$$

Axiom 3(a).

If $A_1, A_2,...,A_n,...,...$ is any countable sequence of mutually disjoint events in $\mathbf{B}$, then

$$P(\bigcup_{i=1}^{\infty} A_i/B) = \sum_{i=1}^{\infty} P(A_i/B) \tag{2.7-14}$$

The proofs of these axioms will be left to the reader.

Consider now the sequence of disjoint events $\{B_n\}$. The sequence may be finite or countably infinite. Assume that

$$P(\bigcup_n B_n) = \sum_n P(B_n) = 1 \tag{2.7-15}$$

The event A is an arbitrary event in $\mathbf{B}$ and $P(B_n) > 0$ for every n. The probability of event A may be written as

$$P(A) = P[A \cap (\bigcup_n B_n)] + P[A \cap (\bigcup_n B_n)^c] \tag{2.7-16}$$

But the probability $P(\bigcup_n B_n)^c$ is zero by Eq. (2.7-15). Consequently, this last equation becomes

$$P(A) = P[A \cap (\bigcup_n B_n)] = P(\bigcup_n AB_n) = \sum_n P(AB_n) \tag{2.7-17}$$

or, finally,

$$P(A) = \sum_n P(AB_n) = \sum_n P(A/B_n)P(B_n) \tag{2.7-18}$$

This last expression is frequently called the *Theorem of Total Probabilities*.

2.8 *Independence* - Consider the events A and B, both assumed to have probabilities different from zero. If

$$P(A/B) = P(A) \qquad (2.8\text{-}1)$$

then A and B are said to be *statistically independent*. Note that Eq. (2.8-1) and Eq. (2.7-12) imply that

$$P(B/A) = P(B) \qquad (2.8\text{-}2)$$

which could have been taken equally well as the definition of independence. This concept can be extended as before to cases where more than two variables are involved. Note that Eq. (2.8-1) or (2.8-2) implies that Eq. (2.7-12) can be rewritten for the independent case as

$$P(AB) = P(A)P(B) \qquad (2.8\text{-}3)$$

which also could have been taken as the definition of independence.

Three events $A, B,$ and C are mutually independent if

$$P(AB) = P(A)P(B)$$
$$P(AC) = P(A)P(C) \qquad (2.8\text{-}4)$$
$$P(BC) = P(B)P(C)$$

and if

$$P(ABC) = P(A)P(B)P(C) \qquad (2.8\text{-}5)$$

The last condition is necessary since the first three do not insure that such events as AB and C are independent as they would be if

$$P\{(AB)(C)\} = P(AB)P(C) = P(A)P(B)P(C) \qquad (2.8\text{-}6)$$

Example 2.7

Two dice are thrown. The events $A, B,$ and C are taken to be

event A - "odd face on first die"

event B - "odd face on second die"

event C - "sum of the faces odd"

The probabilities for fair dice are

$$P(A) = 1/2 \quad P(A/B) = 1/2 \quad P(A/C) = 1/2$$
$$P(B) = 1/2 \quad P(B/A) = 1/2 \quad P(B/C) = 1/2$$
$$P(C) = 1/2 \quad P(C/A) = 1/2 \quad P(C/B) = 1/2$$

However, A, B, and C cannot occur simultaneously and it is apparent that the events AB and C are not independent.

For N events A_i to be statistically independent, it is necessary that, for all combinations $1 \leq i \leq j \leq k \leq \cdots \leq N$,

$$P(A_i A_j) = P(A_i)P(A_j)$$
$$P(A_i A_j A_k) = P(A_i)P(A_j)P(A_k)$$
.
.   (2.8-7)
.

$$P(A_1 A_2 ... A_N) = P(A_1)P(A_2)...P(A_N)$$

Note that the probability of the event $A \cup B$ can be written as

$$P(A \cup B) = P(A) + P(B) - P(AB) \qquad (2.8\text{-}8)$$

since

$$A \cup B = A + B - AB \qquad (2.8\text{-}9)$$

If A and B are mutually exclusive, then $AB = \phi$ and

$$P(A \cup B) = P(A) + P(B)$$

as previously discussed.

The dependency relationships can be extended to the case where more than two events can occur. Consider the events A, B, and C. If B is replaced in Eq. (2.8-8) by $B \cup C$, the result is

$$P(A \cup B \cup C) =$$

(2.8-10)

$$P(A) + P(B) + P(C) - P(AB) - P(AC) - P(BC) + P(ABC)$$

where use has been made of the identity

$$A(B \cup C) = AB + AC - ABC \qquad (2.8\text{-}11)$$

Here $A \cup B \cup C$ means that at least one of the events A, B, or C has occurred and ABC means that all of the events A, B, and C have occurred. In a similar way, Eq. (2.8-10) may be extended to the union of any number of events.

2.9 *Random Variables, Discrete and Continuous* - In Section
2.7 we developed the concept of a probability space which completely describes the outcome of a random experiment. Fundamental to the specification is a set of elementary outcomes or
points $\omega \in S$ where S is the sample space associated with the probability space (S, B, P). These elementary outcomes are the build-
ing blocks from which the events are constructed. As was
emphasized earlier, the elementary outcomes can be any objects
whatsoever. Certainly, they need not be numbers. In the tossing
of a coin, for example, the outcomes will usually be "heads" or
"tails". On the other hand, if a single die is tossed, one possible
set of outcomes are the numbers 1,2,3,4,5, and 6 corresponding to
the number of dots on the upper face of the die.

It is usually convenient to associate a real number with each
of the elementary outcomes (or elements) ω of the sample space S.
The result is a mapping of the space S to the real line R in the
manner indicated in Fig. 2.3. The assignment of the real numbers
amounts to defining a real valued function on the elements of the

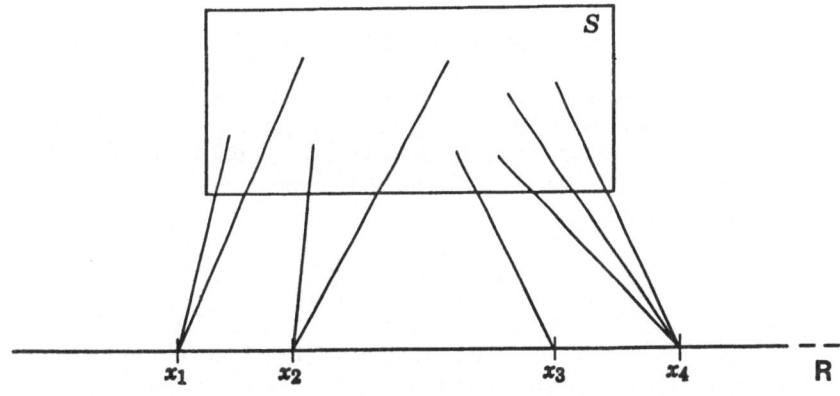

Fig. 2.3 - A mapping from the basic space S to the real line R
produced by an arbitrary random variable

sample space; this function is called a *random variable*. It is
desired that this random variable assume each of its possible
values with a definite probability; therefore we are led to the following equivalent definitions of a random variable:

Definition 1(a): A *random variable X* is a real-and single-valued function defined on the points (elementary outcomes) of a probability space. To each point, there corresponds a real number, the *value* of the random variable at that point. For every real number x, the set $\{X=x\}$, the set of points at which X assumes the value x, is an event. Also, for every pair of real numbers x_1 and x_2, the sets $\{x_1<X<x_2\}$, $\{x_1\leq X\leq x_2\}$, $\{x_1<X\leq x_2\}$, $\{x_1\leq X<x_2\}$, $\{X<x_1\}$, $\{X\leq x_1\}$, $\{X<x_2\}$, and $\{X\leq x_2\}$ are events.

Definition 1(b): A *random variable X* on a probability space $(S,\mathbf{B},P)$ is a real-and single-valued function whose domain is S and which is **B** *-measurable;* that is, for every real number x,

$$\{\omega\in S \mid X(\omega) \leq x\} \in \mathbf{B}.$$

Several comments should be made about these definitions. In Definition 1(a), if S is a finite space, then the condition of the last sentence is automatically satisfied; that is, every set of points is an event. In terms of Definition 1(b), all subsets of a finite probability space are **B** *-measurable.* The notation used in Definition 1(b) is somewhat cumbersome and it will be convenient to express the set in curly brackets by $\{X\leq x\}$; that is, we shall use the following notation:

$$\{X\leq x\} = \{\omega \in S \mid X(\omega) \leq x\} \tag{2.9-1}$$

and, for any Borel set **B**,

$$\{X \in \mathbf{B}\} = \{\omega \in S \mid X(\omega)\in \mathbf{B}\} \tag{2.9-2}$$

One of the simplest examples of a random variable is the *indicator* or *indicator function* I_A of an event A. If $A \in \mathbf{B}$ (that is, if A is an event), then I_A is a function taking on the value unity at all points of A and the value zero at all other points given by A^c; thus

$$I_A(\omega) = \begin{cases} 1, & \omega \in A \\ 0, & \omega \notin A \end{cases} \tag{2.9-3}$$

Every event has an indicator and every function assuming only the values unity and zero is the indicator of some event. The indicator for the event A is shown schematically in Fig. 2.4. For an arbitrary function f, the product $I_A f$ is another function taking on the values f on A and zero elsewhere.

It is clear that the concept of random variable has been introduced in order to map each element of the abstract sample space S onto the real line R. The domain of X is S and the range is R. Thus subsets of the sample space become sets of points on the real line. To every set A in S there corresponds a set in R, called the *image* of A and denoted by $X(A)$. Also, for every set T in R, there exists in S the *inverse image* $X^{-1}(T)$ where

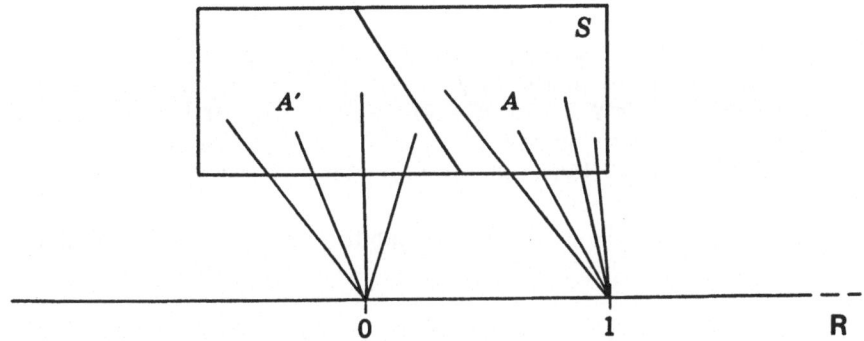

Fig. 2.4 - A mapping from the basic space S to the real line R
produced by the indicator function $I_A(\bullet)$

$$X^{-1}(T) = \{\omega \in S \mid X(\omega) \in T\} \tag{2.9-4}$$

The function X^{-1} is called the *inverse function* of X and maps the
real line R into the abstract space S. Since X was defined to be
single-valued, the inverse function X^{-1} maps disjoint sets T_1 and
T_2 in R into disjoint sets in S as illustrated in Fig. 2.4. Thus,
the inverse functions preserve all set operations, and, if a class of
sets $\mathbf{F}$ in R is closed under a given set operation, the class $X^{-1}(\mathbf{F})$
in S is closed under the same operation. The class $X^{-1}(\mathbf{F})$ is
defined by

$$X^{-1}(\mathbf{F}) = \{X^{-1}(T) \in S \mid T \in \mathbf{F}\} \tag{2.9-5}$$

Given a probability space $(S,\mathbf{B},P)$, it is clear that the random variable X defined on this space induces on its range space R
a new probability space $(R,\mathbf{F},\mathbf{P}_X)$ where the new probability
measure P_X is defined by

$$P_X(T) = P[X^{-1}(T)] \tag{2.9-6}$$

It is also clear that, to every event $A \in S$ with associated probability $P(A)$, there corresponds an event $T \in R$ with equal probability $P_X(T)$, that is,

$$P_X(T) = P(A) \quad , \quad X(A) = T \text{ or } X^{-1}(T) = A \tag{2.9-7}$$

Thus it is no longer necessary to consider the basic space $(S, \mathbf{B}, P)$; the new space $(R, \mathbf{F}, P_X)$ is equivalent and its sample points $X(x)$ are real numbers.

Let us now consider several rather simple examples of random variables.

Example 2.9

A random experiment consists of tossing a coin three times. Assume that the coin is weighted so that the probability of a head on each toss is $2/3$ and of a tail is $1/3$. The random variable X will be taken to be the number of heads produced in the three tosses. Thus X takes on the four values $x_0 = 0$, $x_1 = 1$, $x_2 = 2$, and $x_3 = 3$.

The sample space and probabilities $P(x_i)$ may be tabulated as:

ω	$X(\omega)$	$P(\{\omega\})$	
HHH	3	8/27	$P(X=3) = 8/27$
HHT	2	4/27	
HTH	2	4/27	$P(X=2) = 12/27$
THH	2	4/27	
HTT	1	2/27	
THT	1	2/27	$P(X=1) = 6/27$
TTH	1	2/27	
TTT	0	1/27	$P(X=0) = 1/27$

Example 2.10

A marksman is given four cartridges and is told to fire at a target until he has hit it or has used up the four cartridges. The probability of a hit on each shot is p and of a miss is $q = 1-p$, independently of other shots. Let the number of used cartridges be the random variable X taking on values 1,2,3, and 4. Find the function $P\{X=x\}$, $x=1,2,3,4$; this is the *(probability) density function* of the random variable X.

The probabilities are easily tabulated; we have:

$P\{X=1\} = p$: by definition.
$P\{X=2\} = qp = (1-p)p$: miss with first shot; hit with second.
$P\{X=3\} = q^2p = (1-p)^2p$:miss with first two shots; hit with third.
$P\{X=4\} = q^3 = (1-p)^3$:miss with first three shots; doesn't matter whether fourth shot is hit or miss.

Note the sum of these probabilities must be unity since the associated events exhaust the sample space; that is,

$$P(S) = \sum_{x=1}^{4} P\{X=x\} = P\{X=1\}+P\{X=2\}+P\{X=3\}+P\{X=4\}$$

$$= p + (1-p)p + (1-p)^2p + (1-p)^3 = 1$$

Suppose now that the supply of cartridges is unlimited and the marksman fires at the target until he hits? In this case the probabilities are

$$P\{X=n\} = (1-p)^{n-1}p \quad , \quad n=1,2,...$$

Again the sum must be unity:

$$\sum_{n=1}^{\infty} P\{X=n\} = \sum_{n=1}^{\infty} (1-p)^{n-1}p = \frac{p}{1-(1-p)} = 1$$

———————

Example 2.11

A fair coin is tossed; two possible outcomes, "heads" and "tails", exist each with probability one-half. A random variable X is defined by assigning the value unity to the occurrence of a head and the value zero to the occurrence of a tail; thus

$$P\{X=x\} = \begin{cases} 1/2, & x=0 \\ 1/2, & x=1 \\ 0, & \text{elsewhere} \end{cases}$$

Another well-defined event is the event $X \leq x$ for $-\infty < x < \infty$ with probability

$$P\{X \leq x\} = \begin{cases} 0, & x<0 \\ 1/2, & 0 \leq x < 1 \\ 1, & x \geq 1 \end{cases}$$

The function $P\{X=x\}$ and $P\{X \leq x\}$ are plotted in Fig. 2.5.

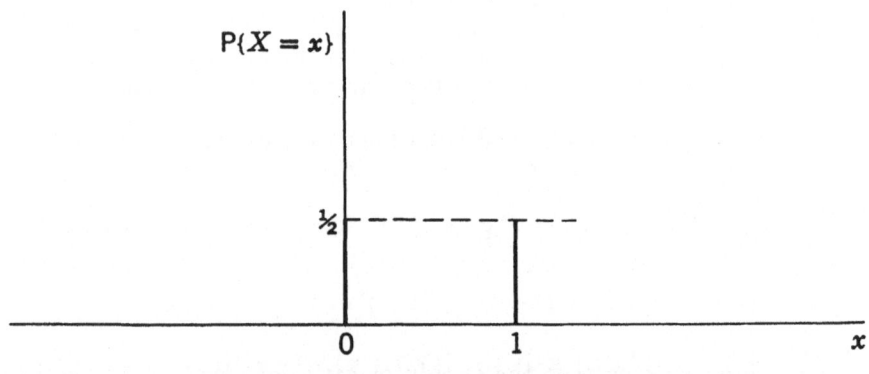

(a) The density function

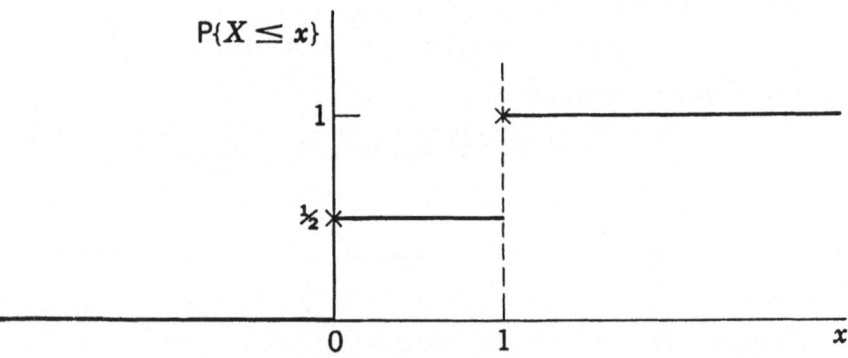

(b) The (cumulative) distribution function

Fig. 2.5 - A coin-tossing experiment.

The first is called the *(probability) density function* and the second is called the *(cumulative) distribution function* of the random variable X.

Discrete Random Variables

In the examples just considered, the random variable has taken on only a discrete set of values. In such cases we say that we are dealing with a *discrete random variable,* sometimes denoted as X_i where the subscript i indexes the possible values of X_i. More precisely, *a random variable X will be said to be discrete* if there is a denumerable sequence of distinct numbers x_i such that

$$P\{\cup_i [X=x_i]\} = \sum_i P\{X=x_i\} = 1 \qquad (2.9\text{-}8)$$

It is clear that, if X is discrete, it can be written in terms of indicators as

$$X = \sum_i x_i \, I_{[X=x_i]} \qquad (2.9\text{-}9)$$

Example 2.12

Let A, B, and C be three mutually exclusive and exhaustive events. Let a random variable X be defined which associates the number 1 with event A, the number 2 with event B, and the number 3 with with event C. This random variable may be written as

$$X(\omega) = 1 \times I_A(\omega) + 2 \times I_B(\omega) + 3 \times I_C(\omega)$$

where $\omega \in S$. For example, when $\omega \in B$, $X(\omega) = 2$.

In accordance with the definition, a discrete random variable takes on only a denumerable number of possible values x_i, $i=1,2,\dots$. As mentioned in Examples 2.10 and 2.11, the *density function* $f_X(x_i)$ of a discrete random variable X is determined by this set of numbers x_i and their associated probabilities; that is,

$$f_X(x_i) = P\{X=x_i\} \quad , \quad i=1,2,\dots \qquad (2.9\text{-}10)$$

It is clear that

$$f_X(x_i) \geq 0 \qquad (2.9\text{-}11)$$

and that

$$\sum_i f_X(x_i) = 1 \qquad (2.9\text{-}12)$$

As in Example 2.11, it will sometimes be convenient to treat the function f as though it were defined for all real x. In this case we will write

$$f_X(x) = P\{X=x\} \qquad (2.9\text{-}13)$$

and consider that $f_X(x)$ is zero whenever x is not one of the set of numbers $\{x_i\}$.

The probabilities of such events as $\{a<X<b\}$ are easily calculated since they are given by the sum

$$P\{a<X<b\} = \sum_{a<x_i<b} f_X(x_i) \qquad (2.9\text{-}14)$$

In particular, the probability of the event $\{X \leq x\}$ will be denoted by $F_X(x)$ and is given by

$$F_X(x) = P\{X \le x\} = \sum_{x_i \le x} f_X(x_i) \qquad (2.9\text{-}15)$$

As previously mentioned, the function $F_X(x)$ is called the *(cumulative) distribution function* (c.d.f.) of the random variable X. Various properties of this function will be developed later.

A discrete random variable whose range consists of a finite set of values is sometimes called a *simple random variable*. If the range consists of a set which is denumerably infinite, the random variable is sometimes called *elementary*. Thus a discrete random variable is either simple or elementary. Note that an indicator function is an example of a simple random variable.

Continuous Random Variables

Not all random variables are discrete. For example, X may assume a continuum of values in some interval as the following example indicates.

Example 2.13

Consider the *uniform distribution* where the range of the random variable is the finite interval (a,b). Let (x_1, x_2) be any interval inside (a,b). The probability that the continuous random variable X lies in this interval is assumed to be proportional to the length of the interval; that is,

$$P\{x_1 < X < x_2\} = \lambda(x_2 - x_1) \ , \quad a \le x_1 < x_2 \le b$$

The probability that X lies outside the interval (a,b) is zero. By Axiom 1 of Section 2.7, $P(A) \ge 0$ for any event A; consequently $\lambda \ge 0$. By Axiom 2, we have

$$P(S) = 1 = P\{a < X < b\} = \lambda(b - a) = 1$$

or

$$\lambda = \frac{1}{b-a}$$

Therefore the probability that X lies in the interval (x_1, x_2) is

$$P\{x_1 < X < x_2\} = \frac{x_2 - x_1}{b - a} \ , \quad a \le x_1 < x_2 \le b$$

Note that this last expression can be rearranged as

$$\frac{P\{x_1 < X < x_2\}}{x_2 - x_1} = \frac{1}{b - a} \ , \quad a \le x_1 < x_2 \le b$$

In this form it is called the *(probability) density function* for the uniformly distributed random variable X and is shown in Fig. 2.6(a).

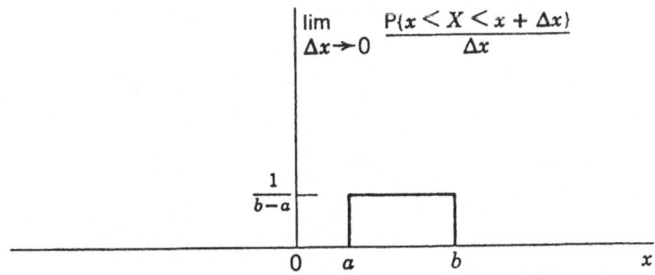

(a) The density function.

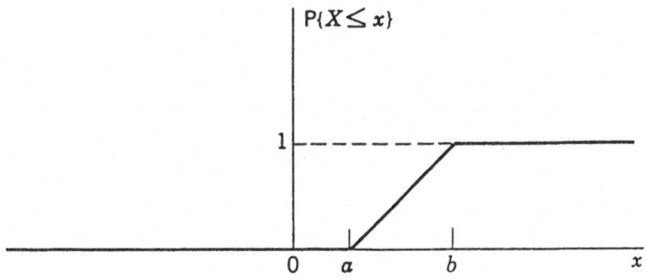

(b) The (cumulative) distribution function.

Fig. 2.6 - The uniform distribution.

Let us calculate now the probability $P\{X \leq x\}$ where x is any real number. If $x < a$, it is clear that this probability is zero; in the same way, if $x > b$, the probability must be unity. For $a \leq x \leq b$, we have

$$P\{X \leq x\} = P\{a < X \leq x\} = \frac{x-a}{b-a} \ , \ \ a \leq x \leq b$$

The probability $P\{X \leq x\}$ is called the *(cumulative) distribution function* for the uniformly distributed random variable X and is plotted in Fig. 2.6(b).

A random variable X will be said to be continuous if $P\{X = x\} = 0$ for all x and if a function $f_X(x)$ exists such that, for all x,

$$P\{x < X < x + \Delta x\} = \int_{x}^{x+\Delta x} f_X(y) \ dy \tag{2.9-16}$$

The function $f_X(x)$ is called the *density function* of the continuous random variable X. From an alternative point of view, if $f_X(x)$ is continuous at x, it could be defined as

$$f_X(x) = \lim_{\Delta x \to 0} \frac{P\{x < X < x + \Delta x\}}{\Delta x} \qquad (2.9\text{-}17)$$

It follows from Eq. (2.9-16) and the Mean-Value Theorem that, if $f_X(x)$ is continuous at x,

$$P\{x < X < x + dx\} \approx f_X(x)\ dx \qquad (2.9\text{-}18)$$

The quantity $f_X(x)dx$ will be called the *probability differential* $dF_X(x)$. Now the probability of the event $\{x_1 < X < x_2\}$ is

$$P\{x_1 < X < x_2\} = \int_{x_1}^{x_2} f_X(x)\ dx = \int_{x_1}^{x_2} dF_X(x) \qquad (2.9\text{-}19)$$

If $f_X(x)$ exists; that is, if $f_X(x) < \infty$, Eq. (2.9-16) implies that the probability of the event $\{X = x\}$ is zero since

$$P\{X = x\} = \lim_{\Delta x \to 0} P\{x < X < x + \Delta x\} = \lim_{\Delta x \to 0} f_X(x)\Delta x = 0 \quad (2.9\text{-}20)$$

It should be kept in mind that this last expression does not preclude the occurrence of the event $\{X = x\}$. The probability of the null event is zero but the converse is not true.

The probability of the event $\{X \leq x\}$ may be written as

$$P\{X \leq x\} = \int_{-\infty}^{x} f_X(x)\ dx = \int_{-\infty}^{x} dF_X(x) \qquad (2.9\text{-}21)$$

or, from the Fundamental Theorem of the Calculus*,

$$f_X(x) = \frac{d}{dx} F_X(x) \qquad (2.9\text{-}22)$$

where $F_X(x)$ is the distribution function of X.

It should be clear from the previous discussion that any function $f_X(x)$ is a density function for some continuous random variable X if

$$f_X(x) \geq 0 \qquad (2.9\text{-}23)$$

---

*Let

$$G(u) = \int_{a(u)}^{b(u)} H(x, u)\, dx$$

Then the reader will recall that the derivative of this definite integral is given by

$$\frac{dG}{du} = \int_a^b \frac{\partial H}{\partial u}\, dx + H(b, u)\frac{db}{du} - H(a, u)\frac{da}{du}$$

if $H(x, u)$ is a continuous function of x and u.

and

$$\int_{-\infty}^{\infty} f_X(x)dx = 1 \qquad (2.9\text{-}24)$$

This last expression follows from Eq. (2.9-19) on noting that

$$P\{-\infty < X < \infty\} = 1 = \int_{-\infty}^{\infty} f_X(x)dx$$

It is clear that, for the continuous random variable X, a continuum of values is permissible over some interval or set of intervals. In this case, the probability of a single selected value of X occurring is zero, as shown by Eq. (2.9-20). The situation is analogous to considering the area under a simple curve. The area under any point is zero but the area under any segment of non-zero length may be non-zero. We are led to consider not the probability that a random variable X takes on some value x but rather that it lies in some interval $a \leq X \leq b$, that is, $P\{a \leq X \leq b\}$.

Thus, it is apparent that a one-to-one correspondence does not exist between the point values of a continuous random variable and the underlying events $A,B,C,...$ even if an infinite number of events are considered since the points in a continuum are not denumerable. The problem has already been resolved, however, in an elementary fashion. Divide the real line into two parts by the number x and consider values of the random variable $X \leq x$ and $X > x$. With an arbitrary event A associate the values of the random variable $X \leq x$. Thus the probability $P(A)$ of the event A becomes the probability $P\{X \leq x\}$. This probability has already been denoted by $F_X(x)$ and is the distribution function of X. If this function is given for all values of x, $-\infty < x < \infty$, then the values of the random variable have been defined.

2.10 *Distribution Functions and Densities* - For the random variable X, we have defined the cumulative distribution function $F_X(x)$ by Eq. (2.9-1); that is,

$$F_X(x) = P\{X \leq x\}$$

A number of basic properties of $F_X(x)$ follow directly from its definition:

(1) $F_X(x)$ is *non-decreasing* (monotonicity); that is, $F_X(x+h) \geq F_X(x)$ if $h > 0$.

> The function $F_X(x)$ is a probability by definition. Let us associate the event A with the probability $P(X \leq x)$ and the event B with the probability $P(X \leq x + h)$ where $h > 0$. Then $B \supseteq A$ and $P(B) \geq P(A)$, or $F_X(x+h) \geq F_X(x)$.

(2) $F_X(x)$ *is right-continuous;* that is,

$$\lim_{\substack{h \to 0 \\ h > 0}} F_X(x + h) = F_X(x)$$

From the definitions of events A and B in (1), we have

$$P(x < X \le x + h) = F_X(x + h) - F_X(x)$$

or

$$\lim_{\substack{h \to 0 \\ h > 0}} P(x < X \le x + h) = P(\phi) = 0$$

or

$$0 = \lim_{\substack{h \to 0 \\ h > 0}} F_X(x + h) - F_X(x)$$

(3) $\lim_{x \to -\infty} F_X(x) = 0$; $\lim_{x \to \infty} F_X(x) = 1$

If we assume that the random variable can take on only finite values*, then

$$\lim_{a \to -\infty} F_X(a) = P(\phi) = 0$$

and

$$\lim_{a \to \infty} F_X(a) = P(S) = 1$$

We shall assume that the random variable assumes only finite values and shall use the notation:

$$\lim_{x \to -\infty} F_X(x) = F_X(-\infty)$$

$$\lim_{x \to \infty} F_X(x) = F_X(\infty)$$

$$\lim_{\substack{h \to 0 \\ h > 0}} F_X(x + h) = F_X(x^+) = F_X(x + 0)$$

$$\lim_{\substack{h \to 0 \\ h > 0}} F_X(x - h) = F_X(x^-) = F_X(x - 0)$$

---

*Suppose that the random variable can assume the values ∞ and $-\infty$ with non-zero probability. Then we have

$$\lim_{x \to \infty} F_X(x) < F_X(\infty) = 1$$

$$\lim_{x \to -\infty} F_X(x) = F_X(-\infty) > 0$$

This situation could occur when a random variable has a non-zero probability of assuming the value zero; its logarithm would have a non-zero probability of assuming the value $-\infty$.

There are a number of corollaries to these properties which are either obvious or follow immediately from the previous discussion:

(1) $0 \leq F_X(x) \leq 1$ (2.10-1)

(2) $F_X(b) - F_X(x) = P(x < X \leq b)$ $, b > x$ (2.10-2)

(3) $F_X(x) - F_X(x^-) = P(X = x)$ (2.10-3)

(4) $F_X(x^-) = P(X < x)$ (2.10-4)

It is apparent from Corollary (3) that the c.d.f. is continuous for a continuous random variable X. For a discrete random variable X_i, the c.d.f. is constant except for jumps at the points x_i, these jumps being equal to the probability $P(x_i)$. These characteristics are illustrated in Example 2.11 and Figure 2.5 of Section 2.9. The crosses on the graph are used to indicate right-continuity.

It can be shown [3] that if a function $F_X(x)$ of a real variable x satisfies the basic properties (1), (2), and (3), then there is a random variable X (or X_i) for which $F_X(x)$ is the c.d.f. In other words we could have started with these three properties as the definition of a cumulative distribution function [4].

It will often be convenient to use the symbol X to denote either the continuous or the discrete random variable. It will be apparent from the context as to which situation exists and whether the sum of Eq. (2.10-6) or the integral of Eq. (2.10-8) is to be used. If we are willing to use a somewhat more general concept of the integral (the Riemann-Stieltjes integral) then both equations can be expressed in a single form. In addition the case is easily handled where the random variable is mixed; that is, where it assumes a continuum of values but also has a non-zero probability of assuming some distinct values. From the applied point of view, this last situation is easily handled using the Dirac delta-function. The Riemann-Stieltjes integral and the Dirac delta-function are discussed briefly in Appendices A and B respectively.

Example 2.14

The following function $f_X(x)$ is to be tested as a density function:

$$f_X(x) = \begin{cases} e^x & , x \leq 0 \\ 0 & , x > 0 \end{cases}$$

It is apparent that $f_X(x)$ is non-negative. In addition

$$\int_{-\infty}^{\infty} f_X(x)\,dx = \int_{-\infty}^{0} e^x\,dx = 1$$

Thus this $f_X(x)$ is an example of a continuous density function and may be plotted as shown in Fig. 2.7.

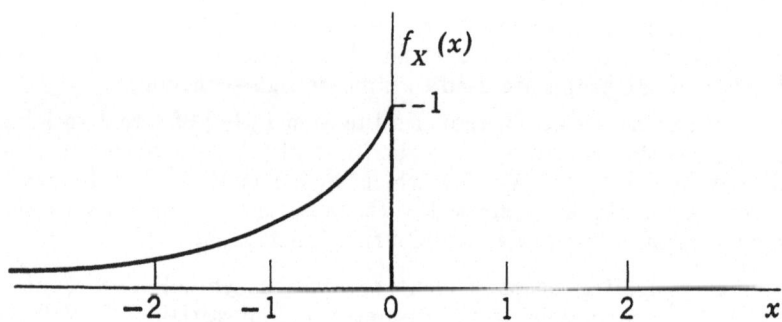

Fig. 2.7 - A continuous density function

Example 2.15

Let us define a function $b(k;n,p)$ as

$$b(k;n,p) = \binom{n}{k} (p)^k (1-p)^{n-k}$$

for $k=0,1,2...,n$. In this case let $n=2$ and $p=1/2$ so that

$$b(k;2,1/2) = \frac{2!}{(2-k)!k!} (1/2)^n$$

Again it is apparent that $b(k;2,1/2)$ is non-negative and that

$$\sum_{k=0}^{2} b(k;2,1/2) = 1/4 + 2/4 + 1/4 = 1$$

Thus the function $b(k;2,1/2)$ [or, in general, $b(k;n,p)$] is a discrete density. The general form is called the *binomial distribution* and will be discussed in more detail in Section 2.16. The plot of $b(k;2,1/2)$ is shown in Fig. 2.8.

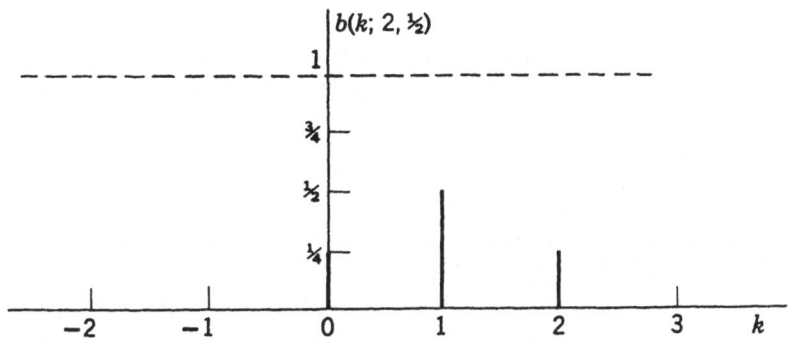

Fig. 2.8 - A discrete density function

It is clear from the previous discussion and from Section 2.9 that the c.d.f. $F_X(x)$ and the d.f. $f_X(x)$ are related, in the discrete case, by

$$F_X(x) = P(X \leq x) = \sum_{x_i \leq x} f_X(x_i) \qquad (2.10\text{-}5)$$

and, in the continuous case, by

$$F_X(x) = \int_{-\infty}^{x} f_X(y)dy \qquad (2.10\text{-}6)$$

We prove that the function $F_X(x)$ defined by either Eq. (2.10-5) or (2.10-6) is a c.d.f. by showing that it has the basic properties (1), (2), and (3) developed at the beginning of this section. The proof will be given only for the continuous case; extension to the discrete case should be apparent.

First Property

(1) $F_X(x)$ *is non-decreasing*

$$F_X(x+h) = F_X(x) + \int_{x}^{x+h} f_X(y)dy \quad , \quad h > 0$$

Since $f_X(x)$ is non-negative, the last term of this expression must be non-negative and

$$F_X(x+h) \geq F_X(x) \quad , \quad h > 0$$

Second Property

(2) $F_X(x)$ *is right-continuous*

$$F_X(x^+) - F_X(x) = \lim_{\substack{h \to 0 \\ h > 0}} \int_x^{x+h} f_X(y)\,dy$$

$$= \lim_{\substack{h \to 0 \\ h > 0}} h f_X(x) \text{ (by the Mean–Value Theorem)} = 0$$

Third Property

(3) *Behavior at infinity*

$$F_X(-\infty) = 0 \quad ; \quad F_X(\infty) = 1$$

$$F_X(-\infty) = \lim_{x \to -\infty} \int_{-\infty}^x f_X(y)\,dy = 0$$

$$F_X(+\infty) = \lim_{x \to \infty} \int_{-\infty}^x f_X(y)\,dy = 1$$

The properties of $F_X(x)$ corresponding to Eqs. (2.10-1), (2.10-2), (2.10-3), and (2.10-4) are easily derived in a similar fashion. The derivations are left to the reader.

2.11 *The Transformation of Random Variables* - We consider now the problem that arises when a random variable X is transformed to a new random variable Y through a functional relationship $y = h(x)$. The usual problem is, given the distribution of X, to find the distribution of Y.

We begin with the idea of a *monotone function*. A function $h(x)$ is *monotonically increasing* if $h(x_2) > h(x_1)$ when $x_2 > x_1$. Similarly a function $h(x)$ is *monotonically decreasing* if $h(x_2) < h(x_1)$ when $x_2 > x_1$. It is apparent that for monotone functions, there is a one-to-one correspondence between $h(x)$ and x; that is, for each value of x, there is one, and only one, value of $h(x)$, and, for each value of $h(x)$, there is one, and only one, value of x.

Suppose $y = h(x)$ is monotonically increasing and differentiable. Let the random variable X have a density $f(x)$ and a c.d.f. $F(x)$ while the random variable Y has a density $g(y)$ and a c.d.f. $G(y)$. If $h(a) = b$, then it is clear that

$$P(Y \leq b) = P(X \leq a) \tag{2.11-1}$$

or

$$G(b) = F(a) \qquad (2.11\text{-}2)$$

We may differentiate Eq. (2.11-2) with respect to b to yield

$$\frac{dG(b)}{db} = \frac{dF(a)}{da}\frac{da}{db} \qquad (2.11\text{-}3)$$

Since x and y are related one-to-one, a unique inverse exists for h and

$$a = h^{-1}(b) \qquad (2.11\text{-}4)$$

where the notation h^{-1} is used to denote the relationship

$$h(a) = h[h^{-1}(b)] = b \qquad (2.11\text{-}5)$$

Thus Eq. (2.11-3) becomes

$$g(b) = f[h^{-1}(b)]\frac{d[h^{-1}(b)]}{db}$$

or, replacing b by y,

$$g(y) = f[h^{-1}(y)]\frac{d[h^{-1}(y)]}{dy} \qquad (2.11\text{-}6)$$

As a matter of convenience, we shall frequently write Eq. (2.11-6) in the form

$$g(y) = f(x)\frac{dx}{dy} \qquad (2.11\text{-}7)$$

but it will be understood that Eq. (2.11-6) is what we have actually in mind.

Example 2.16

Let $y = e^x$ and let $f(x)$ be uniformly distributed in $(0,1)$. Then, from Eq. (2.11-7), we have

$$g(y) = f(x)e^{-x} = e^{-x} \quad , \quad 0 \le x \le 1$$

or

$$g(y) = 1/y \quad , \quad 0 \le x \le 1$$

For $0 \le x \le 1$, it follows that $1 \le y \le 2.718...$; hence

$$g(y) = \begin{cases} 1/y \ , & 1 \le y \le 2.718... \\ \\ 0 & , \quad \text{elsewhere} \end{cases}$$

It is clear that $g(y) \ge 0$ and that $\int_{1}^{2.718...} (1/y)\, dy = 1$; therefore $g(y)$ is a density. The two densities are plotted in Fig. 2.9.

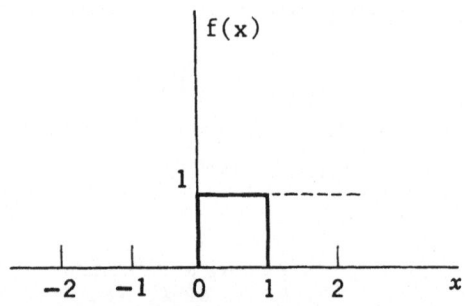

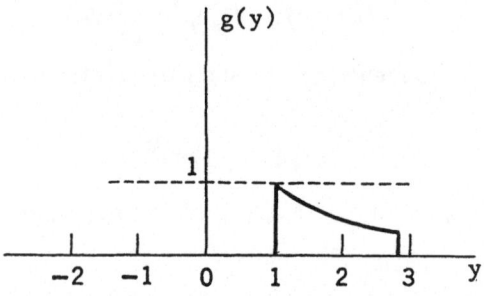

Fig. 2.9 - A plot of the densities $f(x)$ and $g(y)$ of
Example 2.16

Example 2.17

Let $y = x^3$ and let $f(x)$ be uniformly distributed in $(0,1)$.
We find the density $g(y)$ directly from the c.d.f.

$$P(Y \leq b) = P(X^3 \leq b) = P(X \leq b^{1/3})$$

or

$$P(X \leq b^{1/3}) = \begin{cases} 0 & , \ b < 0 \\ b^{1/3} & , \ 0 \leq b \leq 1 \\ 1 & , \ b > 1 \end{cases}$$

or

$$g(b) = \frac{dP(X \le b^{1/3})}{db} = \begin{cases} 0 & , \ b < 0 \\ \dfrac{1}{3} b^{-2/3} & , \ 0 \le b \le 1 \\ 0 & , \ b > 1 \end{cases}$$

We plot $G(y)$ and $g(y)$ in Fig. 2.10. It is left to the interested reader to show that the same result is obtained when Eq. (2.11-7) is used.

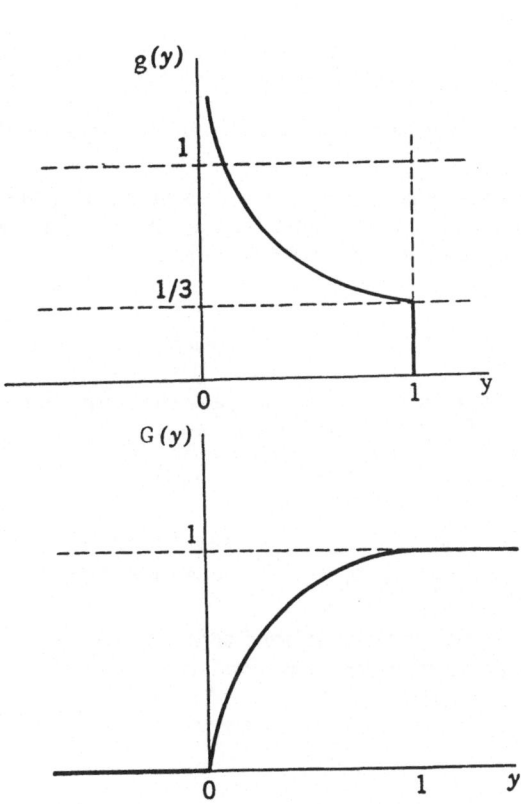

Fig. 2.10 - A cumulative distribution function and the corresponding density function of Example 2.17

We take up next the case where $g = h(x)$ is *monotonically decreasing*. Now we have, for $b = h(a)$, that

$$P(Y \le b) = P(X \ge a)$$

or

$$G(b) = 1 - F(a)$$

or, on differentiating,

$$g(b) = -f[h^{-1}(b)]\frac{d[h^{-1}(b)]}{db}$$

However, the derivative $\frac{d[h^{-1}(b)]}{db}$ is always negative, and we can write

$$g(b) = f[h^{-1}(b)] \left| \frac{d[h^{-1}(b)]}{db} \right| \qquad (2.11\text{-}8)$$

This expression is identical to Eq. (2.11-6) and, hence, we can write in general

$$g(y) = f(x)\left| \frac{dx}{dy} \right| \qquad (2.11\text{-}9)$$

when $y = h(x)$ is a one-to-one continuously differentiable function of x.

For the case where $y = h(x)$ is not monotone, we must consider the problem in more general terms. Now several values of x may correspond to a single value of y, and the inverse h^{-1} is multivalued. We now write, for $b = h(a)$,

$$P(Y \le b) = P(set \ of \ x \ such \ that \ y \le b)$$

and we must take into account the fact that, for each value of x, a set of values of y may exist. Thus

$$G(b) = \int_{\{x/y \le b\}} f(x)dx$$

where the notation $\{x/y \le b\}$ means simply that we integrate over all x for which $y \le b$. We now illustrate this procedure with several examples.

Two common transformations that will be encountered later are the *square-law transformation* and the *absolute-value transformation*. For the former

$$y = x^2 \qquad (2.11\text{-}10)$$

and

$$G(b) = \int_{-\sqrt{b}}^{\sqrt{b}} f(x)dx \qquad (2.11\text{-}11)$$

or, differentiating* with respect to b and replacing b by y, we have

---

*see footnote on page 32.

$$g(y) = \frac{f(\sqrt{y}) + f(-\sqrt{y})}{2\sqrt{y}} \qquad (2.11\text{-}12)$$

For the absolute-value transformation

$$y = |x| \qquad (2.11\text{-}13)$$

and

$$G(b) = \int_{-b}^{b} f(x)dx \qquad, \; b \geq 0 \qquad (2.11\text{-}14)$$

or

$$g(y) = f(y) + f(-y) \qquad, \; y \geq 0 \qquad (2.11\text{-}15)$$

Example 2.18

Let $y = x^2$ and let $f(x)$ be given by

$$f(x) = \frac{1}{\sqrt{2\pi}} e^{-x^2/2} \quad, \quad -\infty < x < \infty$$

This density is called the *normal* or *Gaussian* density and will be encountered extensively in noise problems. It will be discussed in more detail in Section 2.18. We have, from Eq. (2.11-12),

$$g(y) = \frac{1}{\sqrt{2\pi}\,2\sqrt{y}} \left[e^{-y/2} + e^{-y/2} \right], \; 0 < y < \infty$$

or

$$g(y) = \frac{1}{\sqrt{2\pi}} y^{-1/2} e^{-y/2} \; , \; 0 < y < \infty$$

This density is called the *chi-squared distribution* with one degree of freedom [5]. Both it and the normal distribution are plotted in Fig. 2.11.

2.12 *Expectation* - In the random experiments that motivate our study of probability, a given outcome cannot be predicted but its probability of occurrence is presumed to be available. We seek now some way of determining the "average" behavior of these outcomes.

Suppose that we are offered the opportunity of playing a game with the following payoff: The probability of winning one dollar is $p_1 = 0.3$, of winning three dollars is $p_2 = 0.1$, and of losing one dollar is $p_3 = 0.6$. A reasonable question to ask is whether or not it is profitable to play the game. In other words, what is the average winnings or *expectation* per game? Assume that the game is played n times and we

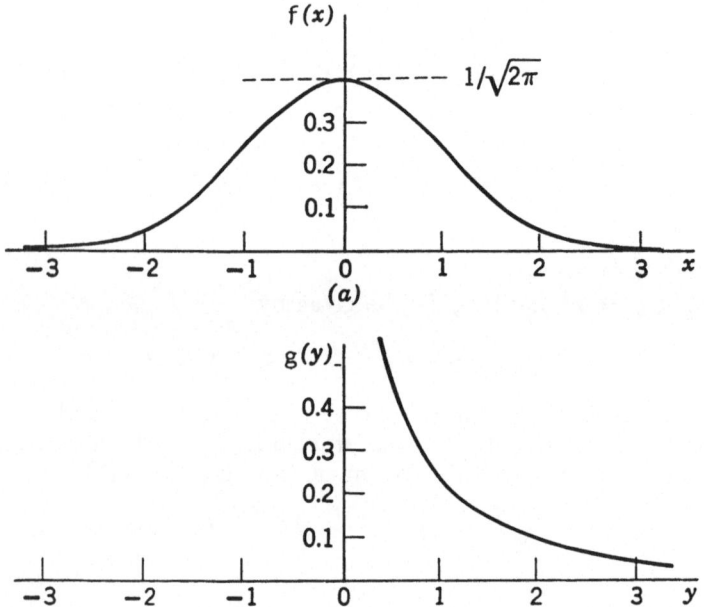

Fig. 2.11 - The normal and the chi-squared distribution

win one dollar	m_1 times
win three dollars	m_2 times
lose one dollar	m_3 times

where $n = m_1 + m_2 + m_3$. In n games, the total winnings will be $m_1 + 3m_2 - m_3$ and the average winnings per game will be

$$E = \frac{m_1}{n} + 3\frac{m_2}{n} - \frac{m_3}{n} \qquad (2.12\text{-}1)$$

or, approximately,

$$E \approx p_1 + 3p_2 - p_3 \qquad (2.12\text{-}2)$$

If the random variable X_i was defined, for this game, as the dollars won on each game, then

$$E \approx \sum_i x_i\, p\,(x_i) \qquad (2.12\text{-}3)$$

would be a formula defining this expectation.

We generalize from this simple example. If X_i is a discrete random variable, we define the *expectation of the function* $h(X_i)$ by

$$E[h(X_i)] = \sum_i h(x_i) f_X(x_i) \tag{2.12-4}$$

where $f_X(x_i)$ is the density function associated with the random variable and the x_i are the values of the discrete random variable for which $f_X(x_i)$ is defined. In the same way, if X is a continuous random variable, we define the *expectation of the function* $h(X)$ by

$$E[h(X)] = \int_{-\infty}^{\infty} h(x) f_X(x) dx \tag{2.12-5}$$

where $f_X(x)$ is the density function associated with the random variable X.

As previously mentioned and as discussed in Appendix A for the case where the random variable can take on both discrete and continuous characteristics, we make use of the Stieltjes integral to write the expectation of $h(X)$ as

$$E[h(X)] = \int_{-\infty}^{\infty} h(x) dF_X(x) \tag{2.12-6}$$

where $F_X(x)$ is the c.d.f. corresponding to the density $f(x)$. Of course this last equation includes Eqs. (2.12-4) and (2.12-5) as special cases and could be used as the general definition of the expectation of $h(X)$. However, it will often be convenient to use the other forms when there is no confusion as to the nature of the random variable.

The expectation has a number of rather obvious properties:

(1) $E[h_1(X) + h_2(X)] = E[h_1(X)] + E[h_2(X)]$ $\qquad$ (2.12-7)

(2) $E[ch(X)] = cE[h(X)]$ $\qquad$ (2.12-8)

These two are *linearity* properties and are a consequence of the fact that the expectation is a *linear operator*. In addition; for c a constant,

(3) $E[c] = c;$ $\quad$ $E[1] = 1$ $\qquad$ (2.12-9)

Also

(4) $E[h(X)] \geq 0$ if $h(X) \geq 0$ $\qquad$ (2.12-10)

These properties are obtained in an elementary way from the definition of expectation.

2.13 *Moments* - One of the principal problems which will arise later is that of characterizing an arbitrary distribution or density function. The *moments* of a distribution will turn out to be one of the most convenient ways for such characterization.

For the discrete case, we define the *r-th moment about the origin* for a given density function $f(x_i)$ as

$$u_r' = E(X_i^r) = \sum_i x_i^r f(x_i) \qquad (2.13\text{-}1)$$

and, for the continuous case, with density function $f(x)$,

$$u_r' = E(X^r) = \int_{-\infty}^{\infty} x^r f(x)dx \qquad (2.13\text{-}2)$$

In particular, the first moment about the origin is called the *mean*. It is often written as u and, for the continuous case, we have

$$u = u_1' = E(X) = \int_{-\infty}^{\infty} x \, f(x)dx \qquad (2.13\text{-}3)$$

This is just the centroid of the area under the curve $f(x)$.

The *r-th moment about the mean* is defined for the discrete case and for the density function $f(x_i)$ as

$$u_r = E\{(X_i - u)^r\} = \sum_i (x_i - u)^r f(x_i) \qquad (2.13\text{-}4)$$

For the continuous case, , with density function $f(x)$, we have

$$u_r = E\{(X - u)^r\} = \int_{-\infty}^{\infty} (x - u)^r f(x)dx \qquad (2.13\text{-}5)$$

In particular, the second moment about the mean is called the *variance*, is usually written as σ^2, and is given for the continuous case by

$$\sigma^2 = u_2 = E\{(X - u)^2\} = \int_{-\infty}^{\infty} (x - u)^2 f(x)dx \qquad (2.13\text{-}6)$$

Moments have a number of interesting properties. Some of the most obvious and most useful are listed below:

1. *The first moment about the mean is zero.*

$$u_1 = E(X-u) = E(X) - u = u - u = 0$$

$$u_1 = 0 \qquad (2.13\text{-}7)$$

2. *The second moment about the origin is equal to the variance plus the square of the mean.*

$$\sigma^2 = E\{(X - u)^2\} = E(X^2) - 2uE(X) + u^2$$

$$\sigma^2 = u_2' - u^2$$

$$u_2' = \sigma^2 + u^2 \qquad (2.13\text{-}8)$$

3. *The least second moment is the variance.* We find the value of b for which $E\{(X - b)^2\}$ is a minimum. Expanding, we have

$$E\{(X-b)^2\} = E(X^2) - 2bu + b^2$$

We take the partial derivative with respect to b and equate the result to zero:

$$\frac{\partial}{\partial b}[E(X^2) - 2bu + b^2] = -2u + 2b = 0$$

$$b = u$$

Check for minimum: the second derivative is

$$\frac{\partial}{\partial b}[-2u + 2b] = 2 \quad \text{positive!}$$

4.   *There is a general relationship between moments about the mean (central moments) and moments about the origin.* From the binomial expansion

$$(x - u)^r = \sum_{s=0}^{r} (-1)^s \binom{r}{s} x^{r-s} u^s$$

we write

$$E\{(X - u)^r\} = \sum_{s=0}^{r} (-1)^s \binom{r}{s} u^s E\{X^{r-s}\}$$

or

$$u_r = \sum_{s=0}^{r} (-1)^s \binom{r}{s} u^s u_{r-s}' \qquad (2.13\text{-}9)$$

For the case $r=2$, we have

$$u_2 = \sigma^2 = u_2' - 2u^2 + u^2$$

or

$$u_2' = \sigma^2 + u^2$$

as was obtained in 2.

It is apparent that the mean u is a measure of the central tendency or center of gravity of a density. On the other hand, the variance σ^2 is a measure of the dispersion of the density about the mean. A density with a small variance tends to be concentrated about the mean. The positive square root σ of the variance is often used instead of the variance and is called the *standard deviation*.

One of the most commonly encountered transformations of random variables is the *linear transformation*

$$Y = a + bX \qquad (2.13\text{-}10)$$

where a and b are constants and X and Y are random variables. In this case the means and variances of X and Y are simply related. Let u_y be the mean of Y and u_x the mean of X. Then

$$u_y = E(Y) = a + b\ u_x$$

$$u_y = a + b\ u_x \qquad (2.13\text{-}11)$$

In the same way, the variance of Y is

$$\sigma_y^2 = E\{(Y - u_y)^2\} = E\{(a + bX - a - bu_x)^2\}$$

$$= b^2 E\{(X - u_x)^2\} = b^2 \sigma_x^2 \qquad (2.13\text{-}12)$$

2.14 *The Chebychev Inequality* - As mentioned in the previous section, the variance σ^2, or the standard deviation σ, is closely related to the idea of the dispersion of the distribution about the mean. This relationship is shown clearly by the following very useful and very general inequality:

The Chebychev Inequality: For any random variable X with density $f_X(x)$ with finite mean u and variance σ^2, we have

$$P\{|X - u| \geq \lambda\sigma\} \leq \frac{1}{\lambda^2} \qquad (2.14\text{-}1)$$

for any real $\lambda > 0$.

Proof: Define a random variable Y by

$$Y = \begin{cases} 0 & \text{for } |X - u| < \lambda\sigma \\ \lambda^2\sigma^2 & \text{for } |X - u| \geq \lambda\sigma \end{cases} \qquad (2.14\text{-}2)$$

The expected value of Y is given by

$$E(Y) = 0[P(Y=0)] + \lambda^2\sigma^2[P(Y=\lambda^2\sigma^2)]$$

or

$$E(Y) = \lambda^2\sigma^2 P\{|X - u| \geq \lambda\sigma\} \qquad (2.14\text{-}3)$$

It is apparent from Eq. (2.14-2) that

$$Y \leq (X - u)^2$$

or that

$$E(Y) \leq E\{(X - u)^2\} = \sigma^2$$

Substituting from Eq. (2.14-3) for $E(Y)$, we obtain

$$\lambda^2\sigma^2 P\{|X - u| \geq \lambda\sigma\} \leq \sigma^2$$

which is the inequality desired.

2.15 *Generating Functions* - We consider now two functions which are closely related to the moments of a distribution. These will be useful later in the representation and manipulation of moment relationships. They are the *moment generating function* $M(t)$ and the *characteristic function* $M(jv)$.

The moment generating function (m.g.f.)$M(t)$ is defined for a random variable X with density function $f(x)$ as

$$M(t) = E(e^{tX})$$ (2.15-1)

where t is a real variable defined on $(-\infty, \infty)$. Since e^{tx} has the power series expansion

$$e^{tx} = 1 + tx + \frac{t^2 x^2}{2!} + \cdots + \frac{t^n x^n}{n!} + \cdots$$ (2.15-2)

we may write Eq. (2.15-1) as

$$M(t) = 1 + tE(X) + \frac{t^2}{2!} E(X^2) + \cdots + \frac{t^n}{n!} E(X^n) + \cdots$$ (2.15-3)

or as

$$M(t) = 1 + tu_1' + \frac{t^2}{2} u_2' + \cdots + \frac{t^n}{n!} u_n' + \cdots$$ (2.15-4)

It is apparent that the moments u_1' are easily found as follows:

$$M(0) = 1$$ (2.15-5)

$$\frac{dM(t)}{dt} \Big|_{t=0} = u_1'$$ (2.15-6)

$$\cdot \quad \cdot$$
$$\cdot \quad \cdot$$
$$\cdot \quad \cdot$$

$$\frac{d^n M(t)}{dt^n} \Big|_{t=0} = u_n'$$ (2.15-7)

It can be shown that the m.g.f. uniquely specifies the distribution $f(x)$; that is, no two distinct distributions correspond to the same m.g.f.

The *characteristic function* $M(jv)$ is defined for a random variable X with density function $f(x)$ as

$$M(jv) = E(e^{jvX})$$ (2.15-8)

where $j = \sqrt{-1}$ and v is a real variable defined on $(-\infty, \infty)$. For the continuous random variable X, this becomes

$$M(jv) = \int_{-\infty}^{\infty} f(x) e^{jvx} \, dx$$ (2.15-9)

and for the discrete random variable x_i,

$$M(jv) = \sum_i f(x_i) e^{jvx_i}$$ (2.15-10)

Equation (2.15-9) is the Fourier transform of $f(x)$ [actually, the inverse Fourier transform]. Consequently, from the theory of Fourier integrals [6], this equation may be inverted to give

$$f(x) = \frac{1}{2\pi} \int_{-\infty}^{\infty} M(jv)e^{-jvx}\, dv \qquad (2.15\text{-}11)$$

The function $M(jv)$ is uniquely determined by $f(x)$ through Eq. (2.15-9) and $f(x)$ is uniquely determined by $M(jv)$ through Eq. (2.15-11).

An additional advantage of the characteristic function is that it always exists as follows from the fact that $|e^{jvx}| = 1$. Thus

$$|M(jv)| \leq 1 = M(j0)$$

Equation (2.15-8) may be expanded as in the case of the m.g.f. to yield

$$M(jv) = 1 + jvu_1' + \frac{(jv)^2}{2!} u_2' + \cdots + \frac{(jv)^n}{n!} u_n' + \cdots (2.15\text{-}12)$$

or

$$\frac{d^n M(jv)}{dv^n} \Big|_{v=0} = (j)^n u_n' \qquad (2.15\text{-}13)$$

A number of specific examples of both the characteristic function and the moment generating function will be given in the next sections where certain special distributions that are commonly encountered are considered in detail.

2.16 *The Binomial Distribution* - This distribution $b(k;n,p)$ has been given previously as Example 2.15 of Section 2.10 and a special case is plotted in Fig. 2.8. The importance of the binomial distribution in our studies lies in the fact that both the Poisson distribution and the normal (or Gaussian) distribution can be considered to be limiting forms of the binomial distribution. We proceed now to derive the form of this distribution.

In many of the simplest situations where probability concepts are to be applied, it happens that there is a sequence of repeated independent trials with only two possible outcomes on each trial and with the probabilities of these two outcomes fixed throughout the trials. Such sequences are called *Bernoulli trials*. Perhaps the commonest example is a sequence of coin-tossings, where a typical sequence might be

HTTHTHH.....

It is conventional to call one of the two outcomes S (success) with probability of occurrence given by p and the other outcome F (failure) with probability $q = 1-p$. Since the individual trials are independent by definition, the probabilities multiply and the probability of any given sequence is the product obtained when S is replaced by p and F by $q = 1-p$. Thus, for the previous example, if H is considered a success, the probability P of the sequence is

$$P \;=\; p\,(1-p\,)(1-p\,)p\,(1-p\,)pp....$$

or, for a sequence of length n,

$$P \;=\; p^{\,k}(1-p\,)^{n-k} \tag{2.16-1}$$

where k is the number of successes and $(n-k)$ is the number of failures.

Example 2.19

Consider the successive tosses of a true die. Let throwing a two be considered success and all other outcomes failure. Then the probability P of the Bernoulli sequence $F\,2FFF\,2$, where F stands for any number 1,3,4,5,6, is

$$P \;=\; (\frac{1}{6})^2(\frac{5}{6})^4$$

Suppose that we are interested in the probability of a given number of successes (say k successes) in n Bernoulli trials. One way to obtain k successes would be to have the sequence

$$\underbrace{SSS........S}_{k} \qquad \underbrace{FFF........F}_{n-k}$$

where the first k trials are successes and the last $(n-k)$ are failures. The probability P of this sequence has already been given by Eq. (2.16-2) and is

$$P \;=\; p^{\,k}(1-p\,)^{n-k} \tag{2.16-2}$$

It is apparent that any other sequence containing exactly k successes will be acceptable. In fact there are available n cells or slots into which k successes can be put (The remaining cells each have a failure placed in them in each case). The number of different ways in which this can be done is just the number of combinations of n things taken k at a time or

$$\begin{pmatrix} n \\ k \end{pmatrix} \;=\; \frac{n\,!}{(n-k\,)!k\,!} \tag{2.16-3}$$

Each of these $\begin{pmatrix} n \\ k \end{pmatrix}$ sequences has a probability given by Eq. (2.16-2). Consequently the probability of exactly k successes in n Bernoulli trials is

$$b\,(k\,;n\,,p\,) \;=\; \begin{pmatrix} n \\ k \end{pmatrix} p^{\,k}(1-p\,)^{n-k} \tag{2.16-4}$$

where p is the probability of success in a single trial.

The quantity $b(k;n,p)$ is called the *binomial distribution*. It is the k-*th* term in the binomial expansion of $(p+q)^n$

$$(p+q)^n = \sum_{k=0}^{n} \binom{n}{k} p^k q^{n-k} = \sum_{k=0}^{n} b(k;n,p) \qquad (2.16\text{-}5)$$

It is apparent that $q=1-p$ and, consequently, that

$$[p + (1-p)]^n = \sum_{k=0}^{n} b(k;n,p) = 1 \qquad (2.16\text{-}6)$$

Since, in addition, $b(k;n,p)$ is non-negative, it is a density function.

The mean u of the binomial distribution may be found as follows:

$$u = E(K) = \sum_{k=0}^{n} k\, b(k;n,p) \qquad (2.16\text{-}7)$$

or

$$u = \sum_{k=0}^{n} k\, \frac{n(n-1)\,\cdots\,(n-k+1)}{k(k-1)\,\cdots\,(3)(2)(1)}\, p^k (1-p)^{n-k} \qquad (2.16\text{-}8)$$

Since the term for which $k=0$ is zero, we may write

$$u = np \sum_{k=1}^{n} \binom{n-1}{k-1} p^{k-1}(1-p)^{(n-1)-(k-1)} \qquad (2.16\text{-}9)$$

If we change the index of summation by placing $i=k-1$, then Eq. (2.16-9) becomes

$$u = np \sum_{i=0}^{n-1} \binom{n-1}{i} p^i (1-p)^{(n-1)-i} \qquad (2.16\text{-}10)$$

or

$$u = np(p + 1 - p)^{n-1} = np \qquad (2.16\text{-}11)$$

so that we conclude that the mean u is given by np. This result is intuitively reasonable. In a large number of trials n with probability p of success on each trial, we would expect approximately np successes.

If other moments of the binomial distribution are desired, it will be profitable to calculate the m.g.f. and obtain the moments by differentiation. The m.g.f. $M(t)$ is given by

$$M(t) = E(e^{tK}) = \sum_{k=0}^{n} e^{tk} \binom{n}{k} p^k (1-p)^{n-k} \qquad (2.16\text{-}12)$$

or

$$M(t) = \sum_{k=0}^{n} \binom{n}{k} (pe^t)^k (1-p)^{n-k} \qquad (2.16\text{-}13)$$

$$M(t) = (pe^t + 1-p)^n \qquad (2.16\text{-}14)$$

We note that $M(0)=1$ as required by Eq. (2.15-4). Also the mean u can be found from

$$u = \frac{dM(t)}{dt}\Big|_{t=0} = n(p + 1 - p)^{n-1}p = np$$

as was found previously by direct calculation. The second moment about the origin is

$$u_2' = \frac{d^2M(t)}{dt^2}\Big|_{t=0}$$

We have

$$\frac{d^2M(t)}{dt^2} = n(n-1)(pe^t + 1 - p)^{n-2}p^2 e^{2t} + n(pe^t + 1 - p)^{n-1}pe^t$$

or

$$u_2' = n(n-1)p^2 + np = \sigma^2 + u^2 \qquad (2.16\text{-}15)$$

Consequently the variance σ^2 is given by

$$\sigma^2 = u_2' - n^2p^2 = np(1-p) \qquad (2.16\text{-}16)$$

For large n, it will often be convenient to work with the normalized random variable

$$Y = K/n \qquad (2.16\text{-}17)$$

which cannot exceed unity. For example, if k is the number of heads observed in n tosses of a coin, then y is the proportion of heads. The mean of Y is

$$E(Y) = E(K/n) = np/n = p \qquad (2.16\text{-}18)$$

and the variance is

$$\sigma_y^2 = np(1-p)/n^2 = p(1-p)/n \qquad (2.16\text{-}19)$$

This last expression indicates that, as n becomes large, the distribution of Y clusters more and more about the mean p. A question of some interest is: How close to the mean p is Y for large n? Since Y is a random variable, the question can be answered only in a probabilistic sense. We have, from the Chebychev inequality,

$$P\{|Y - p| \geq \epsilon\} \leq \sigma_y^2/\epsilon^2 \qquad (2.16\text{-}20)$$

for any arbitrary $\epsilon > 0$. Since σ_y^2 is given by Eq. (2.16-19), we write, for $\epsilon > 0$,

$$\lim_{n\to\infty} P\{|Y - p| \geq \epsilon\} \leq \lim_{n\to\infty} \frac{p(1-p)}{n\epsilon^2} = 0 \qquad (2.16\text{-}21)$$

Thus, in a probabilistic sense, the random variable Y can be made to approach the mean p arbitrarily closely by making the number of trials n large enough. This relationship is called *Bernoulli's Theorem* and is a special case of the *Law of Large Numbers* [6]. The type of convergence given by Eq. (2.16-21) is called *convergence in probability;* that is, we say that Y converges to p in probability. This subject will be discussed further in Section 3.5.

2.17 *The Poisson Distribution* - As previously mentioned, this distribution can be obtained as a limiting approximation to the binomial distribution. Suppose the situation exists where n is very large and p is very small but the product $np = \lambda$ is of reasonable size. More precisely, we want n to approach infinity and p to approach zero in such a way that the product λ exists and is non-zero. For this case we seek an approximation to the binomial distribution in a form more amenable to calculation and manipulation.

Consider the term $b(0;n,p)$:

$$b(0;n,p) = \begin{pmatrix} n \\ 0 \end{pmatrix} p^0(1-p)^n = (1-p)^n \qquad (2.17\text{-}1)$$

Replace p by λ/n so that

$$\lim_{n \to \infty} b(0;n,p) = \lim_{n \to \infty} (1 - \frac{\lambda}{n})^n = e^{-\lambda} \qquad (2.17\text{-}2)$$

since

$$\lim_{x \to \infty} (1 + \frac{1}{x})^x = e \qquad (2.17\text{-}3)$$

Thus, for large n, the term $b(0;n,p)$ is given approximately by $e^{-\lambda}$ where $\lambda = np$. In the same way

$$b(1;n,p) = \begin{pmatrix} n \\ 1 \end{pmatrix} p(1-p)^{n-1} = \frac{np}{1-p} b(0;n,p) \qquad (2.17\text{-}4)$$

and

$$\lim_{n \to \infty} b(1;n,p) = \lim_{n \to \infty} \frac{\lambda}{1-\lambda/n} e^{-\lambda} = \lambda e^{-\lambda} \qquad (2.17\text{-}5)$$

We continue and find that

$$\lim_{n \to \infty} b(2;n,p) = \frac{\lambda^2}{2} e^{-\lambda} \qquad (2.17\text{-}6)$$

Finally, if we proceed far enough, we see that

$$\lim_{n \to \infty} b(k;n,p) \overset{\Delta}{=} p(k;\lambda) = \frac{\lambda^k}{k!} e^{-\lambda} \quad , \quad k = 0,1,2,... \qquad (2.17\text{-}7)$$

if p approaches zero in such a way that $np = \lambda$. For large n, Eq. (2.17-7) is called the *Poisson approximation to the binomial distribution* and, in the limit, is the *Poisson distribution*.

We note that

$$\sum_{k=0}^{\infty} p(k;\lambda) = e^{-\lambda} \sum_{k=0}^{\infty} \frac{\lambda^k}{k!} \qquad (2.17\text{-}8)$$

or, from the power series expansion of e^{λ}

$$\sum_{k=0}^{\infty} p(k;\lambda) = e^{-\lambda} e^{\lambda} = 1 \qquad (2.17\text{-}9)$$

Thus, since $p(k;\lambda)$ is also non-negative, it is a density function if defined for $k = 0,1,2,...$

We would suspect that the mean of the Poisson distribution is given by $\lambda = np$. By direct calculation,

$$E(K) = \sum_{k=0}^{\infty} k \frac{\lambda^k}{k!} e^{-\lambda} = \lambda e^{-\lambda} \sum_{k=1}^{\infty} \frac{\lambda^{k-1}}{(k-1)!} \qquad (2.17\text{-}10)$$

Again, we let $i = k - 1$ and find that

$$E(K) = \lambda e^{-\lambda} \sum_{i=0}^{\infty} \frac{\lambda^i}{i!} = \lambda e^{-\lambda} e^{\lambda} = \lambda \qquad (2.17\text{-}11)$$

as expected. The second moment may be found in a similar fashion:

$$E[K(K-1)] = \sum_{k=0}^{\infty} k(k-1) \frac{\lambda^k}{k!} e^{-\lambda} = \lambda^2 e^{-\lambda} \sum_{k=2}^{\infty} \frac{\lambda^{k-2}}{(k-2)!} \qquad (2.17\text{-}12)$$

or

$$E(K^2) - E(K) = \lambda^2 e^{-\lambda} e^{\lambda} = \lambda^2 \qquad (2.17\text{-}13)$$

Equation (2.17-13) may be rewritten as

$$u_2{}' - u = \lambda^2 = u_2{}' - \lambda$$

or, since $u_2{}' = \sigma^2 + u^2$

$$\sigma^2 = \lambda^2 - \lambda^2 + \lambda = \lambda \qquad (2.17\text{-}14)$$

The variance and the mean are both equal to λ.

These moments could have been found from the m.g.f. which is given by

$$M(t) = E(e^{tK}) = \sum_{k=0}^{\infty} e^{tk} e^{-\lambda} \frac{\lambda^k}{k!} \qquad (2.17\text{-}15)$$

and

$$M(t) = e^{-\lambda} \sum_{k=0}^{\infty} \frac{(\lambda e^t)^k}{k!} = e^{-\lambda} e^{\lambda e^t}$$

or, finally,

$$M(t) = e^{\lambda(e^t - 1)} \qquad (2.17\text{-}16)$$

is the moment generating function for the Poisson distribution.
We have

$$M(0) = 1 \qquad\qquad (2.17\text{-}17)$$

$$\frac{dM(t)}{dt}\Big|_{t=0} = e^{\lambda(e^t - 1)}\lambda e^t\Big|_{t=0} = \lambda \qquad\qquad (2.17\text{-}18)$$

$$\frac{d^2M(t)}{dt^2} = e^{\lambda(e^t - 1)}\lambda^2 e^{2t} + e^{\lambda(e^t - 1)}\lambda e^t$$

$$u_2' = \lambda^2 + \lambda \qquad\qquad (2.17\text{-}19)$$

as before.

2.18 *The Normal or Gaussian Distribution* - We can show [6]
that if X is binomially distributed, that is, if

$$f_X(x) = \left(\begin{array}{c} n \\ x \end{array}\right) p^x (1 - p)^{n-x} \qquad x = 0,1,2,\dots \qquad (2.18\text{-}1)$$

then, as n becomes large, the density of X approaches

$$f_X(x) = \frac{1}{\sqrt{np(1-p)}\sqrt{2\pi}}\, e^{-\frac{(x-np)^2}{2np(1-p)}} \qquad\qquad (2.18\text{-}2)$$

in the sense that the ratio of Eq. (2.18-1) and (2.18-2) approaches
unity. In the limit as $n \to \infty$ the random variable X can be considered to be continuous $(-\infty,\infty)$ and*

$$\lim_{n \to \infty} f_X(x) = \frac{1}{\sqrt{2\pi\sigma^2}}\, e^{-\frac{1}{2}\left(\frac{x-u}{\sigma}\right)^2} \quad,\quad -\infty < x < \infty \quad (2.18\text{-}3)$$

where

$$u = np \qquad \sigma^2 = np(1-p)$$

As discussed by Feller [1], the approximation is amazingly good for
small n $(n \geq 5)$, provided p is not too far from $1/2$.

Equation (2.18-3) is called the *normal density or distribution*
(also called the *Gaussian distribution)* with mean u and variance
σ^2. For the moment, we will take $u = 0$ and $\sigma^2 = 1$ and consider
the *unit normal distribution*

$$f_X(x) = \frac{1}{\sqrt{2\pi}}\, e^{-x^2/2} \quad,\quad -\infty < x < \infty \qquad (2.18\text{-}4)$$

This density has already been plotted in Fig. 2.11. It is non-
negative and we can show that the area under the curve is equal
to unity; that is, that

---

*As we have defined them, both u and σ^2 are infinite in the limit. It is neces-
sary, of course, for Eq.(2.18-3) to be renormalized so that u and σ^2 are finite.
See [1] and [9] for detailed discussion.

$$\frac{1}{\sqrt{2\pi}} \int_{-\infty}^{\infty} e^{-x^2/2} \, dx = 1 \tag{2.18-5}$$

Let us denote the integral of Eq. (2.18-5) by J and write

$$J^2 = \frac{1}{2\pi} \int_{-\infty}^{\infty} e^{-x^2/2} dx \int_{-\infty}^{\infty} e^{-y^2/2} dy$$

$$= \frac{1}{2\pi} \int_{-\infty}^{\infty} \int_{-\infty}^{\infty} e^{-(x^2+y^2)/2} dx dy$$

We may change to polar coordinates by the substitution

$$r^2 = x^2 + y^2$$

and

$$r \, dr \, d\theta = dx \, dy$$

to obtain

$$J^2 = \frac{1}{2\pi} \int_{0}^{2\pi} \int_{0}^{\infty} r \, e^{-r^2/2} dr \, d\theta \tag{2.18-6}$$

This expression may be integrated directly to yield

$$J^2 = \frac{1}{2\pi} \int_{0}^{2\pi} \left\{ -e^{-r^2/2} \right\}_{0}^{\infty} d\theta \tag{2.18-7}$$

$$J^2 = \frac{1}{2\pi} \int_{0}^{2\pi} d\theta = 1$$

Since J must be positive, we conclude that $J=1$ and that Eq. (2.18-5) is valid. It is apparent that the same result is obtained for the general form of Eq. (2.18-3), since, after the substitution

$$y = \frac{x - u}{\sigma} \tag{2.18-8}$$

Eq. (2.18-5) is obtained as the integral of the general normal distribution.

The cumulative distribution function for the unit normal distribution is given by

$$F_X(x) = \frac{1}{\sqrt{2\pi}} \int_{-\infty}^{x} e^{-y^2/2} dy \tag{2.18-9}$$

Since the integrand is even,

$$F_X(-x) = 1 - F_X(x) \tag{2.18-10}$$

For the case $x = -x = 0$, we have

$$F_X(0) = 1/2 \tag{2.18-11}$$

Since the normal c.d.f. cannot be evaluated directly, it is necessary to refer to tables which have been obtained by some numerical technique. One often finds tables, not of the c.d.f., but of the *error integral* or *error function* $\theta(t)$ instead, where

$$\theta(t) = erf \ t = \frac{2}{\sqrt{\pi}} \int_0^t e^{-y^2} dy \qquad (2.18\text{-}12)$$

Let us make the substitution

$$y = x/\sqrt{2} \qquad (2.18\text{-}13)$$

Then Eq. (2.18-12) becomes

$$\theta(t) = \frac{2}{\sqrt{2\pi}} \int_0^{\sqrt{2}t} e^{-x^2/2} dx \qquad (2.18\text{-}14)$$

It is apparent from Eq. (2.18-11) that

$$\theta(t) = 2F_X(\sqrt{2}t) - 1 \qquad (2.18\text{-}15)$$

relates the c.d.f. for the unit normal distribution to the error function. As a matter of convenience, Table 2.1 gives the values of the density $f_X(x)$ and the c.d.f. $F_X(x)$ for a range of values of x.

Table 2.1 - The Unit Normal Distribution.

x	$f_X(x)$	$F_X(x)$
0.0	0.3989	0.5000
0.2	0.3910	0.5793
0.4	0.3683	0.6554
0.6	0.3332	0.7257
0.8	0.2897	0.7881
1.0	0.2420	0.8413
1.2	0.1942	0.8849
1.4	0.1497	0.9192
1.6	0.1109	0.9452
1.8	0.0790	0.9641
2.0	0.0540	0.9773
2.2	0.0355	0.9861
2.4	0.0224	0.9918
2.6	0.0136	0.9953
2.8	0.0080	0.9974
3.0	0.0044	0.9987
3.5	0.0009	0.9998
4.0	0.0001	0.9999+
4.5	0.00002	0.9999+

A more complete listing is given in Table 1 in the appendix. For the normal distribution with non-zero mean u and variance σ^2, the linear change of variable of Eq. (2.18-8) may be employed and the data of Table 2.1 used.

Example 2.20

Find the c.d.f. $F_1(1.2)$ for the normal distribution with mean of 0.6 and variance of 9.0.

We want to find

$$F_1(1.2) = \frac{1}{\sqrt{2\pi 9}} \int_{-\infty}^{1.2} e^{-\frac{1}{2}(\frac{x-0.6}{3})^2} dx$$

We let

$$y = \frac{x-0.6}{3} \quad , \quad dx = 3\,dy$$

Then

$$F_1(1.2) = \frac{1}{\sqrt{2\pi}} \int_{-\infty}^{\frac{1.2-0.6}{3}} e^{-y^2/2} dy$$

$$= F_X\left(\frac{1.2-0.6}{3}\right) = F_X(0.2) = 0.5793$$

from Table 2.1. Thus

$$F_1(1.2) = 0.5793$$

The moments for the unit normal distribution are found most conveniently from the m.g.f. $M_X(t)$ where

$$M_X(t) = E\{e^{tX}\} = \frac{1}{\sqrt{2\pi}} \int_{-\infty}^{\infty} e^{tx} e^{-x^2/2} dx \qquad (2.18\text{-}16)$$

This integral is typical of many integrals connected with the normal distribution. The general procedure to follow is to complete the square in the variable of integration and then to make a linear change in variable. We have

$$M_X(t) = e^{t^2/2} \frac{1}{\sqrt{2\pi}} \int_{-\infty}^{\infty} e^{-(x-t)^2/2} dx \qquad (2.18\text{-}17)$$

We make the substitution

$$y = x - t$$

and obtain

$$M_X(t) = e^{t^2/2} \frac{1}{\sqrt{2\pi}} \int\limits_{-\infty}^{\infty} e^{-y^2/2} dy \qquad (2.18\text{-}18)$$

or

$$M_X(t) = e^{t^2/2} \qquad (2.18\text{-}19)$$

for the result. In the same way, we find the characteristic function $M_X(jv)$ to be

$$M_X(jv) = E\{e^{jvx}\} = \frac{1}{\sqrt{2\pi}} \int\limits_{-\infty}^{\infty} e^{jvx} e^{-x^2/2} dx \qquad (2.18\text{-}20)$$

or

$$M_X(jv) = e^{-v^2/2} \qquad (2.18\text{-}21)$$

From Eq. (2.18-19) we write

$$M_X(t) = 1 + \frac{t^2}{2} + \frac{1}{2!}(\frac{t^2}{2})^2 + \ldots + \frac{1}{n!}(\frac{t^2}{2})^n + \cdots \qquad (2.18\text{-}22)$$

The n-th moment about the origin $u_n{}'$ is the coefficient of $t^n/n!$ in this power series expansion of $M_X(t)$. There are no odd powers of t in Eq. (2.18-22) and for n even, the $n/2$–th term is

$$\frac{1}{(n/2)!} (\frac{t^2}{2})^{n/2} = \frac{n(n-1)\ldots(\frac{n}{2}+1)}{8^{n/2}} \frac{t^n}{n!} \qquad (2.18\text{-}23)$$

Consequently, the moments are

for n odd

$$u_n{}' = 0$$

for n even

$$u_n{}' = u_n = \frac{n(n-1)..(\frac{n}{2}+1)}{2^{n/2}} \qquad (2.18\text{-}24)$$

In particular,

$$u_1{}' = u = 0 \qquad (2.18\text{-}25)$$
$$u_2 = \sigma^2 = 1 \qquad (2.18\text{-}26)$$

Let us consider again the linear transformation of Eq. (2.18-8) where Y is a random variable with unit normal distribution. Then X is given by

$$X = u + \sigma Y \qquad (2.18\text{-}27)$$

and the density of X is easily obtained as

$$f_X(x) = \frac{1}{\sqrt{2\pi\sigma^2}} e^{-\frac{1}{2}(\frac{x-u}{\sigma})^2}, \quad -\infty < x < \infty \qquad (2.18\text{-}28)$$

We have immediately from Eq. (2.18-27) that

$$E(X) = u_z = u + \sigma E(Y) = u \qquad (2.18\text{-}29)$$

and

$$E\{(X - u_z)^2\} = \sigma^2 \sigma_y^2 = \sigma^2 \qquad (2.18\text{-}30)$$

as expected for the normal distribution of Eq. (2.18-28) with mean u and variance σ^2. In the same way, the characteristic function $M_X(jv)$ is

$$M_X(jv) = e^{juv - \sigma^2 v^2/2} \qquad (2.18\text{-}31)$$

2.19 *Limit Theorems* - In this section we will consider very briefly three limit theorems of fundamental importance in the study of sequences of random variables and of sums of random variables. These three theorems are

(1) Bernoulli's Theorem

(2) The Law of Large Numbers

(3) The Central Limit Theorem

Bernoulli's Theorem

This theorem has already been discussed at the end of Section 2.16. The theorem states that:

"The relative frequency Y_n with which an event occurs in n Bernoulli trials converges in probability to the probability p of occurrence in a single trial; that is,

$$\lim_{n \to \infty} P\{\,|\,Y_n - p\,|\, \geq \epsilon\} = 0 \qquad (2.19\text{-}1)$$

for any $\epsilon > 0$; $Y_n = K/n$ where K is the number of successes in n trials."

This theorem was proved in Section 2.16 by an application of Chebychev's Inequality. It is a special case of the Law of Large Numbers.

The Law of Large Numbers

This theorem states that:

"The average Y_n of n identically distributed independent random variables X_i with the same mean u and variance σ^2 converges in probability to the mean u."

Let $X_1, X_2, \ldots, X_n$ be independent random variables with the same distribution and with mean u and variance σ^2 (They could be n independent observations of the same random variable). Define the average Y_n as

$$Y_n = (X_1 + X_2 + \cdots + X_n)/n \qquad (2.19\text{-}2)$$

If Chebychev's Inequality is applied to the random variable Y_n , then

$$P\{\,|\,Y_n - u_y\,|\, \geq \lambda \sigma_y\} \leq \frac{1}{\lambda^2} \quad , \quad \lambda > 0 \qquad (2.19\text{-}3)$$

where u_y is the mean and σ_y^2 is the variance of Y_n. If we let $\epsilon = \lambda \sigma_y$, then Eq. (2.19-3) becomes

$$P\{\,|\,Y_n - u_y\,|\, \geq \epsilon\} \leq \sigma_y^2/\epsilon^2 \qquad (2.19\text{-}4)$$

We now find the mean u_y and the variance σ_y^2 for substitution in Eq. (2.19-4). The mean of Y_n is

$$u_y = \frac{1}{n} \sum_{i=1}^{n} E(X_i) = \frac{n\,u}{n} = u \qquad (2.19\text{-}5)$$

and the variance is

$$\sigma_y^2 = E\{(Y_n - u_y)^2\} = \frac{1}{n^2} E\left\{\left[\sum_i (X_i - u)\right]^2\right\} \qquad (2.19\text{-}6)$$

This last expression is more conveniently written as

$$\sigma_y^2 = \frac{1}{n^2} E\left\{\sum_{i=1}^{n}(X_i-u) \sum_{j=1}^{n}(X_j-u)\right\} \qquad (2.19\text{-}7)$$

The double sum contains n terms where the indices are equal; that is, for $i=j$, and $(n^2 - n)$ terms where $i \neq j$. For the cases where $i=j$, each term is

$$E\{(X_i - u)^2\} = \sigma^2 \qquad (2.19\text{-}8)$$

and there are n of these. For the cases where $i \neq j$, each term is

$$E\{(X_i - u)(X_j - u)\} = E\{X_i X_j\} - u^2 \qquad (2.19\text{-}9)$$

Since X_i and X_j are independent, we can write

$$E\{X_i X_j\} = E\{X_i\}E\{X_j\} = u^2 \qquad (2.19\text{-}10)$$

(This subject will be discussed in more detail in Section 2.20). Consequently, all of the cross terms of Eq. (2.19-7) are zero and σ_y^2 becomes

$$\sigma_y^2 = \frac{1}{n^2} n \sigma^2 = \frac{\sigma^2}{n} \qquad (2.19\text{-}11)$$

Now, if Eqs. (2.19-5) and (2.19-11) are substituted in Eq. (2.19-4), the result, in the limit of large n , is

$$\lim_{n \to \infty} P\{\,|\,Y_n - u\,|\, \geq \epsilon\} \leq \frac{\sigma^2}{n\,\epsilon} = 0 \qquad (2.19\text{-}12)$$

since $\epsilon > 0$. Thus we say that Y_n converges in probability to u as was to be proved.

The Central Limit Theorem

This is a very general limit theorem, which states that sums of independent random variables tend to be normally distributed even though the summands are not. More specifically:

"Let $X_1, X_2, ..., X_n$ be a series of independent random variables having arbitrary distributions, means $u_1, u_2, ..., u_n$, and variances $\sigma_1^2, \sigma_2^2, ..., \sigma_n^2$ respectively. We form the new (normalized) random variables Y_n where

$$Y_n = \frac{\sum_{i=1}^{n} X_i - \sum_{i=1}^{n} u_i}{\left[\sum_{i=1}^{n} \sigma_i^2\right]^{1/2}} \quad , \quad n = 1, 2, ... \qquad (2.19\text{-}13)$$

so that the mean of Y_n is zero and its variance is unity. Then, under very general conditions, the distribution of Y_n approaches in the limit the unit normal distribution".

The necessary conditions under which this theorem holds are difficult to state [6], but a *sufficient* set of conditions are that two numbers m and M exist such that

$$\sigma_i^2 > m > 0 \qquad (2.19\text{-}14)$$

and

$$E\{\,|\,X_i - u_i\,|^3\} < M \qquad (2.19\text{-}15)$$

for all $i = 1, 2, ...$. These conditions insure that no one term in the sum dominates.

It is relatively easy to prove the Central Limit Theorem for the special case where the X_i are independent random variables with the same distribution and with mean u and variance σ^2. The proof consists of showing that the moment-generating function for Y_n approaches in the limit the moment-generating function $e^{t^2/2}$ for the unit normal distribution. We proceed as follows:

We form Y_n from Eq. (2.19-13). For our special case it can be written as

$$Y_n = \sum_{i=1}^{n} \frac{X_i - u}{\sqrt{n}\,\sigma} \qquad (2.19\text{-}16)$$

Let $M(t)$ be the moment-generating function for the random variable $(X_i - u)/\sigma$; that is,

$$M(t) = E\{e^{t(X_i - u)/\sigma}\} \qquad (2.19\text{-}17)$$

Then, if the symbol $M_n(t)$ is used to denote the m.g.f. for Y_n, it follows from Eq. (2.19-16) that

$$M_n(t) = E\{e^{tY_n}\} = E\left\{e^{\frac{t}{\sqrt{n}}\sum_i \frac{X_i - u}{\sigma}}\right\} \qquad (2.19\text{-}18)$$

Since the X_i are independent, the expectation of the product is equal to the product of the expectations (see Section 2.20) and Eq. (2.19-18) becomes

$$M_n(t) = \prod_{i=1}^{n} M(\frac{t}{\sqrt{n}}) = \left[M(\frac{t}{\sqrt{n}}) \right]^n \qquad (2.19\text{-}19)$$

It has been pointed out previously that the m.g.f. $M(t)$ may be expanded as

$$M(t) = 1 + t u_1' + \frac{t^2}{2!} u_2' + \cdots + \frac{t^n}{n!} u_n' + \cdots \qquad (2.19\text{-}20)$$

or

$$M(\frac{t}{\sqrt{n}}) = \qquad (2.19\text{-}21)$$

$$1 + \frac{t}{\sqrt{n}} u_1' + \frac{t^2}{2n} u_2' + \cdots + \frac{t^n}{n!(n)^{n/2}} u_n' + \cdots$$

Since the terms $(X_i - u)/\sigma$ have zero means and variances of unity, Eq. (2.19-21) becomes

$$M(\frac{t}{\sqrt{n}}) = 1 + 0 + \frac{t^2}{2n} + \cdots \qquad (2.19\text{-}22)$$

If n is taken large enough, then the function $M(\frac{t}{\sqrt{n}})$ is given by

$$M(\frac{t}{\sqrt{n}}) = 1 + \frac{t^2}{2n} + higher\ order\ terms \qquad (2.19\text{-}23)$$

and it follows from Eq. (2.19-19) that the m.g.f. for Y_n is, in the limit of large n,

$$\lim_{n \to \infty} M_n(t) = \lim_{n \to \infty} \left[1 + \frac{t^2}{2n} \right]^n = e^{t^2/2} \qquad (2.19\text{-}24)$$

which was to be proved.

The Central Limit Theorem holds under much weaker conditions and even, in certain cases, for dependent random variables; however, the interested reader is referred elsewhere for further discussion [6].

2.20 *Bivariate Distributions* - Thus far we have been concerned with single random variables and their distributions. We call these *univariate* cases. In the last section, we encountered for the first time the expression $E\{X_i X_j\}$; here we were interested in statistical relations between two random variables or in the *bivariate* case. In general, when we are concerned with more than one random variable, we talk about *multivariate* statistics, distributions, etc. For our purposes, it will be sufficient to consider bivariate problems. The extension to more than two random variables will usually be obvious (or too difficult for this treatment).

The joint or bivariate cumulative distribution function $F_{X_1,X_2}(x_1,x_2)$ for two real numbers x_1 and x_2 is defined on the random variables X_1 and X_2 as the joint probability

$$F_{X_1,X_2}(x_1,x_2) = P\{X_1 \leq x_1 \text{ and } X_2 \leq x_2\} \qquad (2.20\text{-}1)$$

The random variables may be discrete or continuous. It will usually be apparent from the context as to which is the case. This c.d.f. $F_{X_1,X_2}(x_1,x_2)$ has the following properties:

(1) $F_{X_1,X_2}(x_1,x_2)$ is monotone non-decreasing in both variables; that is, for h and $k > 0$,

$$F_{X_1,X_2}(x+h,y) \geq F_{X_1,X_2}(x,y)$$
$$F_{X_1,X_2}(x,y+k) \geq F_{X_1,X_2}(x,y)$$

(2) For any $h \geq 0$ and $k \geq 0$,

$$F_{X_1,X_2}(x+h,y+k) - F_{X_1,X_2}(x+h,y) - F_{X_1,X_2}(x,y+k) + F_{X_1,X_2}(x,y) \geq 0$$

(3) If *either* x_1 or x_2 approaches $-\infty$, then $F_{X_1,X_2}(x_1,x_2) \to 0$.

(4) If *both* x_1 and x_2 approach $+\infty$, then $F_{X_1,X_2}(x_1,x_2) \to 1$.

(5) $P\{X_1 \leq x_1 \text{ and no condition on } X_2\} = F_{X_1,X_2}(x_1,\infty) = F_{X_1}(x_1)$

where $F_{X_1}(x_1)$ is the c.d.f. of X_1.

(6) $P\{\text{no condition on } X_1 \text{ and } X_2 \leq x_2\} = F_{X_1,X_2}(\infty,x_2) = F_{X_2}(x_2)$

where $F_{X_2}(x_2)$ is the c.d.f. of X_2.

For the discrete case, we consider the *bivariate* or *joint density* $f_{X_1,X_2}(x_{i2},x_{j2})$ as a function defined on the discrete random variables X_{i1} and X_{j2} and satisfying

$$(1) \quad f_{X_1,X_2}(x_{i1},x_{j2}) \geq 0 \qquad (2.20\text{-}2a)$$

$$(2) \quad \sum_{i,j} f_{X_1,X_2}(x_{i1},x_{j2}) = 1 \qquad (2.20\text{-}2b)$$

$$(3) \quad \sum_{i} f_{X_1,X_2}(x_{i1},x_{j2}) = f_{X_2}(x_{j2}) \qquad (2.20\text{-}3a)$$

$$(4) \quad \sum_{j} f_{X_1,X_2}(x_{i1},x_{j2}) = f_{X_1}(x_{i1}) \qquad (2.20\text{-}3b)$$

It is apparent that each bivariate density is related to a bivariate c.d.f. by

$$F_{X_1,X_2}(a_1,a_2) = \sum_{\substack{x_{i1} \leq a_1 \\ x_{j2} \leq a_2}} f_{X_1,X_2}(x_{i1},x_{j2}) \qquad (2.20\text{-}4)$$

For the continuous case, we consider the *bivariate* or *joint density* $f_{X_1,X_2}(x_1,x_2)$ as a function defined on the continuous random variables X_1 and X_2 and satisfying

(1) $f_{X_1,X_2}(x_1,x_2) \geq 0$ \hfill (2.20-5a)

(2) $\int\limits_{-\infty}^{\infty} \int\limits_{-\infty}^{\infty} f_{X_1,X_2}(x_1,x_2) dx_1 dx_2 = 1$ \hfill (2.20-5b)

(3) $\int\limits_{-\infty}^{\infty} f_{X_1,X_2}(x_1,x_2) dx_1 = f_{X_2}(x_2)$ \hfill (2.20-6a)

(4) $\int\limits_{-\infty}^{\infty} f_{X_1,X_2}(x_1,x_2) dx_2 = f_{X_1}(x_1)$ \hfill (2.20-6b)

It is apparent again that each bivariate density is related to a bivariate c.d.f. by

$$F_{X_1,X_2}(a_1,a_2) = \int\limits_{-\infty}^{a_2} \int\limits_{-\infty}^{a_1} f_{X_1,X_2}(x_1,x_2) dx_1 dx_2 \qquad (2.20\text{-}7a)$$

and that, for continuous and differentiable $F(x_1,x_2)$,

$$f_{X_1,X_2}(x_1,x_2) = \frac{\partial^2}{\partial x_1 \partial x_2} \left[F_{X_1,X_2}(x_1,x_2) \right] \qquad (2.20\text{-}7b)$$

As in the univariate case, an expectation operator can be defined. For discrete random variables, the expectation of the function $h(X_1,X_2)$ is given by

$$E\{h(X_{i1},X_{j2})\} = \sum_{i,j} h(x_{i1},x_{j2}) f_{X_1,X_2}(x_{i1},x_{j2}) \qquad (2.20\text{-}8)$$

and, in the continuous case, by

$$E\{h(X_1,X_2)\} = \int\limits_{-\infty}^{\infty} \int\limits_{-\infty}^{\infty} h(x_1,x_2) f_{X_1,X_2}(x_1,x_2) dx_1, dx_2 \qquad (2.20\text{-}9)$$

As a matter of convenience, we shall use the notation for the continuous random variable in the future even when we deal with the discrete case.

The expectation operator has the same properties in the bivariate case as were developed in Section 2.12 for univariate random variables:

(1) $E\{h_1(X_1,X_2) + h_2(X_1,X_2)\} =$ (2.20-10)

$$E\{h_1(X_1,X_2)\} + E\{h_2(X_1,X_2)\}$$

(2) $E\{ch(X_1,X_2)\} = cE\{h(X_1,X_2)\}$ (2.20-11)

(3) $E(c) = c$ (2.20-12)

(4) $E\{h(X_1,X_2)\} \geq 0$ if $h(X_1,X_2) \geq 0$ (2.20-13)

We come now to the concept of *independence*, which was originally defined in Section 2.8. In terms of our present notation, the definition of Eq. (2.8-3) is as follows: Two random variables X_1 and X_2 are independent if

$$P\{X_1 \leq x_1 \text{ and } X_2 \leq x_2\} = P\{X_1 \leq x_1\}P\{X_2 \leq x_2\} \quad (2.20\text{-}14)$$

for all x_1 and x_2. In other words, Eq. (2.20-14) states that

$$F_{X_1,X_2}(x_1,x_2) = F_{X_1}(x_1)F_{X_2}(x_2) \quad (2.20\text{-}15)$$

where $F_{X_1,X_2}(\bullet,\bullet)$ is the joint c.d.f. of X_1 and X_2, $F_{X_1}(\bullet)$ is c.d.f. of X_1, and $F_{X_2}(\bullet)$ is the c.d.f. of X_2. Eq. (2.20-15) implies that the bivariate density can be factored in the *independent* case, or that

$$f_{X_1,X_2}(x_1,x_2) = f_{X_1}(x_1)f_{X_2}(x_2) \quad (2.20\text{-}16)$$

where $f_{X_1,X_2}(\bullet,\bullet)$ is the bivariate density of X_1 and X_2, $f_{X_1}(\bullet)$ is the (univariate) density of X_1 and $f_{X_2}(\bullet)$ is the (univariate) density of X_2.

A fifth property of the expectation can be written when X_1 and X_2 are independent:

(5) $E\{h_1(X_1)h_2(X_2)\} = E\{h_1(X_1)\}E\{h_2(X_2)\}$ (2.20-17)

The proof is straightforward, since Eq. (2.20-17) can be rewritten from Eq. (2.20-16) as

$$\int_{-\infty}^{\infty}\int_{-\infty}^{\infty} h_1(x_1)h_2(x_2)f_{X_1,X_2}(x_1,x_2)dx_1dx_2$$

$$= \int_{-\infty}^{\infty} h_1(x_1)f_{X_1}(x_1)dx_1 \int_{-\infty}^{\infty} h_2(x_2)f_{X_2}(x_2)dx_2$$

In particular, if the random variables X_1 and X_2 are independent, then

$$E(X_1X_2) = E(X_1)E(X_2) \quad (2.20\text{-}18)$$

This was the relationship used several times in Section 2.19.

We proceed now to define moments for bivariate distributions. We define the rs-th moment about the origin u_{rs}' for the distribution $f_{X_1,X_2}(x_1,x_2)$ as

$$u_{rs}' = E\{X_1^r X_2^s\}$$
$$= \int_{-\infty}^{\infty} \int_{-\infty}^{\infty} x_1^r x_2^s f_{X_1,X_2}(x_1,x_2) dx_1 dx_2 \qquad (2.20\text{-}19)$$

In particular, the mean of X_1 is

$$u_{10}' = u_{x_1} = E(X_1) \qquad (2.20\text{-}20)$$

and the mean of X_2 is

$$u_{01}' = u_{x_2} = E(X_2) \qquad (2.20\text{-}21)$$

As in the univariate case, we find it convenient to define also the rs-th moment about the mean as

$$u_{rs} = E\{(X_1-u_{x_1})^r (X_2-u_{x_2})^s\} \qquad (2.20\text{-}22)$$

In particular, the variance of X_1 is

$$u_{20} = \sigma_{x_1}^2 = E\{(X_1 - u_{x_1})^2\} \qquad (2.20\text{-}23)$$

and the variance of X_2 is

$$u_{02} = \sigma_{x_2}^2 = E\{(X_2-u_{x_2})^2\} \qquad (2.20\text{-}24)$$

In addition, there is a third possibility, the second mixed moment u_{11} called the *covariance* and sometimes written as $\sigma_{x_1 x_2}$

$$u_{11} = \sigma_{x_1 x_2} = E\{(X_1-u_{x_1})(X_2 - u_{x_2})\} \qquad (2.20\text{-}25)$$

We note that the covariance is zero if X_1 and X_2 are independent, since, in that case,

$$u_{11} = E(X_1-u_{x_1})E(X_2-u_{x_2}) = 0 \qquad (2.20\text{-}26)$$

The reader will recall that, for the univariate case, the variance σ^2 and the second moment were related by

$$u_2' = \sigma^2 + u^2$$

where u was the mean of the distribution. In the bivariate case a similar relationship can be derived by expanding Eq. (2.20-25):

$$\sigma_{x_1 x_2} = E(X_1 X_2) - u_{x_1}u_{x_2} - u_{x_2}u_{x_1} + u_{x_1}u_{x_2}$$

or

$$u_{11}' = \sigma_{x_1 x_2} + u_{x_1}u_{x_2} \qquad (2.20\text{-}27)$$

We can normalize the covariance and define the *correlation coefficient* $\rho_{x_1 x_2}$ by

$$\rho_{x_1 x_2} = \frac{\sigma_{x_1 x_2}}{\sigma_{x_1}\sigma_{x_2}} \qquad (2.20\text{-}28)$$

It can be shown that this quantity is bounded in magnitude by unity; that is,

$$-1 \leq \rho \leq 1 \qquad (2.20\text{-}29)$$

Let us consider the non-negative function

$$[(X_1 - u_{x_1})a + (X_2 - u_{x_2})b]^2 \geq 0 \qquad (2.20\text{-}30)$$

where a and b are real variables. Taking the expectation of this function, we have

$$\sigma_{x_1}^2 a^2 + 2u_{11} ab + \sigma_{x_2}^2 b^2 \geq 0 \qquad (2.20\text{-}31)$$

The left side of this equation is a homogeneous quadratic form in a and b and could be written as

$$Aa^2 + B\ 2ab + Cb^2$$

where A and C are non-negative. The condition that this expression be non-negative is that it have no real roots, or that $AC - B^2$ be non-negative; that is,

$$\sigma_{x_1}^2 \sigma_{x_2}^2 - u_{11}^2 \geq 0 \qquad (2.20\text{-}32)$$

whence

$$\rho^2 = \frac{u_{11}^2}{\sigma_{x_1}^2 \sigma_{x_2}^2} \leq 1 \qquad (2.20\text{-}33)$$

as was to be proved.

If X_1 and X_2 are linearly related, then we have equality in Eq. (2.20-29). Let

$$X_1 = bX_2$$

where b is an arbitrary real non-zero number. It is clear that

$$\rho_{x_1 x_2} = \frac{E\{(bX_2 - bu_{x_2})(X_2 - u_{x_2})\}}{|b|\,\sigma_{x_2}\sigma_{x_2}}$$

or

$$\rho_{x_1 x_2} = \begin{cases} 1 \text{ if } b > 0 \\ -1 \text{ if } b < 0 \end{cases} \qquad (2.20\text{-}34)$$

We consider now the mean and variance of a linear combination of random variables. Let Y be defined by

$$Y = \sum_i a_i X_i \qquad (2.20\text{-}35)$$

where the X_i are random variables with mean u_{x_i} and variances $\sigma_{x_i}^2$. The mean of Y is

$$u_y = \sum_i a_i u_{x_i}$$

as has been previously established. The variance, on the other hand, is more complicated because of the presence of cross-terms which may not be zero. We have

$$\sigma_y^2 = E\{(Y-u_y)^2\} = E\{[\sum_i a_i (X_i-u_{x_i})]^2\} \qquad (2.20\text{-}36)$$

or

$$\sigma_y^2 = \sum_i a_i^2 \sigma_{x_i}^2 + \sum_i \sum_j a_i\, a_j\, \sigma_{x_i x_j} \qquad (2.20\text{-}37)$$
$$i \neq j$$

If the X_i are all uncorrelated; that is, if all

$$\sigma_{x_i x_j} = 0 \quad , \quad i \neq j$$

then Eq. (2.20-37) reduces to

$$\sigma_y^2 = \sum_i a_i^2 \sigma_{x_i}^2 \qquad (2.20\text{-}38)$$

A sufficient condition for this to be true is that the X_i be independent as already discussed in conjunction with Eq. (2.20-17). However, random variables may be uncorrelated but not independent; thus, although independence implies Eq. (2.20-38), the converse is not necessarily true.

Suppose we are given the bivariate density $f_{X_1 X_2}(x_1, x_2)$. The probability of X_1 assuming some value less than or equal to a_1, independent of the value of X_2 is sometimes called the *marginal* c.d.f. of X_1 and is given by

$$F_{X_1 X_2}(a_1, \infty) = F_{X_1}(a_1) = \int_{-\infty}^{a_1} \int_{-\infty}^{\infty} f_{X_1 X_2}(x_1, x_2)\, dx_2\, dx_1 \qquad (2.20\text{-}39)$$

In the same way, we define the marginal c.d.f. of X_2 as

$$F_{X_1 X_2}(\infty, a_2) = F_{X_2}(a_2) = \int_{-\infty}^{a_2} \int_{-\infty}^{\infty} f_{X_1 X_2}(x_1, x_2)\, dx_1\, dx_2 \qquad (2.20\text{-}40)$$

The joint characteristic function $M_{X_1 X_2}(jv_1, jv_2)$ and the joint moment generating function $M_{X_1 X_2}(t_1, t_2)$ are defined in a manner analogous to that of the univariate case:

$$M_{X_1 X_2}(jv_1, jv_2) = E\{e^{j(v_1 X_1 + v_2 X_2)}\}$$

$$= \int_{-\infty}^{\infty} \int_{-\infty}^{\infty} e^{j(v_1 x_1 + v_2 x_2)} f_{X_1 X_2}(x_1, x_2)\, dx_1\, dx_2 \qquad (2.20\text{-}41)$$

and

$$M_{X_1,X_2}(t_1,t_2) = E\{e^{t_1X_1 + t_2X_2}\} \qquad (2.20\text{-}42)$$

As in one dimension we can invert Eq. (2.20-41) to give

$$f_{X_1,X_2}(x_1,x_2) \qquad (2.20\text{-}43)$$

$$= \frac{1}{(2\pi)^2} \int_{-\infty}^{\infty} \int_{-\infty}^{\infty} e^{-j(v_1x_1 + v_2x_2)} M_{X_1,X_2}(jv_1,jv_2)\,dv_1\,dv_2$$

The problem of functional transformations is considerably more complicated for bivariate distributions although the principles are the same as in Section 2.11. Suppose that

$$y_1 = h_1(x_1,x_2) \qquad (2.20\text{-}44)$$

and

$$y_2 = h_2(x_1,x_2) \qquad (2.20\text{-}45)$$

are both one-to-one continuously differentiable transformations of (x_1,x_2) into (y_1,y_2). It is clear that

$$P\left\{\begin{array}{c} a_1 \le X_1 \le a_1 + da_1 \\ \text{and} \\ a_2 \le X_2 \le a_2 + da_2 \end{array}\right\} = P\left\{\begin{array}{c} b_1 \le Y_1 \le b_1 + db_1 \\ \text{and} \\ b_2 \le Y_2 \le b_2 + db_2 \end{array}\right\} \qquad (2.20\text{-}46)$$

where $h_1(a_1,a_2) = b_1$ and $h_2(a_1,a_2) = b_2$. In other words, we have

$$f_{X_1,X_2}(x_1,x_2)\,dx_1\,dx_2 = g_{Y_1,Y_2}(y_1,y_2)\,dy_1\,dy_2 \qquad (2.20\text{-}47)$$

where $f_{X_1,X_2}(x_1,x_2)$ is the joint density of the X_i and $g_{Y_1,Y_2}(y_1,y_2)$ is the joint density of the Y_i. This expression may be written as

$$f_{X_1,X_2}(x_1,x_2) = g_{Y_1,Y_2}(y_1,y_2)\,|\,J(y/x)\,| \qquad (2.20\text{-}48)$$

where $J(y/x)$ is the *Jacobian* of the transformation of Eqs. (2.20-44) and (2.20-45) and is given by the determinant

$$J(y/x) = \begin{vmatrix} \dfrac{\partial y_1}{\partial x_1} & \dfrac{\partial y_1}{\partial x_2} \\ \dfrac{\partial y_2}{\partial x_1} & \dfrac{\partial y_2}{\partial x_2} \end{vmatrix} \qquad (2.20\text{-}49)$$

This is just a problem in coordinate transformation and is discussed in more detail in Appendix C, where the general n-dimensional case is considered. Note that, in the univariate case, we have

$$J(y/x) = \frac{dy}{dx} \qquad (2.20\text{-}50)$$

which agrees with the results in Section 2.11. As pointed out in Appendix C, Eq. (2.20-48) may be written as

$$g_{Y_1,Y_2}(y_1,y_2) = f_{X_1,X_2}(x_1,x_2) \, | \, J(x/y) \, | \qquad (2.20\text{-}51)$$

Example 2.21

Suppose $Y_1 = X_1 + X_2$ where X_1 and X_2 are independent random variables with densities $f_{X_1,X_2}(x_1,x_2) = f_{X_1}(x_1)f_{X_2}(x_2)$. Find the density $g_1(y_1)$.

Let $Y_2 = X_2$ for convenience and form

$$J(y/x) = \begin{vmatrix} 1 & 1 \\ 0 & 1 \end{vmatrix} = 1$$

Then $g_{Y_1,Y_2}(y_1,y_2) = f_{X_1}(x_1)f_{X_2}(x_2)$ or

$$g_{Y_1}(y_1) = \int_{-\infty}^{\infty} g_{Y_1,Y_2}(y_1,y_2)dy_2$$

$$g_{Y_1}(y_1) = \int_{-\infty}^{\infty} f_{X_1}(y_1-y_2)f_{X_2}(y_2)dy_2$$

This last expression is called the *convolution* of f_1 and f_2 and is sometimes written as

$$g_{Y_1}(y_1) = f_{X_1}(y_1)*f_{X_2}(y_1)$$

In the general case where the transformations are not one-to-one, we start with Eq. (2.20-47) and have

$$F_{Y_1,Y_2}(b_1,b_2) = \int\int_{\{(x_1,x_2)/(y_1 \leq b_1, y_2 \leq b_2)\}} f_{X_1,X_2}(x_1,x_2)dx_1dx_2 \qquad (2.20\text{-}52)$$

where $F_{Y_1,Y_2}(b_1,b_2)$ is the joint c.d.f. of the Y's and integration is over that (x_1,x_2) region where $y_1 \leq b_1$ and $y_2 \leq b_2$. Choosing the proper regions of integration may be difficult. An excellent and comprehensive discussion with many examples is given in Reference [7], Chapter 5.

2.21 *The Bivariate Normal Distribution* - The normal distribution is of such importance in problems to be considered later that we devote this section and the next to the discussion of some of the elementary properties of the bivariate and multivariate normal distributions.

Suppose we are given the random variables X_1 and X_2 with densities $f_{X_1}(x_1)$ and $f_{X_2}(x_2)$ respectively. It is not, in general, a simple problem to construct an appropriate joint density $f_{X_1,X_2}(x_1,x_2)$. For the case where X_1 and X_2 are independent, we have immediately

$$f_{X_1,X_2}(x_1,x_2) = f_{X_1}(x_1)f_{X_2}(x_2) \qquad (2.21\text{-}1)$$

For normally distributed random variables with means u_1 and u_2 and variances σ_1^2 and σ_2^2, this becomes

$$f_{X_1,X_2}(x_1,x_2) = \frac{1}{2\pi\sigma_1\sigma_2} e^{-\frac{1}{2}\left[\frac{(x_1-u_1)^2}{\sigma_1^2} + \frac{(x_2-u_2)^2}{\sigma_2^2}\right]} \qquad (2.21\text{-}2)$$

The characteristic function $M_{X_1,X_2}(jv_1,jv_2)$ is given by Eq. (2.20-41). This expression may be integrated by completing the square, first in one variable of integration and then in the other as in previous examples. Let us rewrite Eq. (2.20-41) with the substitution of Eq. (2.21-2):

$$M_{X_1,X_2}(jv_1,jv_2) = E\{e^{jv_1X_1}\}E\{e^{jv_2X_2}\} \qquad (2.21\text{-}3)$$

or

$$M_{X_1,X_2}(jv_1,jv_2) = e^{[j(u_1v_1 + u_2v_2) - \frac{1}{2}(\sigma_1^2v_1^2 + \sigma_2^2v_2^2)]} \qquad (2.21\text{-}4)$$

from Eq. (2.18-31). An examination of Eq. (2.21-4) shows that the second part of the exponent is a quadratic form with the cross-terms missing. Let us form a new characteristic function given by

$$M_{X_1,X_2}(jv_1,jv_2) = e^{j\sum\limits_{i=1}^{2} u_iv_i - \frac{1}{2}\sum\limits_{i=1}^{2}\sum\limits_{j=1}^{2}\sigma_{ij}v_iv_j} \qquad (2.21\text{-}5)$$

where the notation means that $\sigma_{ii} = \sigma_i^2$ and $\sigma_{ij}\,(i \neq j)$ is the covariance of X_i and X_j. Here we have added the cross-terms. We will now find the joint density $f_{X_1,X_2}(x_1,x_2)$ from Eq. (2.20-43).

The algebra is considerably simplified if we take the means to be zero so that

$$u_1 = u_2 = 0$$

A simple linear change of variable later will allow for non-zero means. We now consider Eq. (2.20-43) with the substitution of $M_{X_1,X_2}(jv_1,jv_2)$ from Eq. (2.21-5). We complete the square in the exponent, first in v_1 and then in v_2. The exponent becomes

$$-\frac{1}{2}\sigma_1^2\left[(v_1 + \frac{\sigma_{12}^2v_2}{\sigma_1^2} + j\frac{x_1}{\sigma_1^2})^2 + \frac{\sigma_1^2\sigma_2^2-\sigma_{12}^2}{\sigma_1^4}(v_2 - j\frac{\sigma_{12}x_1-\sigma_1^2x_2}{\sigma_1^2\sigma_2^2-\sigma_{12}^2})^2\right]$$
$$-\frac{1}{2\sigma_1^2}\left[x_1^2 + \frac{(\sigma_{12}x_1-\sigma_1^2x_2)^2}{\sigma_1^2\sigma_2^2-\sigma_{12}^2}\right]$$

Let us make the change of variables

$$w = \sigma_1 \{v_1 + \frac{\sigma_{12}v_2}{\sigma_1^2} + j\ \frac{x_1}{\sigma_1^2})$$

and

$$z = \left(\frac{\sigma_1^2\sigma_2^2 - \sigma_{12}^2}{\sigma_1^2}\right)^{1/2} \left(v_2 - j\ \frac{\sigma_{12}x_1 - \sigma_1^2 x_2}{\sigma_1^2\sigma_2^2 - \sigma_{12}^2}\right)$$

After these changes, the exponent becomes

$$-\frac{1}{2}\ w^2 - \frac{1}{2}z^2 - \frac{1}{2}\ \frac{\sigma_2^2 x_2^2 - 2\sigma_{12}x_1 x_2 + \sigma_1^2 x_2^2}{\sigma_1^2\sigma_2^2 - \sigma_{12}^2}$$

and Eq. (2.20-43) can be written as

$$f_{X_1,X_2}(x_1,x_2) = \qquad\qquad (2.21\text{-}6)$$

$$\frac{1}{2\pi\sqrt{\sigma_1^2\sigma_2^2 - \sigma_{12}^2}}\ e^{-\frac{1}{2}d}\left[\frac{1}{\sqrt{2\pi}}\int_{-\infty}^{\infty} e^{-\frac{1}{2}w^2}\,dw\right]\left[\frac{1}{\sqrt{2\pi}}\int_{-\infty}^{\infty} e^{-\frac{1}{2}z^2}\,dz\right]$$

where, for convenience, we have used d to denote

$$d = \frac{\sigma_2^2 x_1^2 - 2\sigma_{12}x_1 x_2 + \sigma_1^2 x_2^2}{\sigma_1^2\sigma_2^2 - \sigma_{12}^2}$$

The two terms in brackets in Eq. (2.21-6) are each unity since they are each the integral of the unit normal density. Let us now introduce the correlation coefficient defined by Eq. (2.20-20) or, in our notation,

$$\rho = \frac{\sigma_{12}}{\sigma_1\sigma_2} \qquad\qquad (2.21\text{-}7)$$

Equation (2.21-6) can be rewritten as

$$f_{X_1,X_2}(x_1,x_2) = \frac{1}{2\pi\sigma_1\sigma_2(1-\rho^2)^{1/2}}\ e^{-\frac{1}{2(1-\rho^2)}[\frac{x_1^2}{\sigma_1^2} - 2\rho\frac{x_1 x_2}{\sigma_1\sigma_2} + \frac{x_2^2}{\sigma_2^2}]} \qquad (2.21\text{-}8)$$

which is the bivariate normal distribution with means zero, variances σ_1^2 and σ_2^2, and correlation coefficient ρ. Note that when $\rho = 0$, the joint density factors into the product of two univariate normal densities. Thus *two uncorrelated normal random variables are independent*, although zero correlation does not imply independence in the general non-normal case.

If we replace x_1 by $x_1 - u_1$ and x_2 by $x_2 - u_2$, then Eq. (2.21-8) becomes the bivariate normal distribution with means u_1 and u_2.

2.22 *The Multivariate Normal Distribution* - Our purpose in this section is to generalize the normal distribution to the multivariate case. At this point it will be convenient to introduce some matrix concepts in order to simplify the notation. As we proceed we will use the bivariate case as an illustration.

Let us denote the random variables $X_1, X_2, ..., X_n$ by the column matrix (also called a column vector) $\mathbf{X}$ where

$$\mathbf{X} = \begin{bmatrix} X_1 \\ X_2 \\ \cdot \\ \cdot \\ \cdot \\ X_n \end{bmatrix} \qquad (2.22\text{-}1)$$

The *transpose* $\mathbf{X}^c$ of the matrix is obtained by interchanging rows and columns so that $\mathbf{X}^c$ is the row matrix (row vector) given by

$$\mathbf{X}^c = [X_1, X_2, ..., X_n] \qquad (2.22\text{-}2)$$

We will define a *covariance matrix* Λ by

$$\Lambda = [\sigma_{ij}] = \begin{bmatrix} \sigma_{11} & \sigma_{12} & \cdot & \cdot & \cdot & \sigma_{1n} \\ \sigma_{21} & \sigma_{22} & \cdot & \cdot & \cdot & \sigma_{2n} \\ \cdot & & & & & \\ \cdot & & & & & \\ \cdot & & & & & \\ \sigma_{n1} & \sigma_{n2} & \cdot & \cdot & \cdot & \sigma_{nn} \end{bmatrix} \qquad (2.22\text{-}3)$$

with determinant $|\Lambda|$ or $|\sigma_{ij}|$. Here the element σ_{ij} is given by

$$\sigma_{ij} = E\{(X_i - u_i)(X_j - u_j)\} \qquad (2.22\text{-}4)$$

and $\sigma_{ij} = \sigma_{ji}$. Note that it might be more correct (but less conventional) to refer to Λ as the variance-covariance matrix. In the same way we define the column matrix $\mathbf{u}$ of the means of $X_1, X_2, ..., X_n$ by

$$\mathbf{u} = \begin{bmatrix} u_1 \\ u_2 \\ \cdot \\ \cdot \\ \cdot \\ u_n \end{bmatrix} \qquad (2.22\text{-}5)$$

In the bivariate case, the covariance matrix is

$$\Lambda = \begin{bmatrix} \sigma_{11} & \sigma_{12} \\ \sigma_{21} & \sigma_{22} \end{bmatrix} = \begin{bmatrix} \sigma_1^2 & \sigma_1\sigma_2\rho \\ \sigma_1\sigma_2\rho & \sigma_2^2 \end{bmatrix} \qquad (2.22\text{-}6)$$

and the characteristic function $M_{X_1 X_2}(jv_1, jv_2)$ of Eq. (2.21-5) can be written as

$$M_X(j\mathbf{v}) = e^{j\mathbf{u}^c\mathbf{v} - \frac{1}{2}\mathbf{v}^c\Lambda\mathbf{v}} \qquad (2.22\text{-}7)$$

where $\mathbf{v}$ is the column matrix

$$\mathbf{v} = \begin{bmatrix} v_1 \\ v_2 \end{bmatrix}$$

or, in general,

$$\mathbf{v} = \begin{bmatrix} v_1 \\ v_2 \\ . \\ . \\ . \\ v_n \end{bmatrix} \qquad (2.22\text{-}8)$$

Not only is Eq. (2.22-7) more compact than Eq. (2.21-5) but it represents the n-variate characteristic function [the case where the index of summation in Eq. (2.21-5) takes on the values $1,2,...,n$].

Let us denote the *inverse* of the matrix Λ [or (σ_{ij})] by Λ^{-1} [or (σ^{ij})]. For the bivariate case, the determinant of Λ is

$$|\Lambda| = |\sigma_{ij}| = \sigma_1^2\sigma_2^2(1 - \rho^2) \qquad (2.22\text{-}9)$$

and the inverse Λ^{-1} is

$$\Lambda^{-1} = [\sigma^{ij}] = \begin{bmatrix} \dfrac{\sigma_2^2}{|\sigma_{ij}|} & -\dfrac{\sigma_1\sigma_2\rho}{|\sigma_{ij}|} \\[2ex] -\dfrac{\sigma_1\sigma_2\rho}{|\sigma_{ij}|} & \dfrac{\sigma_1^2}{|\sigma_{ij}|} \end{bmatrix}$$

$$\qquad (2.22\text{-}10)$$

$$= \begin{bmatrix} \dfrac{1}{\sigma_1^2(1-\rho^2)} & \dfrac{-\rho}{\sigma_1\sigma_2(1-\rho^2)} \\[2ex] \dfrac{-\rho}{\sigma_1\sigma_2(1-\rho^2)} & \dfrac{1}{\sigma_2^2(1-\rho^2)} \end{bmatrix}$$

The determinant of this inverse is

$$|\sigma^{ij}| = \frac{1}{|\sigma_{ij}|} = \frac{1}{\sigma_1^2\sigma_2^2(1-\rho^2)} \qquad (2.22\text{-}11)$$

The bivariate normal density [Eq. (2.21-8) with the substitution $(x_i - u_i)$ for x_i] can now be written in matrix notation as

$$f_{\mathbf{x}}(\mathbf{x}) = \frac{|\sigma^{ij}|^{1/2}}{2\pi} e^{-\frac{1}{2}(\mathbf{x} - \mathbf{u})^c A^{-1}(\mathbf{x} - \mathbf{u})} \qquad (2.22\text{-}12)$$

It can be shown that, for the n-variate case, the joint density $f_{\mathbf{x}}(\mathbf{x})$ corresponding to the characteristic function $M_{\mathbf{x}}(j\mathbf{v})$ of Eq. (2.22-7) is given by

$$f_{\mathbf{x}}(\mathbf{x}) = \frac{|\sigma^{ij}|^{1/2}}{(2\pi)^{n/2}} e^{-\frac{1}{2}(\mathbf{x}-\mathbf{u})^c A^{-1}(\mathbf{x}-\mathbf{u})} \qquad (2.22\text{-}13)$$

This will be taken to be the *n-variate normal density*. A more complete discussion in terms of the moment generating function $M_{\mathbf{x}}(\mathbf{t})$ rather than the characteristic function is given in Reference [8], Chapter 9.

As a matter of interest, we note that the n-variate normal moment generating function can be written as

$$M_{\mathbf{x}}(\mathbf{t}) = e^{\mathbf{u}^c \mathbf{t} + \frac{1}{2}\mathbf{t}^c A \mathbf{t}} \qquad (2.22\text{-}14)$$

where $\mathbf{t}$ is a column matrix

$$\mathbf{t} = \begin{bmatrix} t_1 \\ t_2 \\ . \\ . \\ . \\ t_n \end{bmatrix} \qquad (2.22\text{-}15)$$

Since this m.g.f. is defined by

$$M_{\mathbf{x}}(\mathbf{t}) = E\{e^{\mathbf{t}^c \mathbf{X}}\} = E\{e^{\sum\limits_{i=1}^{n} t_i X_i}\} \qquad (2.22\text{-}16)$$

it follows by direct differentiation that

$$E\{X_i^a X_j^b X_k^c \cdots\} = \frac{\partial^{a+b+c+\cdots} M_{\mathbf{x}}(\mathbf{t})}{\partial t_i^a \partial t_j^b \partial t_k^c \cdots} \Big|_{\substack{t_m = 0 \\ \text{all } m}} \qquad (2.22\text{-}17)$$

Thus, for the normal distribution, in the univariate case,

$$\frac{\partial M_{X_1}(t_1)}{\partial t_1} \Big|_{t_1 = 0} = E(X_1) \qquad (2.22\text{-}18)$$

in the 2-variate case,

$$\frac{\partial^2 M_{X_1, X_2}(t_1, t_2)}{\partial t_1 \partial t_2} \Big|_{t_1 = t_2 = 0} = E(X_1 X_2) \qquad (2.22\text{-}19)$$

in the 3-variate case,

$$\frac{\partial^3 M_{X_1,X_2,X_3}(t_1,t_2,t_3)}{\partial t_1 \partial t_2 \partial t_3}\bigg|_{t_1=t_2=t_3=0} = E(X_1 X_2 X_3)$$

(2.22-20)

$$= u_1 E(X_2 X_3) + u_2 E(X_3 X_1) + u_3 E(X_1 X_2) - 2u_1 u_2 u_3$$

in the 4-variate case

$$\frac{\partial^4 M_{X_1,X_2,X_3,X_4}(t_1,t_2,t_3,t_4)}{\partial t_1 \partial t_2 \partial t_3 \partial t_4}\bigg|_{t_1=t_2=t_3=t_4=0} = E(X_1 X_2 X_3 X_4)$$

(2.22-21)

$$= E(X_1 X_2)E(X_3 X_4) + E(X_1 X_3)E(X_2 X_4)$$
$$+ E(X_1 X_4)E(X_2 X_3) - 2u_1 u_2 u_3 u_4$$

These relationships are useful in treating problems involving non-linear operations on normal random variables.

2.23 *Linear Transformations on Normal Random Variables* - One of the interesting and useful properties of normal random variables is that they are invariant to linear transformations. In other words, a linear transformation on a normal random variable, or a set of normal random variables, results in a normal random variable, or a set of normal random variables.

Let us represent an arbitrary linear transformation on the set of normal random variables $X_1, X_2,...,X_n$ by

$$\mathbf{Y} = \mathbf{A}\mathbf{X} \tag{2.23-1}$$

where **Y** is a column matrix

$$\mathbf{Y} = \begin{bmatrix} Y_1 \\ Y_2 \\ \cdot \\ \cdot \\ \cdot \\ Y_n \end{bmatrix} \tag{2.23-2}$$

and **A** is a matrix

$$\mathbf{A} = \begin{bmatrix} a_{11} & a_{12} & \cdots & a_{1n} \\ a_{21} & a_{22} & \cdots & a_{2n} \\ \cdot & & & \\ \cdot & & & \\ \cdot & & & \\ a_{n1} & a_{n2} & \cdots & a_{nn} \end{bmatrix} \tag{2.23-3}$$

Thus, in the bivariate case, we have

$$Y_1 = a_{11}X_1 + a_{12}X_2$$
$$Y_2 = a_{21}X_1 + a_{22}X_2$$

The column matrix of the mean of $\mathbf{Y}$ is given by

$$\mathbf{u}_y = E(\mathbf{Y}) = \mathbf{A}\,\mathbf{u} \qquad (2.23\text{-}4)$$

and the covariance matrix, on direct substitution of Eq. (2.23-1), by

$$\mathbf{\Lambda}_y = E\{(\mathbf{Y}-\mathbf{u}_y)(\mathbf{Y}-\mathbf{u}_y)^c\} = E\{\mathbf{A}(\mathbf{X}-\mathbf{u})(\mathbf{X}-\mathbf{u})^c\,\mathbf{A}^c\} \qquad (2.23\text{-}5)$$

Since

$$\mathbf{\Lambda} = E\{(\mathbf{X}-\mathbf{u})(\mathbf{X}-\mathbf{u})^c\} \qquad (2.23\text{-}6)$$

then Eq. (2.23-5) becomes

$$\mathbf{\Lambda}_y = \mathbf{A}\,\mathbf{\Lambda}\,\mathbf{A}^c \qquad (2.23\text{-}7)$$

We have already written (Eq. 2.22-7) the characteristic function of $\mathbf{X}$ as

$$M_X(j\mathbf{v}) = e^{j\mathbf{u}^c\mathbf{v} - \frac{1}{2}\mathbf{v}^c\mathbf{\Lambda}\mathbf{v}} \qquad (2.22\text{-}7)$$

Suppose we make the substitution

$$\mathbf{v} = \mathbf{A}^c\,\mathbf{w} \qquad (2.23\text{-}8)$$

where $\mathbf{A}$ is the linear transformation matrix and $\mathbf{w}$ is a column matrix defined by Eq. (2.22-8); that is

$$\mathbf{w} = \begin{bmatrix} w_1 \\ w_2 \\ \cdot \\ \cdot \\ \cdot \\ w_n \end{bmatrix} \qquad (2.23\text{-}9)$$

With this substitution Eq. (2.22-7) becomes

$$M_X(j\mathbf{v}) = e^{j(\mathbf{A}\mathbf{u})^c\mathbf{w} - \frac{1}{2}\mathbf{w}^c(\mathbf{A}\mathbf{\Lambda}\mathbf{A}^c)\mathbf{w}} \qquad (2.23\text{-}10)$$

or, with the use of Eqs. (2.23-4) and (2.23-7),

$$M_X(j\mathbf{v}) = e^{j\mathbf{u}_y^c\mathbf{w} - \frac{1}{2}\mathbf{w}^c\mathbf{\Lambda}_y\mathbf{w}} \qquad (2.23\text{-}11)$$

However, this is the characteristic function of n normally distributed random variables with mean matrix $\mathbf{u}_y$ and covariance matrix $\mathbf{\Lambda}_y$. We have that the characteristic function of $\mathbf{Y}$ is defined by

$$M_Y(j\,\mathbf{w}) = E\left\{e^{j\,\mathbf{w}^c\,Y}\right\} \qquad (2.23\text{-}12)$$

But, since $Y = AX$ from Eq. (2.23-1) we can rewrite this as

$$M_Y(j\,\mathbf{w}) = E\left\{e^{j\,\mathbf{w}^c\,AX}\right\} = E\left\{e^{j\,\mathbf{v}^c\,X}\right\} \qquad (2.23\text{-}13)$$

or

$$M_Y(j\,\mathbf{w}) = M_X(j\,\mathbf{v}) \qquad (2.23\text{-}14)$$

We compare Eqs. (2.23-11) and (2.23-14) and conclude that **Y** has an n-variate normal characteristic function and, hence, is normally distributed with mean vector $\mathbf{u}_y$ and covariance matrix Λ_y. Thus, we have proved that linear transformation on normal random variables yield normal random variables.

PROBLEMS

1. Indicate by cross-hatching the following sets on the Venn diagram.

   (a) $B \cap C$
   (b) $A \cap B^c$
   (c) $(A \cup B^c) \cap C^c$
   (d) $(A^c \cap B)^c \cup C$

   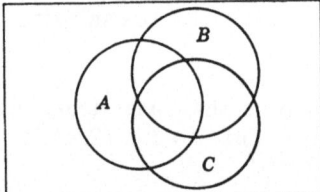

2. Let A be the set of positive even integers, let B be the set of positive integers divisible by 3, and let C be the set of positive odd integers. Describe the following sets:

   (a) $A \cup C$ (d) $(A \cup B) \cap C$
   (b) $A \cap B$ (e) $A \cup (B \cap C)$
   (c) $A \cap C$

3. Using the basic definitions of set equality, intersection, and union prove the distribution law

   $$A \cap (B \cup C) = (A \cap B) \cup (A \cap C)$$

4. If $\begin{pmatrix} n \\ 13 \end{pmatrix} = \begin{pmatrix} n \\ 6 \end{pmatrix}$, what is n? If $\begin{pmatrix} 13 \\ r \end{pmatrix} = \begin{pmatrix} 13 \\ r-3 \end{pmatrix}$, what is r?

5. Three identical coins, marked A, B, and C for identification, are tossed simultaneously. If the order in which the coins fall is noted, how many ways can the coins fall assuming each coin has a "head" and "tail" side.

6. How many distinguishable arrangements are there of ten persons (a) in a row, (b) in a ring?

7. Three dice are rolled. Let A be the event that at least one ace appears and let B be the event that no two dice show the same value. Find $P(A)$, $P(B)$, and $P(AB)$. Are A and B statistically independent? Prove your answer.

8. Consider a game which consists of two successive trials. The first trial has outcomes A or B and the second outcomes C or D. The probabilities for the four possible outcomes of the game are as follows:

Outcome	AC	AD	BC	BD
Probability	1/3	1/6	1/6	1/3

Are A and C statistically independent? Prove your answer.

9. Consider a family with two children of two different ages. Assume that each child is as likely to be a boy as it is to be a girl. What is the conditional probability that both children are boys, given that (a) the older child is a boy, (b) at least one of the children is a boy?

10. The membership of a certain club is composed of 30 men and 20 women. If 40% of the men and 60% of the women play bridge, what is the conditional probability that a member who plays bridge is a man?

11. The network of switches A, B, C, D are connected across the power lines x, y as shown. Each switch has probability p of not closing when operated, and each switch functions independently of the other switches. What is the probability that the circuit from x to y will fail to close when all four switches are operated? Will the addition of bus m change this probability? If so, what is the new probability?

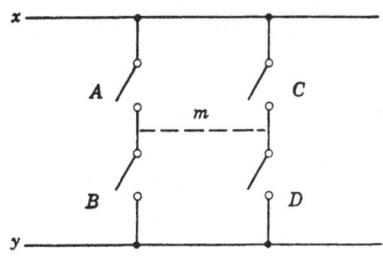

12. The three letters in the word "cat" are permuted in all possible distinct ways and then one permutation is picked at random. What is the probability that it spells a recognizable word in the English language?

13. What is the probability of throwing a "4" before a "7" with two dice?

14. Find the value of A which makes f a density function.

$$f(x) = \begin{cases} Ax & , \quad 0 \le x \le 4 \\ A(8-x) & , \quad 4 \le x \le 8 \\ 0 & , \quad \text{otherwise} \end{cases}$$

Plot the c.d.f. and density function. What is the probability that $X \le 6$?

15. Which of the following are density functions?

(a) $f(x) = \begin{cases} 1/(b-a) & , \quad a < x < b \\ 0 & , \quad \text{otherwise} \end{cases}$

(b) $f(x) = \begin{cases} \dfrac{1}{\pi} e^{-|x|} & , \quad -3 < x < 1 \\ 0 & , \quad \text{otherwise} \end{cases}$

(c) $f(x) = \begin{cases} |x| & , \quad -1 < x < 1 \\ 0 & , \quad \text{otherwise} \end{cases}$

16. Plot the c.d.f. and density function for the random variable X which is (a) the sum of the outcomes of two dice, (b) the product of the outcomes of two dice.

17. Which of the following are discrete density functions:

(a) $p(x) = \begin{cases} 1/12 & , \quad x = 0 \\ 7/12 & , \quad x = 1 \\ 1/4 & , \quad x = 3 \\ 0 & , \quad \text{otherwise} \end{cases}$

(b) $p(x) = \begin{cases} 2^{-|x|} & , \quad x = ... -1, 0, 1, 2, ... \\ 0 & , \quad \text{otherwise} \end{cases}$

$$(c) \quad p(x) = \begin{cases} \dfrac{2}{3}(\dfrac{1}{3})^x & , \quad x = 0,1,2... \\ \\ 0 & , \quad \text{otherwise} \end{cases}$$

18. A random variable N is defined as the number of times a fair coin is tossed until a tail appears.
    (a) Find and sketch the d.f. and c.d.f. associated with this random variable.
    (b) Find the mean of N and its second moment about the origin and about the mean.

19. If the random variable X has a density function

$$f(x) = \begin{cases} 1/2 & , \quad -1 < x \le 0 \\ \\ \dfrac{1}{4}(2-x) & , \quad 0 < x < 2 \\ \\ 0 & , \quad \text{otherwise} \end{cases}$$

and the random variable Y is given by $Y = |X|$, find $g(y)$, the density function of Y.

20. If the random variable X has a density function

$$f(x) = \begin{cases} 1/\pi & , \quad -(\pi/2) < x < \pi/2 \\ \\ 0 & , \quad \text{otherwise} \end{cases}$$

and the random variable Y is given by $Y = \cos X$, find $g(y)$, the density function of Y.

21. A random variable is said to be uniformly distributed in (a,b) if

$$f(x) = \begin{cases} 1/(b-a) & , \quad a < x < b \\ \\ 0 & , \quad \text{otherwise} \end{cases}$$

Find the mean and variance of such a uniformly distributed random variable.

22. If the random variable X is the product of the two outcomes of a pair of dice, what is the mean of X? What is the most probable value of X?

23. Three people take turns tossing a coin, in a fixed order, until a head appears. The man who first tosses a head keeps the coin. Find a set of entrance fees to make the game fair; that is, give equal expected gain for all players.

24. Find the moment generating function and characteristic function for a random variable uniformly distributed in the interval (a,b).

25. Using Table 2.1 plot $P\{|X| \geq \lambda\}$ for $\lambda \leq 3$ for the unit normal distribution [Eq. (2.18-4)]. On the same graph plot the upper bound given by Chebychev's Inequality [Eq. (2.14-1)] and compare the two plots.

26. A book with 100 pages has 100 misprints distributed at random. Assuming that the misprints follow a Poisson distribution with λ equal to the average number of misprints per page, estimate the probability that a page contains at least 2 misprints.

27. An ordinary die is thrown seven times. What is the most probable number of sixes?

28. A submarine launches n torpedos at a ship. Each torpedo has probability 1/2 of hitting the ship independently of the other torpedos. What is the minimum number of torpedos which must be launched so that the probability of at least one hit is at least 0.8?

29. An insurance company writes a policy for A dollars which must be paid if some event E occurs. If E has probability p of occurring, what premium C must the company charge so that the expected profit is (a) 10% of A, (b) 10% of C?

30. The random variable X is uniformly distributed in $(-\frac{\pi}{2}, \frac{\pi}{2})$. If $Y = A \sin X$, find the density, the mean, and variance of Y.

31. A die is rolled 1,000 times. What is the probability that the sum of the outcomes lies between 3,534 and 3,466? You may use the Central Limit Theorem.

32. If X is a random variable with an exponential density function

$$f(x) = Ae^{-2A \, |x|} \quad , \quad -\infty < x < \infty$$

find the n-th moment of X.

33. A certain random variable is normally distributed with an average value of 10 and a standard deviation of unity. What is the probability of the random variable assuming a value (a) less than 11, (b) between 9 and 11?

34. Let X be a random variable with a binomial distribution with $n=100$ and $p=1/5$. *Estimate* the probability that $x \leq 25$.

35. The random variable X is normally distributed with zero mean and variance σ^2. Find the mean of the random variable $Y = |X|$.

36. Let X be the random variable which takes on the values -2,-1,1,2 each with probability 1/4. Let the random variable Y be given by $Y = X^2$. Find the correlation coefficient of X and Y. Are X and Y statistically independent?

37. If the random variables b and c are uniformly distributed on (0,1) and independent, what is the probability that the roots of

$$x^2 + 2bx + c = 0$$

are real?

38. The two random variables X and Y have the joint density function

$$f(x,y) = \begin{cases} e^{-(x+y)} \,, & x \geq 0, y \geq 0 \\ \\ 0 \,, & \text{otherwise} \end{cases}$$

Verify that $f(x,y)$ is a density function. Find the probability of the event that $X < 3$ and $Y < 1$. Are X and Y independent?

39. A red die and a green die are thrown. Let X be the random variable which is the outcome of the red die and let Y be the random variable which is the larger of the two outcomes. Find (a) the joint probability density function of X and Y, (b) the mean of X, and (c) the mean of Y.

40. A marksman find that his shots at a target are distributed about the center of the target in a bivariate normal distribution with zero means, unit variances, and no correlation; that is, the x and y coordinates of any shot are independent random variables with zero means and unit variances. What is the radius of the circle centered on the target such that the probability is 0.95 that a shot falls inside the circle?

41. The exponent in a bivariate normal distribution is

$$-\frac{1}{6}[4(x+1)^2 - 2(x+1)(y-2) + (y-2)^2]$$

What are the means, variances, and correlation coefficient of the random variables X and Y?

42. The random variables X and Y have a bivariate normal distribution. Show that the curves for which the distribution is a constant are ellipses on an x, y coordinate plane.

43. The covariance matrix Λ of three random variables X_1, X_2, X_3 is given by

$$\Lambda = \begin{bmatrix} \sigma_{11} & 0 & \sigma_{13} \\ 0 & \sigma_{22} & 0 \\ \sigma_{31} & 0 & \sigma_{33} \end{bmatrix}$$

The linear transformation $\mathbf{y} = \mathbf{A}\,\mathbf{x}$ is made where

$$\mathbf{A} = \begin{bmatrix} 1 & 0 & 0 \\ 0 & 2 & 0 \\ 1 & 0 & 1 \end{bmatrix}$$

Find the covariance matrix for the random variables Y_1, Y_2, Y_3.

44. Assume that X is a non-negative random variable with density $f(x)$ and mean u. Derive Markov's inequality

$$P\{X \geq c\} \leq u/c \quad , \quad c > 0$$

45. Let us define for the random variable X the r-*th absolute moment* ν_r about the origin by

$$\nu_r = E\{\,|X-u\,|^r\} \quad , \quad u = E\{X\}$$

Prove the general inequality

$$P\{\,|X-u\,| \geq \lambda \nu_r^{1/r}\} \leq 1/\lambda^r$$

for real $\lambda > 0$. For the special case where $r=2$, show that this reduces to the *Chebychev inequality*.

REFERENCES

1. W. Feller, *An Introduction to Probability Theory and Its Applications, Vol. I,* John Wiley & Sons, Inc., New York, 1968.

2. E. Kreyszig, *Introductory Mathematical Statistics,* John Wiley & Sons, Inc., New York, 1970.

3. M. Loeve, *Probability Theory,* Van Nostrand Co., Inc., Princeton, NJ, 1963.

4. E. Lukacs, *Characteristic Functions,* Hafner, New York 1970.

5.. M.G. Kendall and A. Stuart, *The Advanced Theory of Statistics, Vol. I,* Hafner, New York, 1977.

6. H. Cramer, *Mathematical Methods of Statistics,* Princeton University Press, Princeton, NJ, 1946.

7.   G.P. Wadsworth and J.G. Bryan, *Introduction to Probability and Random Variables*, McGraw-Hill, New York, 1960.

8.   A.M. Mood, *Introduction to the Theory of Statistics*, McGraw-Hill, New York, 1950.

9.   R.C. Dubes, *The Theory of Applied Probability*, Prentice-Hall, Inc., Englewood Cliffs, NJ, 1968.

Chapter 3

RANDOM PROCESSES AND SPECTRAL ANALYSIS

3.1 *Definition* - In Sections 2.22 and 2.23, a finite collection of random variables has been called a *vector random variable* or *random vector* and has been denoted by **X** where

$$\mathbf{X} = \begin{bmatrix} X_1 \\ X_2 \\ . \\ . \\ . \\ X_n \end{bmatrix} \qquad (3.1\text{-}1)$$

(Sometimes, when it is necessary to denote explicitly that **X** has n components, it will be convenient to use the notation $\mathbf{X}_n$). Such a random vector **X** is described by the joint distribution function

$$F_n(\mathbf{x}) = P\left\{ X_1 \leq x_1, X_2 \leq x_2, \ldots, X_n \leq x_n \right\} \qquad (3.1\text{-}2)$$

where **x** is the real vector.

$$\mathbf{x} = \begin{bmatrix} x_1 \\ x_2 \\ . \\ . \\ . \\ x_n \end{bmatrix}$$

The random vector **X** is also called a *random or stochastic process*; however, the concept of a random process is basically more general and involves *arbitrary collections* of random variables. More specifically, we have the following definition [1,2,3,4].

Definition 3.1: "A *random process* X_t, or $X(t)$, is an *indexed family* of random variables where the *index* or *parameter* t belongs to some set T; that is $t \in T$."

The set T is called the *parameter set* or *index set* of the process. It may be finite or infinite, denumerable or nondenumerable; it may be an interval or set of intervals on the real line or the whole real line $R_1 = (-\infty, \infty)$. In many applied problems, the index t will be time and the underlying intuitive notion will be that of a random variable developing in time. However, other parameters such as position, temperature, etc. may also enter in a natural manner. Random processes are used as models of phenomena associated with electron emission, noises associated with thermal agitation, atmospheric noise, economic changes, population growth, queues, and many other natural phenomena. There are at least two ways to view an arbitrary random process $X(t)$:

1. As a set of random variables - The definition of a random process was framed in these terms. For each (fixed) value of t, the random variable X exists with realizations x which may be either a denumerable set for discrete random variables or a nondenumerable set for continuous random variables.

2. As a set of functions of t - For each choice of the underlying random variable, a time function $x(t)$ is chosen with domain T. Each such time function is called a *sample function* or a *realization of the process*.

From a physical point of view, it is the sample function which is important since this will be the quantity that will almost always be observed in dealing experimentally with the random process. As will be pointed out later, one of the important practical aspects of the study of random processes is the determination of properties of the random variable $X(t)$ for fixed t on the basis of measurements performed on a single sample function $x(t)$.

A vector *random (or stochastic) process* $\mathbf{X}(t)$ consists of n random processes each with parameter t belonging to the same parameter set T. More formally, we have

$$\mathbf{X}(t) = \left\{ X_1(t), X_2(t), ..., X_n(t) \mid t \in T \right\}$$

In the same way, a *complex random process* $X(t)$ is given by

$$X(t) = X_1(t) + jX_2(t)$$

where $X_1(t)$ and $X_2(t)$ are (real) random processes and $j = \sqrt{-1}$.

A central problem in the study of random processes is their *classification*. Classes of processes are usually defined by imposing suitable restrictions on their n-dimensional distribution functions. In this way, we can define, for example

(1) *stationary processes* whose joint distribution functions are invariant to time translation;

(2) *Gaussian (or normal) processes* whose joint distribution functions are multivariate normal;

(3) *Markov processes* where, given the value of $X(t)$, the values of $X(s), s > t$, do not depend on the values of $X(u), u < t$; in other words, the future behavior of the process, given its present state, is not changed by additional knowledge about its past.

and many other processes.

Another elementary way to classify random process is based on the observation that the index set T can consist of denumerable sets such as

$$T_1 = \{0,1,2,3,...\} \tag{3.1-3}$$

$$T_2 = \{...,-2,-1,0,1,2,...\} \tag{3.1-4}$$

or non-denumerable sets such as

$$T_3 = \{t \mid t \geq 0\} \tag{3.1-5}$$

$$T_4 = \{t \mid -\infty < t < \infty\} \tag{3.1-6}$$

Processes with index sets as T_1 and T_2 are called *discrete parameter processes* while those with index sets such as T_3 and T_4 are called *continuous parameter processes*. Since the random variable $X(t_i)$, any i, can be either *discrete* or *continuous*, it is clear that there is a convenient way to distinguish four classes of random processes:

1. *Discrete random processes with discrete parameter;* also sometimes called *discrete random series* or *discrete random sequences*. Here the random variable takes on only discrete values at any time t and the parameter t assumes only discrete values.

2. *Discrete random processes with continuous parameter;* also sometimes called *discrete random processes*. Here the random variable is discrete but the parameter t assumes a continuum of values.

3. *Continuous random processes with discrete parameter*, or *continuous random series*, or *continuous random sequences*.

4. *Continuous random processes with continuous parameter*, sometimes called just *continuous random processes*.

For added clarity we now illustrate each of these classes by an example.

Example 3.1

Let us toss a coin N times. We define a random variable $X(i)$ by associating the value 1 with the occurrence of a head and 0 with the occurrence of a tail at each toss, and by indexing the number of the toss with i. Thus $X(i) = 1$ if a head occurred on the i-*th* toss and $X(i) = 0$ if a tail occurred.

There are 2^N possible distinct sequences of zeroes and ones of length N. Each sequence is a *realization* or a *sample function* of the random process. We might call the j-*th* sample function $x^{(j)}(i)$, but it will often be convenient to avoid this added complexity and call it simply $x(i)$ when no confusion will result.

We can assume that the probability of a head or of a tail on each toss is one-half and that the outcomes are all independent. The set (or *ensemble*) of the 2^N possible sample functions and this probability law comprise the *discrete random process with discrete parameter* or *discrete random series*. This process is illustrated in Fig. 3.1.

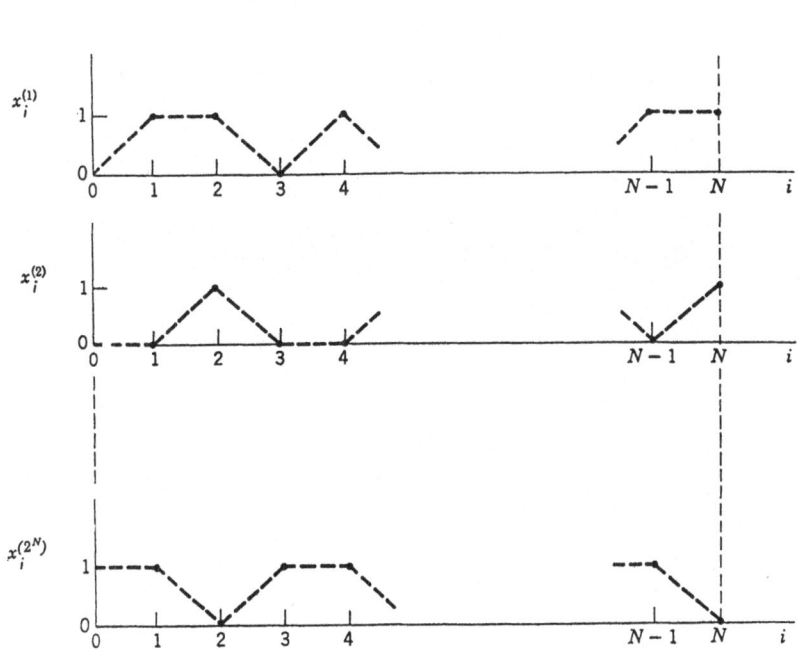

Fig. 3.1 - A discrete random series $X(i)$.

Example 3.2

It is well-known that a noise voltage develops across the ends of a resistor at any temperature above absolute zero. The mean-squared value of this *thermal noise* voltage is given by

$$E(X^2) = 4kT_e R \Delta f \quad (volts)^2$$

where

k = Boltzmann' s Constant = 1.37×10^{-23} watt sec/ $^\circ K$

T_e = Absolute temperature, $^\circ K$

R = Resistance, ohms

Δf = Measurement bandwidth, Hz.

Suppose that we consider the ensemble (set) of all possible resistors having the same given resistance R ohms, maintain all of these resistors at the same temperature T_e, and display the thermal noise voltage of each on an ideal (noiseless) oscilloscope with bandwidth Δf. The result will be something similar to Fig. 3.2 where each resistor is shown together with the thermal noise voltage produced across its terminals.

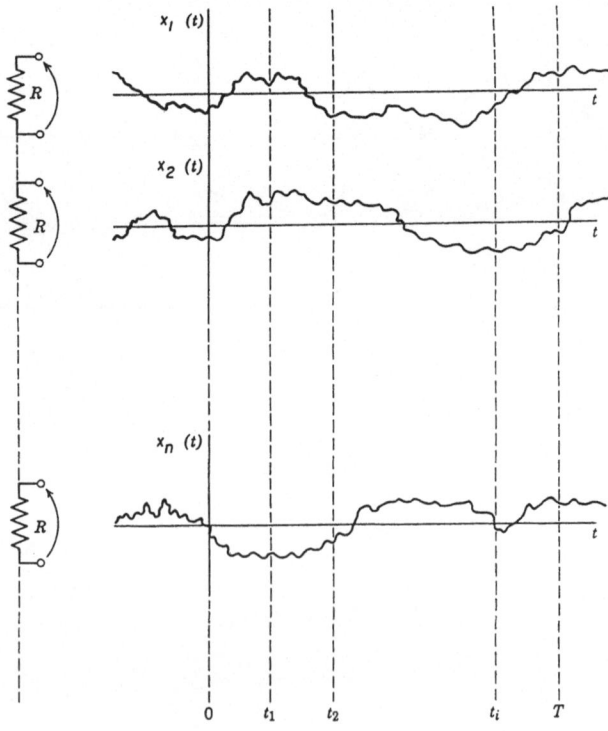

Fig. 3.2 - A physical model for a random process $X(t)$

The (infinite) set of waveforms $x_1(t)$, $x_2(t)$,... comprise a random process $X(t)$, or $\{X(t), -\infty < t < \infty\}$. An inherent part of this random process is the physical foundations from which it arises and from which it derives its statistical properties. Fig 3.2 illustrates a *continuous random process with continuous parameter*. Here the sample space (range of the random variable) is $(-\infty,\infty)$ and the process could be defined for any interval T (or set of intervals) of the parameter t.

Example 3.3

Suppose we sample the random process of Fig. 3.2 at equally spaced intervals t_o as shown in Fig. 3.3. The result is a *continuous random process with discrete parameter*.

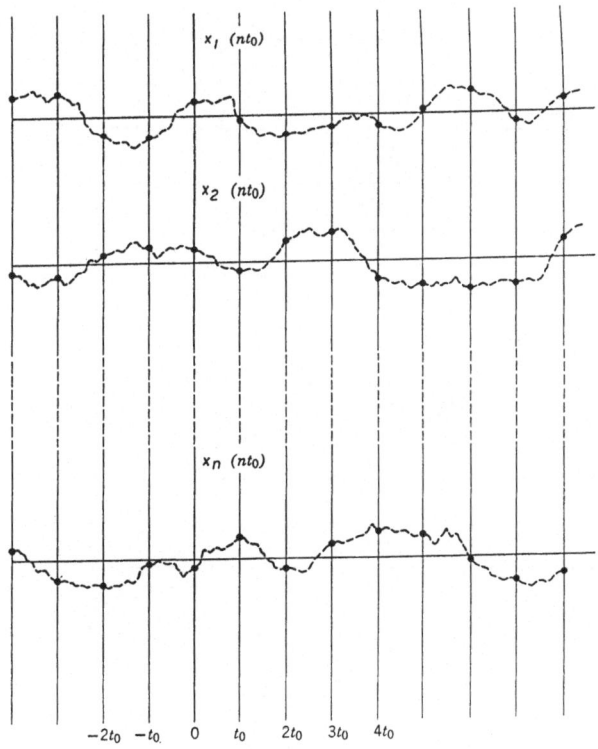

Fig. 3.3 - A continuous random series $X(nt_0)$

The range of the random variable $X(nt_0)$ is the same as that of any $X(t)$ in the previous example, but the parameter time takes on only the discrete values nt_o.

Example 3.4

Let us pass the random process $X(t)$ of Fig. 3.2 through a *hard clipper* or *hard limiter* to produce the process $Y(t)$ defined by

$$Y(t) = \begin{cases} 1 & , \quad X(t) > 0 \\ -1 & , \quad X(t) < 0 \end{cases}$$

This process is illustrated in Fig. 3.4 and is an example of a *discrete random process with continuous parameter*.

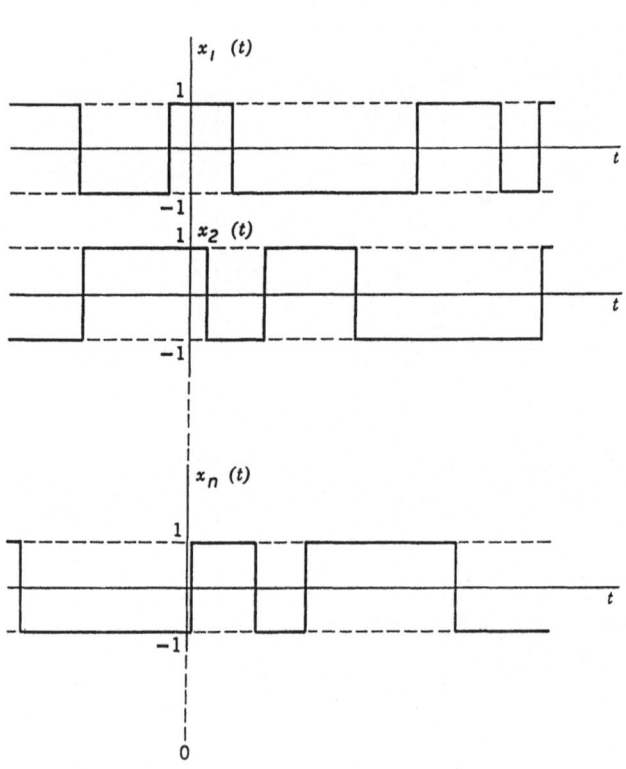

Fig. 3.4 - A discrete random process $Y(t)$

Let (Ω,β,P) be a probability space and let T be any index set. If $\{X(t); t \in T\}$ is a random process defined on Ω, then for any arbitrary finite subset of T, for example, $t_1,t_2,...,t_n$, the random variables $X(t_1)=X_1, X(t_2)=X_2, ..., X(t_n)=X_n$ will have finite-dimensional distribution functions given by

$$F_n (x_1, x_2, ..., x_n ; t_1, t_2, ..., t_n) = \qquad (3.1\text{-}7)$$

$$P\{X_1 \leq x_1, X_2 \leq x_2, ..., X_n \leq x_n\}$$

The collection of all finite-dimensional distribution functions serves as a complete statistical description of the random process. It is intuitively clear that all distribution functions of the type of Eq. (3.1-7) must satisfy the following two conditions which are called the *Kolmogorov consistency conditions:*

$$K1: \lim_{x_n \to \infty} F_n (x_1, ..., x_n ; t_1, \, ... \, , t_n) = \qquad (3.1\text{-}8)$$

$$F_{n-1}(x_1, ..., x_{n-1} ; t_1, ..., t_{n-1})$$

for any subset $\{t_1, ..., t_{n-1}, t_n\}$ of T. In other words, F_{n-1} can be obtained from F_n by removing any restrictions on the random variable X_n.

$$K2: F_n (x_1, ..., x_i, ..., x_j, ..., x_n ; t_1, \, ... \, , t_i, ..., t_j, ..., t_n) = \qquad (3.1\text{-}9)$$

$$F_n (x_1, ..., x_j, ..., x_i, ..., x_n ; t_1, ..., t_j, ..., t_i, ..., t_n)$$

for all pairs (x_i, t_i) and (x_j, t_j) and for all n. In other words, the finite-dimensional distribution function is invariant when x_i and t_i are subject to the same permutation. Conditions $K1$ and $K2$ are the only restrictions on the finite-dimensional distribution functions given by Eq. (3.1-7).

3.2 *Stationarity* - In a general way, we say that a random process is *stationary* if its statistical properties are invariant to time translation. This invariance implies that the underlying physical mechanism producing the process is not changing with time.

Stationary processes are of great importance for two reasons. First, they are frequently encountered in practice or approximated to a high degree of accuracy. (Actually, from the practical point of view, it is not necessary that a process be stationary for all time but only for some observation interval which is long enough to be suitable for a given problem). Second, many of the important properties of stationary processes commonly encountered are described by first and second moments. Consequently, it is relatively easy to develop a simple but useful theory (spectral theory) to describe these processes.

Processes which are not stationary are called *nonstationary*, although they are also sometimes referred to as *evolutionary* processes [1].

Refer again to the random process $X(t)$ of Fig. 3.2 and consider any arbitrary time t_1 in the interval $(0, T)$. We shall denote the univariate density of the random variable $X(t_1) = X_1$ by $f_1(x; t_1)$. This density $f_1(x; t_1)$ is frequently called the *first order amplitude probability density* when it relates to a random process. We define a process to be *stationary to order one* in $(0, T)$ if

$$f_1(x;t_1) = f_1(x;t_1 + h) = f_1(x) \qquad (3.2\text{-}1)$$

for all t_1 and h such that t_1 and $t_1 + h$ are in $(0,T)$. A process stationary to order one has a constant mean value since

$$E\{X(t_1)\} = E\{X(t_1 + h)\} = u \qquad (3.2\text{-}2)$$

Now let us consider any two random variables $X(t_1) = X_1$ and $X(t_2) = X_2$ with joint density function $f_2(x_1,x_2;t_1,t_2)$. Now

$$f_2(x_1,x_2;t_1,t_2)dx_1 dx_2 = P\left\{ \begin{matrix} x(t_1)<X(t_1)\leq x(t_1)+dx(t_1) \\ x(t_2)<X(t_2)\leq x(t_2)+dx(t_2) \end{matrix} \right\} \qquad (3.2\text{-}3)$$

where the right side of Eq. (3.2-3) is the joint probability with which the random process $X(t)$ takes on values in the interval $(x_1,x_1 + dx_1)$ at time t_1 and values in the interval $(x_2,x_2 + dx_2)$ at time t_2. We define a process as *stationary to order two* in $(0,T)$ if

$$f_2(x_1,x_2;t_1,t_2) = f_2(x_1,x_2;t_1 + h,t_2 + h) \qquad (3.2\text{-}4)$$

for all t_1, t_2, and h such that $t_1,t_2,t_1 + h$, and $t_2 + h$ are in $(0,T)$. For a process which is stationary to order two, we have

$$E\{X(t_1)X(t_2)\} = E\{X(t_1+h)X(t_2+h)\} = E\{X_1X_2\} \qquad (3.2\text{-}5)$$

Let us define the function $R(t_1,t_2)$ by

$$R(t_1,t_2) = E\{X(t_1)X(t_2)\} \qquad (3.2\text{-}6)$$

Then, it follows from Eq. (3.2-5) that

$$R(t_1,t_2) = R(t_1 + h, t_2 + h) \qquad (3.2\text{-}7)$$

for a process stationary to order two. Now place $h = -t_1$ so that Eq. (3.2-7) becomes

$$R(t_1,t_2) = R(0,t_2 - t_1) = R(t_2-t_1) \qquad (3.2\text{-}8)$$

The function $R(t_1,t_2)$ is called the *correlation function* or the *autocorrelation function* of the random process $X(t)$. When the process $X(t)$ has zero mean, $R(t_1,t_2)$ is often called the *covariance function* or *autocovariance function*. For processes which are stationary to order two, the autocorrelation function has an argument which depends not on the times t_1 and t_2 but only on their difference t_2-t_1. In fact, a similar argument shows that, in this case, the second order density depends only on the time difference and could be written for convenience as

$$f_2(x_1,x_2;t_1,t_2) = f_2(x_1,x_2;t_2 - t_1) = f_2(x_1,x_2) \qquad (3.2\text{-}9)$$

We proceed in the same way to define a process as *stationary to order n* in $(0,T)$ if

$$f_n(x_1,x_2, \ldots, x_n ;t_1,t_2, \ldots, t_n) = \qquad (3.2\text{-}10)$$

$$f_n(x_1,x_2, \ldots, x_n ;t_1+h ,t_2+h ,...,t_n +h)$$

for all $t_1,...,t_n$, and h such that $t_1,...,t_n$, $t_1 + h ,...,t_n + h$ are in $(0,T)$. It is clear that Eq. (3.2-10) implies

$$E\{X(t_1) \cdots X(t_n)\} = E\{X(t_1+h) \cdots X(t_n +h)\} \qquad (3.2\text{-}11)$$

and that this function depends only on $n-1$ time differences; for example, $t_2 - t_1, ..., t_n - t_1$.

It should be noted that a process which is stationary to order n is stationary to all orders less than n. For example, consider the special case where $n=2$; then Eq. (3.2-9) holds and we can equate and integrate the two densities to give

$$\int_{-\infty}^{\infty} f_2(x_1,x_2;\tau)dx_2 = \int_{-\infty}^{\infty} f_2(x_1,x_2)dx_2 \qquad (3.2\text{-}12)$$

where $\tau = t_2-t_1$ and where x_2 is the variable of integration. But Eq. (3.2-12) may be written as

$$f_1(x_1,t_1) = f_1(x_1) \qquad (3.2\text{-}13)$$

where $f_1(\bullet)$ is the first order amplitude probability distribution. The process is stationary to order one from Eq. (3.2-1).

A process which is stationary to all orders will be called *strictly stationary,* or simply *stationary* if no confusion will exist.

3.3 *Correlation Functions* - Again we consider the random process $X(t)$ and write X_1 for $X(t_1)$ and X_2 for $X(t_2)$ when convenient. We have already defined the *autocorrelation function* $R(t_1,t_2)$ of the process by Eq. (3.2-6)

$$R(t_1,t_2) = E\{X(t_1)X(t_2)\} = E\{X_1,X_2\}$$

if the random process is real. For a complex random process we write

$$R(t_1,t_2) = E\{X(t_1)X^*(t_2)\} \qquad (3.3\text{-}1)$$

where $X^*(t_2)$ is the complex conjugate of $X(t_2)$. As stated previously, we shall be concerned principally with real processes and shall always assume that we are dealing with such processes unless we specifically state otherwise.

For processes which are stationary at least to order two, it was shown in Section 3.2 that the autocorrelation function can be written as $R(t_2-t_1)$. Since we are interested in developing a theory involving only first and second moments, it will be convenient at this point to define another kind of stationarity. A process $X(t)$ will be said to be *stationary in the wide-sense* [4] or *covariance stationary* [1] or *weakly stationary* [1,4] if

(1) $E\{X(t_i)\} = u = constant < \infty$ for all t_i,   (3.3-2)

(2) $E\{X(t_1)X(t_2)\} = R(t_2-t_1) < \infty$ for all t_1 and t_2.   (3.3-3)

It is apparent that a process which is stationary to order two is also wide sense stationary if its second moments exist. However, the converse may not be true.

Example 3.5

     Let $...X_{-1}, X_0, X_1,... = \{X_n\}$ be a sequence of mutually independent random variables each with zero mean and variance σ^2. Then the process $\{X_n\}$ is wide-sense stationary since $E\{X_n\} = 0$ and the autocorrelation function $R(\bullet)$ is given by

$$R(n) = E\{X_m X_{m+n}\} = \begin{cases} \sigma^2, & n = 0 \\ 0, & n \neq 0 \end{cases}$$

However, the process is not strictly stationary (to any order) unless all of the X_i have a common density function.

     The autocorrelation function of a wide-sense stationary random process $X(t)$ has a number of interesting properties which will be of considerable use later. For convenience, we define the time difference $t_2-t_1 = \tau$ so that

$$R(\tau) = E\{X(t)X(t+\tau)\} \qquad (3.3-4)$$

(1) The *mean-square value* of the random process is given by $R(0)$.

$$R(0) = E\{[X(t)]^2\} \qquad (3.3-5)$$

(2) The autocorrelation function of a wide-sense stationary random process is *even*.

$$R(-\tau) = E\{X(t)X(t-\tau)\}$$

Let $t - \tau = y$ so that

$$R(-\tau) = E\{X(y+\tau)X(y)\} = R(\tau)$$
$$R(-\tau) = R(\tau) \qquad (3.3-6)$$

(3)   The autocorrelation function of a wide-sense stationary random process is *a maximum at the origin*.

Form the non-negative quantity

$$E\{[X(t) \pm X(t + \tau)]^2\} \geq 0$$

$$E\{[X(t)]^2\} + E\{[X(t + \tau)]^2\} \pm 2E\{X(t)X(t + \tau)\} \geq 0$$

$$2R(0) \pm 2R(\tau) \geq 0$$

$$R(0) \geq |R(\tau)| \tag{3.3-7}$$

It should be emphasized, however, that the equality may hold.

Example 3.6

Let us consider the random process

$$Y(t) = A \sin(\omega t + \theta)$$

where A, ω, and θ are possible random variables. We form the autocorrelation function

$$R(t, t + \tau) = E\{Y(t)Y(t + \tau)\}$$

On substituting for $Y(t)$, we have

$$R(t, t + \tau) = E\{A^2\sin(\omega t + \theta)\sin[\omega(t + \tau)+\theta]\}$$

We make use of the trigonometric identity

$$\sin x \, \sin y = \frac{1}{2}\cos(x-y) - \frac{1}{2}\cos(x+y)$$

to write

$$R(t, t + \tau) = \frac{1}{2} E\{A^2\cos \omega\tau - A^2\cos(2\omega t + 2\theta + \omega\tau)\}$$

At this point, it is necessary to specify the nature of A, ω, and θ. We consider two cases.

Case I - A and ω are constants and θ is a random variable. Then we write

$$R(t, t + \tau) = \frac{A^2}{2} \cos \omega\tau - \frac{A^2}{2} E\{\cos(2\omega t + 2\theta + \omega\tau)\}$$

Now suppose that θ is uniformly distributed in $(0, 2\pi)$ so that the last term becomes

$$-\frac{A^2}{2} \frac{1}{2\pi} \int_0^{2\pi} \cos(2\omega t + 2\theta + \omega\tau)d\theta = 0$$

In this case

$$R(t, t + \tau) = R(\tau) = \frac{A^2}{2} \cos \omega\tau$$

and now the process $Y(t)$ is wide-sense stationary since $E\{Y(t)\} = 0 < \infty$.

Case II - ω and θ are constants and A is a random variable. Then we write

$$R(t,t+\tau) = \frac{1}{2}[\cos \omega\tau - \cos(2\omega t + 2\theta + \omega\tau)]E\{A^2\}$$

where $E\{A^2\}$ is determined from the distribution of A. In general, in this case, $R(t,t+\tau)$ is not a function of τ only and the process $Y(t)$ is not wide-sense stationary.

In the definition of wide-sense stationarity, we imposed the condition that

$$E\{X(t)\} = u = a \ constant \tag{3.3-8}$$

In this case, the process is easily centered*; that is, a new process $\{X(t) - u\}$ can be defined with zero mean. The autocorrelation function

$$R(\tau) = E\{[X(t) - u][X(t+\tau) - u]\} \tag{3.3-9}$$

of the new process is a true covariance (second moment about the mean) and is properly called either a correlation function or a covariance function.

Suppose that we have two random processes $X(t)$ and $Y(t)$ with autocorrelation functions $R_{xx}(t_1,t_2)$ and $R_{yy}(t_1,t_2)$ respectively. Then we may define the *cross-correlation functions*

$$R_{xy}(t_1,t_2) = E\{X(t_1)Y(t_2)\} \tag{3.3-10}$$

and

$$R_{yx}(t_1,t_2) = E\{Y(t_1)X(t_2)\} \tag{3.3-11}$$

Now the correlation properties of the two random processes may be defined by the correlation matrix

---

* Of course, the process can be centered in any case. If the mean is a function of time; that is, if

$$E\{X(t)\} = u(t) \ , \ all \ t$$

then the process $\{X(t) - u(t)\}$ has zero mean for all t. However, except in special cases, it may not be practical to determine $u(t)$. On the other hand, the constant u may be much easier to calculate or estimate.

$$\mathbf{R}_{zz}(t_1,t_2) = \begin{bmatrix} R_{zz}(t_1,t_2) & R_{zy}(t_1,t_2) \\ R_{yz}(t_1,t_2) & R_{yy}(t_1,t_2) \end{bmatrix} \qquad (3.3\text{-}12)$$

If this matrix can be written as

$$\mathbf{R}_{zz}(t_2\text{-}t_1) = \begin{bmatrix} R_{zz}(t_2\text{-}t_1) & R_{zy}(t_2\text{-}t_1) \\ R_{yz}(t_2\text{-}t_1) & R_{yy}(t_2\text{-}t_1) \end{bmatrix} \qquad (3.3\text{-}13)$$

then we say that the random processes are each wide-sense stationary and, in addition, are *jointly wide sense stationary*. The cross-correlation function is not generally even, as was true for the autocorrelation function, nor does it necessarily have a maximum at the origin. On the other hand, we have

$$R_{zy}(\tau) = E\{X(t)Y(t+\tau)\}$$
$$R_{zy}(-\tau) = E\{X(t)Y(t-\tau)\} = E\{X(t+\tau)Y(t)\}$$
$$R_{zy}(-\tau) = R_{yz}(\tau) \qquad (3.3\text{-}14)$$

which, for $x=y$, becomes Eq. (3.3-6).

Also, suppose we form the non-negative quantity

$$E\{[X(t) \pm Y(t+\tau)]^2\} \geq 0$$

where $X(t)$ and $Y(t)$ are individually and jointly wide-sense stationary. Then, on expanding, we have

$$R_{zz}(0) + R_{yy}(0) \pm 2R_{zy}(\tau) \geq 0$$

or

$$|R_{zy}(\tau)| \leq \frac{1}{2}[R_{zz}(0) + R_{yy}(0)] \qquad (3.3\text{-}15)$$

For $x=y$, this reduces to Eq. (3.3-7). Also, it follows immediately from Eq. (2.20-29) that

$$|R_{zy}(\tau)|^2 \leq R_{zz}(0)R_{yy}(0) \qquad (3.3\text{-}16)$$

Except for Example 3.5, the notation of this section has suggested a continuous parameter random process $\{X(t); -\infty < t < \infty\}$. For the discrete-parameter case, we might write explicitly $\{X_i = X(t_i); t_i = i = 0,\pm1,\pm2,...\}$. Then the autocorrelation function would be given by

$$R(m,n+m) = E\{X_m X_{n+m}\} \qquad (3.3\text{-}17)$$

In the case where the process X_i is at least wide-sense stationary, this becomes

$$R(n) = E\{X_m X_{n+m}\} \qquad (3.3\text{-}18)$$

Similar notational changes might be convenient when treating cross-correlations, etc.

3.4 *Time Averages and Ergodicity* - In practical problems involving random processes, what will generally be available to the observer is not the random process but one of its sample functions or realizations. Thus, in an experimental study of thermal noise, one will work with a given resistor and with the thermal voltage produced by it. In such cases, the quantities that are easily measured are various time averages, and an important question to answer is: Under what circumstances can these time averages be related to the statistical properties of the process?

We define the *time average* of a sample function $x(t)$ of the continuous-parameter random process $X(t)$ by

$$A\{x(t)\} = \lim_{T \to \infty} \frac{1}{2T} \int_{-T}^{T} x(t)dt \qquad (3.4\text{-}1)$$

Other notations are often used instead of $A\{x(t)\}$ and it will sometimes be convenient to use the symbol $\bar{x}(t)$ for this time average. The *time autocorrelation function* $P(\tau)$ [or $P_{xx}(\tau)$] of a sample function $x(t)$ of the random process $X(t)$ is defined by

$$P(\tau) = A\{x(t)x(t+\tau)\} = \lim_{t \to \infty} \frac{1}{2T} \int_{-T}^{T} x(t)x(t+\tau)dt \qquad (3.4\text{-}2)$$

and the *time cross-correlation function* $P_{xy}(\tau)$ of a sample function $x(t)$ of $X(t)$ and a sample function $y(t)$ of the process $Y(t)$ by

$$P_{xy}(\tau) = A\{x(t)y(t+\tau)\} \qquad (3.4\text{-}3)$$

For the discrete-parameter random process X_i, we define the time average of a sample function x_i as

$$A\{x_i\} = \lim_{N \to \infty} \frac{1}{2N+1} \sum_{i=-N}^{N} x_i \qquad (3.4\text{-}4)$$

and the time autocorrelation function as

$$P(m) = A\{x_i\,x_{i+m}\} = \lim_{N \to \infty} \frac{1}{2N+1} \sum_{i=-N}^{N} x_i\,x_{i+m} \qquad (3.4\text{-}5)$$

and similarly for the time cross-correlation function $P_{xy}(m)$.

For random process arising from physical phenomena, it is reasonable to hope that time averages of different sample functions of the same process should be equal to each other. A process which has the property that any time-average of one sample function is equal to the corresponding time average of any other sample function is sometimes called a *regular* random process. The concept of regularity is the time-average analog of the concept of stationarity based on expectations. For a regular process, the time average properties are invariant to shifts over $\omega \in \Omega$.

For a stationary process, the statistical properties are invariant to shifts over $t \in T$.

Suppose a process $X(t)$ is both regular and stationary; then, for example, in the continuous-parameter case, for each sample function $x(t)$,

$$A \{x(t)\} = k = A \{X(t)\} \tag{3.4-6}$$

where k is a constant and where the notation $k = A \{X(t)\}$ simply means that the time-average of every sample function of the process is k; also, for each time t,

$$E \{X(t)\} = u \tag{3.4-7}$$

where u is the (constant) mean. Let us take the expectation of Eq. (3.4-6) and the time average of Eq. (3.4-7) to obtain

$$EA \{X(t)\} = E \{k\} = k \tag{3.4-8}$$

and

$$AE \{X(t)\} = A \{u\} = u \tag{3.4-9}$$

If $X(t)$ is well-behaved enough that the operations of time average and expectation can be interchanged, these last two equations imply that

$$k = u = A \{X(t)\} = E \{X(t)\} \tag{3.4-10}$$

and the time average of the process is equal to the expectation. In the same way, for a well-behaved process which is regular and stationary

$$R(\tau) = E \{X(t)X(t+\tau)\} = A \{X(t)X(t+\tau)\} = P(\tau) \tag{3.4-11}$$

An additional argument to further support this claim that time averages should equal expectations would go as follows: Divide the index set T of the random process $X(t)$ (or X_i) into long intervals each of length T_1. If these intervals are long enough (compared to the time scale of the underlying physical mechanism), then the statistical properties of the process is one interval T_1 should be very similar to those in any other interval. Furthermore, a new random process could be formed in the interval $(0, T_1)$ by using as sample functions the segments of length T_1 from a single sample function of the original process. This new process should be statistically indistinguishable from the original process and its ensemble averages would correspond to time averages of the sample function from the original process.

The foregoing argument is intended as a very crude justification of the condition of *ergodicity* [1,4]. A random process is said to be *ergodic* if time averages of sample functions of the process can be used as approximations to the corresponding ensemble averages or expectations. A sufficient condition is that the random process be regular and stationary and well enough behaved that the order of taking time averages and expectations can be interchanged. However, the subject is extremely complicated and we have really avoided the issue by defining stationary and regular and by a loose use of the term "well-behaved". Nevertheless, in most physical applications, it is assumed that stationary processes are ergodic and that time averages and expectations can be used interchangeably.

3.5 *Convergence of Random Variables* - In the application of the theory of random processes, we shall frequently encounter sequences of random variables of the form

$$X_1, X_2, ..., X_n$$

and we shall be interested in the convergence of such sequences to a random variable X. In Section 2.19, we have already discussed the concept of *convergence in probability*. Although there are a number of types of convergence of random variables which are commonly encountered [4], we shall consider only *convergence in probability* and *convergence in the mean*.

As mentioned earlier, we say that a sequence of random variables X_n [that is, $X_1, X_2, ..., X_n$] *converges in probability* to the random variable X if

$$\lim_{n \to \infty} P\{ \, | X_n - X | \geq \epsilon \} = 0 \qquad (3.5\text{-}1)$$

for arbitrary $\epsilon > 0$. We sometimes write Eq. (3.5-1) as

$$p \lim X_n = X \qquad (3.5\text{-}2)$$

or

$$X_n \to X \quad i.p. \qquad (3.5\text{-}3)$$

We say that the sequence of random variables X_n *converges in the mean* to the random variable X if

$$\lim_{n \to \infty} E\{ \, | X_n - X |^2 \} = 0 \qquad (3.5\text{-}4)$$

This is sometimes written as

$$l.i.m._{n \to \infty} X_n = X \qquad (3.5\text{-}5)$$

where l.i.m. is read "limit in the mean."

If a sequence X_n converges in the mean to X, then it also converges in probability to X although the converse is not necessarily true. The first part of this statement is easily proved using Chebychev's inequality as follows: We have from Section 2.14

$$P\{ \, | Y - u_y | \geq \epsilon \} \leq \sigma_y^2 / \epsilon^2 \qquad (3.5\text{-}6)$$

for $\epsilon > 0$. Let us make the linear change of variable $z = y - u_y$ so that Eq. (3.5-6) becomes

$$P\{\,|\,Z\,|\,\geq \epsilon\} \leq E(Z^2)/\epsilon^2 \qquad (3.5\text{-}7)$$

Now let $|\,Z\,| = |\,X_n - X\,|$ and we have

$$P\{\,|\,X_n - X\,|\,\geq \epsilon\} \leq E\{\,|\,X_n - X\,|^2\}/\epsilon^2 \qquad (3.5\text{-}8)$$

If X_n converges in the mean to X, then, from Eq. (3.5-4), we have

$$\lim_{n \to \infty} P\{\,|\,X_n - X\,|\,\geq \epsilon\} = 0 \qquad (3.5\text{-}9)$$

and X_n also converges in probability to X, as was to be proved. That the converse is not true can be shown by counterexample.

Example 3.7

We now give an illustration that convergence in probability does not imply convergence in the mean. Consider Cauchy's distribution where the density of X_n is

$$f_n(x) = \frac{n}{\pi} \frac{1}{1+n^2 x^2} , \quad -\infty < x < \infty$$

As $n \to \infty$, this density function becomes the Dirac delta function $\delta(x)$ as discussed in Appendix B. In fact, $f_n(x)$ is just Eq. (B.2-15) with n replaced by α. We have

$$\lim_{n \to \infty} P\{\,|\,X_n - 0\,|\,> \epsilon\} = \int_{-\infty}^{-\epsilon} \delta(x)\,dx + \int_{\epsilon}^{\infty} \delta(x)\,dx = 0$$

and X_n converges in probability to zero. On the other hand

$$E\{\,|\,X_n - 0\,|^2\} = E\{\,|\,X_n\,|^2\}$$

$$= \frac{n}{\pi} \int_{-\infty}^{\infty} \frac{x^2}{1+n^2 x^2}\, dx$$

which is infinite even for finite n, so that X_n does not converge in the mean to zero or to any other finite value.

Example 3.8

Let us define the sequence X_n by

$$X_n = \begin{cases} 1 \text{ with probability } a^n, \ a < 1 \\ 0 \text{ with probability } 1 - a^n \end{cases}$$

Then X_n converges in the mean to zero since

$$E\{\,|\,X_n - 0\,|^2\} = a^n \bullet 1 + (1 - a^n) \bullet 0 = a^n$$

and

$$\lim_{n \to \infty} E\{|X_n - 0|^2\} = 0$$

Consequently, X_n also converges in probability to zero.

3.6 *Fourier Transforms* - In this section, we review briefly the concept of the Fourier transform [5,6,7]. It is presumed that the reader will already have been exposed to transform techniques; however, a review of the background material on Fourier series and an introduction to Fourier transforms is given in Appendix D.

Let $f(t)$ be a real-valued function of the real variable t and assume that

$$\int_0^T |f(t)|^2 dt < \infty \qquad \text{for some } (0, T) \qquad (3.6\text{-}1)$$

A function which satisfies Eq. (3.6-1) is said to be of *integrable square* (or square integrable) for $0 \leq t \leq T$. If $f(t)$ is of integrable square for $-\infty < t < \infty$, then from the theory of Fourier transforms [5], there is a function $F(\omega)$ such that

$$\frac{1}{2\pi} \int_{-\infty}^{\infty} |F(\omega)|^2 d\omega = \int_{-\infty}^{\infty} |f(t)|^2 dt \qquad (3.6\text{-}2)$$

and the functions $f(t)$ and $F(\omega)$ are related by

$$f(t) = \underset{A \to \infty}{\text{l.i.m.}} \frac{1}{2\pi} \int_{-A}^{A} F(\omega) e^{j\omega t} d\omega \qquad (3.6\text{-}3)$$

$$F(\omega) = \underset{A \to \infty}{\text{l.i.m.}} \int_{-A}^{A} f(t) e^{-j\omega t} dt \qquad (3.6\text{-}4)$$

Here the notation l.i.m. of Eq. (3.6-4) means "limit in the mean" and is defined by

$$\lim_{A \to \infty} \int_{-\infty}^{\infty} |F(\omega) - \int_{-A}^{A} f(t) e^{-j\omega t} dt|^2 dt = 0 \qquad (3.6\text{-}5)$$

This is the analog in the time domain of the "convergence in the mean" defined in Section 3.5 on an expectation basis for random variables. We call $f(t)$ and $F(\omega)$ a *Fourier-transform pair*. The function $F(\omega)$ is usually called the (direct) Fourier transform of $f(t)$ and $f(t)$ is called the (inverse) Fourier transform of $F(\omega)$.

If the transform $F(\omega)$ is *absolutely integrable*; that is, if

$$\int_{-\infty}^{\infty} |F(\omega)| d\omega < \infty \qquad (3.6\text{-}6)$$

then Eq. (3.6-3) becomes

$$f(t) = \frac{1}{2\pi} \int_{-\infty}^{\infty} F(\omega) e^{j\omega t} d\omega \qquad (3.6\text{-}7)$$

and, if $f(t)$ is absolutely integrable so that

$$\int_{-\infty}^{\infty} |f(t)| \, dt < \infty \qquad (3.6\text{-}8)$$

then Eq. (3.6-4) becomes

$$F(\omega) = \int_{-\infty}^{\infty} f(t) e^{-j\omega t} \, dt \qquad (3.6\text{-}9)$$

Example 3.9

Find the Fourier transform of

$$f(t) = A e^{-b|t|} \quad, \quad -\infty < t < \infty$$

where A and b are positive constants. The function $f(t)$ is both square integrable and absolutely integrable since

$$\int_{-\infty}^{\infty} |A e^{-b|t|}|^2 dt = A^2/b < \infty$$

and

$$\int_{-\infty}^{\infty} |A e^{-b|t|}| \, dt = 2A/b < \infty$$

Thus from Eq. (3.6-9), we have

$$F(\omega) = A \int_{-\infty}^{0} e^{bt} e^{-j\omega t} \, dt + A \int_{0}^{\infty} e^{-bt} e^{-j\omega t} \, dt$$

$$F(\omega) = \frac{2Ab}{b^2 + \omega^2} \quad, \quad -\infty < \omega < \infty$$

The inverse transform $f(t)$ is obtained from Eq. (3.6-7) as

$$f(t) = \frac{1}{2\pi} \int_{-\infty}^{\infty} \frac{2Ab}{b^2 + \omega^2} e^{j\omega t} \, d\omega$$

$$= \frac{Ab}{\pi} \int_{-\infty}^{\infty} \frac{\cos \omega t}{b^2 + \omega^2} \, d\omega$$

This expression may be evaluated by contour integration; however, it is a standard form given in tables of integrals [8] and we may write

$$f(t) = \begin{cases} A e^{bt} & , \ t < 0 \\ \\ A e^{-bt} & , \ t > 0 \end{cases}$$

as we had originally.

For convenience, some of the common Fourier-transform pairs are given below.

Table 3.1

COMMON FOURIER-TRANSFORM PAIRS

$$f(t) = \frac{1}{2\pi} \int_{-\infty}^{\infty} F(\omega)e^{j\omega t}\, d\omega \qquad F(\omega) = \int_{-\infty}^{\infty} f(t)e^{-j\omega t}\, dt$$

No.	$f(t)$	$F(\omega)$
1.	$u(t-a)-u(t-b)$ $a<b$	$\dfrac{e^{-j\omega a}-e^{-j\omega b}}{j\omega}$
2.	$u(t+a)-u(t-a)$ $a>0$	$2a\,\dfrac{\sin a\omega}{a\omega}$
3.	$e^{-bt}\,u(t)$ $b>0$	$\dfrac{1}{b+j\omega}$
4.	$e^{-b|t|}$ $b>0$	$\dfrac{2b}{b^2+\omega^2}$
5.	$\dfrac{1}{b^2+t^2}$ $b>0$	$\dfrac{\pi}{b}\,e^{b|\omega|}$
6.	1	$2\pi\delta(\omega)$
7.	$\delta(t)$	1
8.	$\dfrac{1}{\sqrt{2\pi}}\,e^{-t^2/2}$	$e^{-\omega^2/2}$
9.	$\dfrac{d^n}{dt^n}[f(t)]$ if $f(\infty)=f(-\infty)=0$	$(j\omega)^n F(\omega)$ if $f(\infty)=f(-\infty)=0$

(1) a,b are real

(2) $u(t)$ = unit step function = $\begin{cases} 0, & t<0 \\ 1, & t>0 \end{cases}$

(3) $\delta(t)$ = Dirac delta-function = $\begin{cases} \infty, & t=0 \\ 0, & t\neq0 \end{cases}$, $\int_{-\infty}^{\infty}\delta(t)dt=1$

3.7 *Integrals of Random Processes* - There will be many places in the study of communication theory where integrals of random processes will arise. For example, in Section 3.4, we have already studied time averages. As another example, it may be desirable to obtain the Fourier transform $G(\omega)$ of the random process $X(t)$. Formally, we could write

$$G(\omega) = \int_{-\infty}^{\infty} X(t)e^{-j\omega t}\, dt \tag{3.7-1}$$

It is apparent that $G(\omega)$ is a random variable if it exists; that is, for each sample function $x(t)$, a realization $g(\omega)$ of the random variable $G(\omega)$ results provided the sample function satisfies Eq. (3.6-1). The probability law governing the random variable $G(\omega)$ is determined by the underlying probability laws of the random process and by the transformation of Eq. (3.7-1). As a matter of fact, $G(\omega)$ determined by Eq. (3.7-1) could be considered a new random process with (real) parameter ω instead of t. It is apparent that $G(\omega)$ will be, in general, a complex random process.

Suppose we are given that $Y(t)$ is a random process which is the result of a general linear operation $h(t,\tau)$ on the random process $X(t)$; that is,

$$Y(t) = \int_a^b h(t,\tau)X(\tau)d\tau \tag{3.7-2}$$

where (a,b) is some interval, finite or infinite. It can be shown [4] that this integral exists if

$$\int_a^b E\{\,|\,h(t,\tau)X(\tau)\,|\,\}d\tau = \int_a^b |\,h(t,\tau)\,|\,E\{\,|\,X(\tau)\,|\,\}d\tau < \infty \tag{3.7-3}$$

Furthermore, in this case, we can write

$$E\{Y(t)\} = E\{\int_a^b h(t,\tau)X(\tau)d\tau\} = \int_a^b h(t,\tau)E\{X(\tau)\}d\tau \tag{3.7-4}$$

and the interchange of the order of integration and expectation operation is justified.

Even if the integral of Eq. (3.7-1) does not exist in the usual sense for each sample function $x(t)$ of $X(t)$, it may be possible to define the equality in some stochastic sense as was done in the convergence of sequences of random variables.

For our purposes, it will be sufficient if integrals of the form of Eq. (3.7-1) can be defined as limits in the mean-square sense of the approximating sums formally associated with the integrals. Let the interval $[a,b]$ be partitioned by the set of points

$$a = t_1 < t_2 < \cdots < t_{n+1} = b$$

Let $S_n(\tau)$ be the approximating sum

$$S_n(\tau) = \sum_{i=1}^{n} h(t_i,\tau)X(t_i)(t_{i+1}-t_i) \tag{3.7-5}$$

It is clear that $S_n(\tau)$ is a random variable dependent on the particular partition. Now consider the limiting sum

$$\lim_{\substack{n \to \infty \\ max\,(t_{i+1}-t_i) \to 0}} S_n(\tau) = \lim S_n(\tau) \tag{3.7-6}$$

The integral of Eq. (3.7-1) will be said to *converge in the mean-square sense* (or in quadratic mean) to $Y(\tau)$ if

$$\lim E\{\,|\,Y(\tau) - S_n(\tau)\,|^2\} = 0 \tag{3.7-7}$$

Let us expand this last expression and take the expectation term by term to yield

$$E\{Y(\tau)Y^*(\tau)\} - \lim E\{Y(\tau)S^*_n(\tau)\} - \lim E\{Y^*(\tau)S_n(\tau)\} \tag{3.7-8}$$

$$+ \lim E\{S_n(\tau)S^*_n(\tau)\} = 0$$

If the autocorrelation function of $Y(t)$ is given by

$$R_Y(t,s) = E\{Y(t)\,Y^*(s)\} \tag{3.7-9}$$

then Eq. (3.7-8) can be written as

$$R_Y(\tau,\tau) - \lim \int_a^b \sum_{i=1}^{n} h^*(t_i,\tau)h(t,\tau)E\{X(t)X(t_i)\}(t_{i+1}-t_i)dt$$

$$- \lim \int_a^b \sum_{i=1}^{n} h(t_i,\tau)h^*(t,\tau)E\{X(t)X(t_i)\}(t_{i+1}-t_i)dt$$

$$+ \lim \sum_{i=1}^{n}\sum_{j=1}^{n} h(t_i,\tau)h^*(s_j,\tau)E\{X(t_i)X(s_j)\}(t_{i+1}-t)(s_{j+1}-s_j)$$

$$= 0 \tag{3.7-10}$$

Suppose the ordinary Riemann integral

$$R_Y(\tau,\tau) = \int_a^b \int_a^b h(\tau,t)h^*(\tau,s)R(t,s)dtds \tag{3.7-11}$$

exists where $R(t,s)$ is the autocorrelation function of $X(t)$ and is given by

$$R(t,s) = E\{X(t)X(s)\} \tag{3.7-12}$$

Then Eq. (3.7-10) becomes

$$R_Y(\tau,\tau) - R_Y(\tau,\tau) - R_Y(\tau,\tau) + R_Y(\tau,\tau) = 0$$

Thus the existence of Eq. (3.7-11) insures that the right side of Eq. (3.7-1) converges mean-square to $Y(\tau)$ as given by Eq. (3.7-7).

A random process $X(t)$ was said to be *continuous in mean square* (m.s.) if, for all $t \in T$,

$$\lim_{h \to 0} E\{[X(t+h)-X(t)]^2\} = 0$$

A necessary and sufficient condition that $X(t)$ be continuous in m.s. at some point t_0 is that the autocorrelation function $R(t,s)$ be continuous at $t = s = t_0$. Form

$$E\{[X(t_0+h) - X(t_0)]^2\} = R(t_0+h,t_0+h) - R(t_0+h,t_0)$$
$$(3.7\text{-}13)$$
$$- R(t_0,t_0+h) + R(t_0,t_0)$$

and the sufficiency is evident. The necessity is obtained by considering

$$R(t+h,t+k) - R(t,t) = E\{[X(t+h) - X(t)][X(t+k) - X(t)]\}$$
$$+ E\{[X(t+h) - X(t)]\,X(t)\} + E\{[X(t+k) - X(t)]X(t)\}$$

Suppose that $R(t,s)$ is not continuous at $t = s = t_0$ so that the left-side of this last expression is not zero at $t = t_0$ in the limit as $h,k \to 0$. We have

$$\lim_{\substack{h \to 0 \\ k \to 0}} \; | \, E\{[X(t_0+h) - X(t_0)][X(t_0+k) - X(t_0)]\}$$

$$+ E\{[X(t_0+h) - X(t_0)]\,X(t_0)\} + E\{[X(t_0+k)-X(t_0)]\,X(t_0)\} \,|^2 > 0$$

The Schwarz inequality

$$\sum |f_n\,g_n|^2 \le \sum |f_n|^2 \sum |g_n|^2 \qquad (3.7\text{-}14)$$

may be applied to the last expression to yield

$$\lim_{\substack{h \to 0 \\ k \to 0}} \; | \; 2E\{[X(t_0+h) - X(t_0)]^2\}+E\{[X(t_0+k)-X(t_0)]^2\} \, | \; \times$$

$$\times \; | \; E\{[X(t_0+k) - X(t_0)]^2\}+2E\{X^2(t_0)\} \, | \; > 0$$

It is clear that $X(t)$ cannot be continuous in m.s. at $t=t_0$ unless $R(t,s)$ is continuous at $t = s = t_0$.

The condition of continuity in m.s. of a random process does not say anything very strong about the continuity of the sample functions of the process. It is easy to construct m.s. continuous random processes whose sample functions have an arbitrary number of discontinuities.

The autocorrelation function $R(t,s)$ of the random process $X(t)$ is *nonnegative definite*. Let $t_1, t_2, \ldots, t_n$ be a set of points that belongs to T. Let $f_1, f_2, \ldots, f_n$ be arbitrary real numbers. Form

$$E\{|\sum_{i=1}^{n} f_i X(t_i)|^2\} = \sum_{i=1}^{n} \sum_{j=1}^{n} f_i f_j R(t_i, t_j) \geq 0 \qquad (3.7\text{-}15)$$

In the same way, let $f(t)$ be an arbitrary real function which is square- integrable in $[a,b]$ and let $X(t)$ be a continuous-parameter process such that $[a,b] \epsilon T$. Form

$$E\{|\int_a^b f(t)X(t)dt|^2\} = \int_a^b \int_a^b f(t)f(s)R(t,s)dtds \geq 0 \qquad (3.7\text{-}16)$$

Functions R with the properties of these last two equations are called *nonnegative definite*. If the inequality holds for all $f(t)$, or f_i, then R is called *positive definite* [11].

In this development, we have ignored some of the more subtle conditions which the random processes must satisfy for these results to hold. A more complete discussion requires some background in measure theory and is beyond the scope of this treatment. Nevertheless these results hold for reasonably well-behaved random processes such as those which arise in the physical world.

3.8 *Power Spectra* - As mentioned previously, a considerable theory (spectral theory) involving second moments only has been developed for random processes which are at least wide-sense stationary. In this section we proceed to discuss some of the more elementary aspects of this theory.

We begin with the random process $X(t)$ assumed to be at least wide-sense stationary so that its autocorrelation function can be written as

$$R(\tau) = E\{X(t)X(t+\tau)\} \qquad (3.8\text{-}1)$$

In addition we will often assume that the process has a constant mean of zero so that the variance σ^2 of the process is a constant given by

$$\sigma^2 = R(0) \qquad (3.8\text{-}2)$$

We note in passing that the restriction to zero mean is not necessary. Actually the process could have any mean $u(t)$ and we could work with the new process $Y(t) = X(t) - u(t)$.

We define the *power spectral density* $\phi(\omega)$ of the random process $X(t)$ to be the Fourier transform of the autocorrelation function $R(\tau)$ of the process:

$$\phi(\omega) = \int_{-\infty}^{\infty} R(\tau) e^{-j\omega\tau} d\tau \tag{3.8-3}$$

Then the autocorrelation function is the inverse Fourier transform of $\phi(\omega)$ or

$$R(\tau) = \frac{1}{2\pi} \int_{-\infty}^{\infty} \phi(\omega) e^{j\omega\tau} d\omega \tag{3.8-4}$$

It was pointed out in Section 3.6 that $R(\tau)$ must be absolutely integrable for Eq. (3.8-3) to hold. An example was given in Section 3.3 where the autocorrelation function was of the form

$$R(\tau) = \cos\omega_0\tau \tag{3.8-5}$$

For this case, the power spectral density $\phi(\omega)$ does not properly exist. However, it can be expressed in terms of the Dirac delta-functions discussed in Appendix B since

$$\phi(\omega) = \int_{-\infty}^{\infty} \cos\omega_0\tau \, e^{-j\omega\tau} d\tau$$

$$= \frac{1}{2} \int_{-\infty}^{\infty} e^{-j(\omega-\omega_0)\tau} d\tau + \frac{1}{2} \int_{-\infty}^{\infty} e^{-j(\omega+\omega_0)\tau} d\tau$$

$$= \pi\delta(\omega-\omega_0) + \pi\delta(\omega+\omega_0) \tag{3.8-6}$$

from Pair No. 6 of Table 3.1. Here the power spectral density consists of two delta-functions, one at $\omega=\omega_0$ and the other at $\omega=-\omega_0$. For most of the practical problems which we will encounter, either the power spectral density will be well-defined or it can be written in terms of delta-functions.

Let us define a random variable $f_T(\omega)$ by the integral

$$f_T(\omega) = \int_{-T}^{T} X(t) e^{-j\omega t} dt \tag{3.8-7}$$

It will now be shown [9,10] that $\phi(\omega)$ is actually given by

$$\phi(\omega) = \lim_{T\to\infty} E\left\{ \frac{|f_T(\omega)|^2}{2T} \right\} \tag{3.8-8}$$

Note that $f_T(-\omega) = f^*_T(\omega)$, and form the non-negative function

$$E\left\{\frac{1}{2T} f_T(\omega) f_T(-\omega)\right\} = E\left\{\frac{1}{2T} \int_{-T}^{T} X(t) e^{-j\omega t} dt \int_{-T}^{T} X(s) e^{j\omega s} ds\right\}$$

$$= \frac{1}{2T} \int_{-T}^{T}\int_{-T}^{T} E\{X(t)X(s)\} e^{-j\omega t} dt\, e^{j\omega s} ds \tag{3.8-9}$$

If $X(t)$ is a random process which is at least wide-sense station-ary, then

$$E\{X(t)X(s)\} = R(t-s) \qquad (3.8\text{-}10)$$

Let us now make the following change of variables in the double integral of Eq. (3.8-9)

$$t-s = u \qquad (3.8\text{-}11\text{a})$$

$$s = v \qquad (3.8\text{-}11\text{b})$$

The corresponding regions of integration are shown in Fig. 3.5.

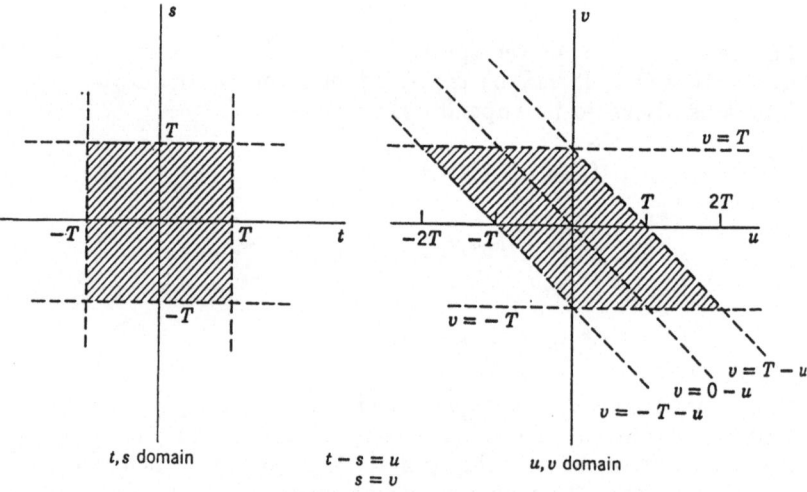

Fig. 3.5 - Regions of integration for Eq. (3.8-12)

If we integrate first with respect to v, we note that we must con-sider two regions of values of u, namely $u < 0$ and $u > 0$. Equa-tion (3.8-9) becomes

$$E\{\frac{1}{2T} f_T(\omega) f_T(-\omega)\} = \frac{1}{2T} \int_{-2T}^{0} R(u)e^{-j\omega u} \int_{-T-u}^{T} dv\,du$$

$$+ \frac{1}{2T} \int_{0}^{2T} R(u)e^{-j\omega u} \int_{-T}^{T-u} dv\,du \qquad (3.8\text{-}12)$$

$$= \int_{-2T}^{0} [1 + \frac{u}{2T}]R(u)e^{-j\omega u}\,du + \int_{0}^{2T}[1-\frac{u}{2T}]R(u)e^{-j\omega u}\,du$$

or, finally,

$$E\left\{\frac{1}{2T}\,f_T(\omega)f_T(-\omega)\right\} = \int_{-2T}^{2T}[1-\frac{|u|}{2T}]R(u)e^{-j\omega u}\,du$$

$$= \phi_T(\omega) \tag{3.8-13}$$

where the definite integral is denoted by $\phi_T(\omega)$. We see immediately that

$$\lim_{T\to\infty} E\left\{\frac{1}{2T}\,f_T(\omega)f_T(-\omega)\right\} = \lim_{T\to\infty}\phi_T(\omega) = \phi(\omega) \tag{3.8-14}$$

where $\phi(\omega)$ is defined by Eq. (3.8-3), as was to be proved.

We now show that the inverse transform relationship of Eq. (3.8-4) is valid. Let us define a function $R_T(u)$ by

$$R_T(u) = \begin{cases} [1-\frac{|u|}{2T}]R(u) & ,\ |u| \le 2T \\ \\ 0 & ,\ |u| > 2T \end{cases} \tag{3.8-15}$$

so that, from Eq. (3.8-13),

$$\phi_T(\omega) = E\left\{\frac{1}{2T}\,f_T(\omega)f_T(-\omega)\right\} = \int_{-\infty}^{\infty} R_T(u)e^{-j\omega u}\,du \ge 0 \tag{3.8-16}$$

We now multiply both right and left sides of this equation by $[1-\frac{|\omega|}{\Omega}]e^{j\omega t}$ and integrate in the variable ω from $-\Omega$ to Ω. We have

$$\int_{-\Omega}^{\Omega}[1-\frac{|\omega|}{\Omega}]e^{j\omega t}\phi_T(\omega)d\omega = \int_{-\Omega}^{\Omega}\int_{-\infty}^{\infty} R_T(u)e^{-j\omega u}\,du\,[1-\frac{|\omega|}{\Omega}]e^{j\omega t}\,d\omega$$

$$= \int_{-\infty}^{\infty} R_T(u)\int_{-\Omega}^{\Omega} e^{-j\omega(u-t)}[1-\frac{|\omega|}{\Omega}]d\omega\,du \tag{3.8-17}$$

The inner integral on the right side is just the Fourier transform of a triangular pulse and is easily calculated so that

$$\int_{-\Omega}^{\Omega}[1-\frac{|\omega|}{\Omega}]e^{j\omega t}\phi_T(\omega)d\omega$$

$$= \int_{-\infty}^{\infty} R_T(u)\left[\frac{\sin\{\frac{1}{2}\Omega(u-t)\}}{\frac{1}{2}\Omega(u-t)}\right]^2 \Omega\,du \tag{3.8-18}$$

As in Section 3 of Appendix B, it can be shown that this transform is also one of the approximating functions for the delta-function. Specifically, we have

$$\lim_{\Omega \to \infty} \left[\frac{\sin\{\frac{1}{2}\Omega(u-t)\}}{\frac{1}{2}\Omega(u-t)\}} \right]^2 \Omega = 2\pi\delta(u-t) \qquad (3.8\text{-}19)$$

where $\delta(u-t)$ is the Dirac delta-function. Thus, in the limit as $\Omega \to \infty$, Eq. (3.8-18) becomes

$$\lim_{\Omega \to \infty} \int_{-\infty}^{\infty} [1 - \frac{|\omega|}{\Omega}] e^{j\omega t} \phi_T(\omega) d\omega \qquad (3.8\text{-}20)$$

$$= \int_{-\infty}^{\infty} \phi_T(\omega) e^{j\omega t} d\omega = 2\pi R_T(t)$$

and, on allowing $T \to \infty$, we have

$$\lim_{T \to \infty} R_T(t) = R(t) = \lim_{T \to \infty} \frac{1}{2\pi} \int_{-\infty}^{\infty} \phi_T(\omega) e^{j\omega t} d\omega$$

$$(3.8\text{-}21)$$

$$= \frac{1}{2\pi} \int_{-\infty}^{\infty} \phi(\omega) e^{j\omega t} d\omega$$

as was to be proved.

Thus for an ergodic (and therefore stationary) random process $X(t)$, the autocorrelation function and the power spectral density are related by

$$P(\tau) = R(\tau) = \frac{1}{2\pi} \int_{-\infty}^{\infty} \phi(\omega) e^{j\omega t} d\omega \qquad (3.8\text{-}22)$$

where

$$P(\tau) = \lim_{T \to \infty} \frac{1}{2T} \int_{-T}^{T} x(t) x(t+\tau) dt \qquad (3.4\text{-}2)$$

$$R(\tau) = E\{X(t)X(t+\tau)\} \qquad (3.8\text{-}1)$$

and

$$\phi(\omega) = \lim_{T \to \infty} E\left\{ \frac{|f_T(\omega)|^2}{2T} \right\} = \int_{-\infty}^{\infty} R(\tau) e^{-j\omega\tau} d\tau \qquad (3.8\text{-}3)$$

The power spectral density $\phi(\omega)$ of the random process $X(t)$ has a number of interesting properties which follow directly from its definition and from the properties of the autocorrelation function $R(\tau)$. For convenience, we list the following:

1. *The integral of the power spectral density is the average power in the random process.*

   This property follows directly from Eq. (3.8-4) since

   $$R(0) = E\{X^2(t)\} = \frac{1}{2\pi} \int\limits_{-\infty}^{\infty} \phi(\omega)d\omega \qquad (3.8\text{-}23)$$

2. *The power spectral density is real.*

   Since $R(\tau)$ is an even function, we can write

   $$\phi(\omega) = \int\limits_{-\infty}^{\infty} R(\tau)[\cos\omega\tau - j\sin\omega\tau]d\tau = \int\limits_{-\infty}^{\infty} R(\tau)\cos\omega\tau \, d\tau$$

   $$(3.8\text{-}24)$$

   $$\phi(\omega) = 2\int\limits_{0}^{\infty} R(\tau)\cos\omega\tau \, d\tau$$

   Since $R(\tau)$ is real for a real random process, then $\phi(\omega)$ is real also.

3. *The power spectral density is non-negative.*

   This property is apparent from Eq. (3.8-13). Thus, when properly normalized, the power spectral density and the autocorrelation function can be related as a density function and its associated characteristic function. Let us write the *normalized autocorrelation function* $\rho(\tau)$ [compare Eq. (2.20-28)] as

   $$\rho(\tau) = \frac{R(\tau)}{R(0)} \qquad (3.8\text{-}25)$$

   Then we may write

   $$\rho(\tau) = \int\limits_{-\infty}^{\infty} \frac{\phi(\omega)}{2\pi R(0)} e^{j\omega\tau}d\omega \qquad (3.8\text{-}26)$$

   or

   $$\rho(v) = E\{e^{j\Omega v}\} \qquad (3.8\text{-}27)$$

   where $\rho(v)$ is the characteristic function corresponding to the density $\Psi(\omega)$ of the random variable Ω where

   $$\Psi(\omega) = \frac{\phi(\omega)}{2\pi R(0)} = \frac{\phi(\omega)}{\int\limits_{-\infty}^{\infty} \phi(\omega)d\omega} \qquad (3.8\text{-}28)$$

   Note that $\Psi(\omega)$ is a true density since

   $$\Psi(\omega) \geq 0 \qquad (3.8\text{-}29)$$

   $$\int\limits_{-\infty}^{\infty} \Psi(\omega)d\omega = 1 \qquad (3.8\text{-}30)$$

Using the Riemann-Stieltjes integral as discussed in Appendix A, we can write Eq. (3.8-26) as

$$\rho(v) = \int_{-\infty}^{\infty} e^{j\omega v} \, dF(\omega) \qquad (3.8\text{-}31)$$

where

$$F(\omega) = \int_{-\infty}^{\omega} \Psi(x) \, dx \qquad (3.8\text{-}32)$$

is a c.d.f. with $F(-\infty) = 0$ and $F(\infty) = 1$. Also we see that $\Psi(\omega)$ is given by

$$\Psi(\omega) = \frac{1}{2\pi} \int_{-\infty}^{\infty} \rho(\tau) e^{-j\omega\tau} d\tau \qquad (3.8\text{-}33)$$

4. *The power spectral density is an even function of ω; that is, $\phi(\omega) = \phi(-\omega)$*

From the definition

$$\phi(-\omega) = \int_{-\infty}^{\infty} R(\tau) e^{j\omega\tau} d\tau \qquad (3.8\text{-}34)$$

After the change of variable $x = -\tau$ and after noting that $R(-x) = R(x)$, we have

$$\phi(-\omega) = \int_{-\infty}^{\infty} R(x) e^{-j\omega x} dx = \phi(\omega) \qquad (3.8\text{-}35)$$

In Section 3.3, we introduced the concept of cross-correlation function. If two processes $U(t)$ and $V(t)$ are wide-sense stationary and jointly wide sense stationary, then the cross-correlation function $R_{uv}(\tau)$ is defined as

$$R_{uv}(\tau) = E\{U(t)V(t+\tau)\} \qquad (3.8\text{-}36)$$

In analogy with Eq. (3.8-3), we define the *cross-spectral density* $\phi_{uv}(\omega)$ as the Fourier transform of the cross-correlation function $R_{uv}(\tau)$:

$$\phi_{uv}(\omega) = \int_{-\infty}^{\infty} R_{uv}(\tau) e^{-j\omega\tau} d\tau \qquad (3.8\text{-}37)$$

It was pointed out earlier that the power spectral density $\phi(\omega)$ was real since $R(\tau)$ was an even function. The cross-spectral density is not necessarily real since the cross-correlation function $R_{uv}(\tau)$ obeys $R_{uv}(\tau) = R_{vu}(-\tau)$. However, direct substitution of this last relation in Eq. (3.8-37) shows that

$$\phi_{uv}(\omega) = \phi_{vu}(-\omega) = \phi^*_{vu}(\omega) \qquad (3.8\text{-}38)$$

where the asterisk denotes the complex conjugate.

Example 3.10

We consider the random process $X(t)$ defined by

$$X(t) = U(t) + V(t)$$

where $U(t)$ and $V(t)$ are individually and jointly wide-sense stationary and have zero means. The autocorrelation function of $X(t)$ is given by

$$R_{xx}(\tau) = E\{[U(t)+V(t)][U(t+\tau) + V(t+\tau)]\}$$

$$R_{xx}(\tau) = R_{uu}(\tau) + R_{vv}(\tau) + R_{uv}(\tau) + R_{vu}(\tau)$$

We find the power spectral density $\phi_{xx}(\omega)$ by taking the Fourier transform term-by-term.

$$\phi_{xx}(\omega) = \phi_{uu}(\omega) + \phi_{vv}(\omega) + \phi_{uv}(\omega) + \phi_{vu}(\omega)$$

if $U(t)$ and $V(t)$ are uncorrelated, then $R_{uv}(\tau)$, $\phi_{uv}(\omega)$, $R_{vu}(\tau)$, and $\phi_{vu}(\omega)$ are zero and the power spectral density of the sum is the sum of the power spectral densities. Also since $\phi_{xx}(\omega)$, $\phi_{uu}(\omega)$, and $\phi_{vv}(\omega)$ are all real, it follows from Eq. (3.8-38) that

$$Re\,[\phi_{uv}(\omega)] = Re\,[\phi_{vu}(\omega)] = \frac{1}{2}[\phi_{uv}(\omega) + \phi^{*}_{uv}(\omega)]$$

$$= \frac{1}{2}\,[\phi_{xx}(\omega) - \phi_{uu}(\omega) - \phi_{vv}(\omega)]$$

We take up now the discrete-parameter or discrete-time case where $\{X_i = X(t_i);\ t_i = i = 0, \pm 1, \pm 2, \cdots\}$ is a sequence of random variables (discrete-parameter random process). We will restrict the parameter set T to be the integers. In some physical applications, this is equivalent to assuming that a continuous-parameter random process is sampled at equi-spaced intervals of time to yield the process X_i. This assumption permits the power spectral density of X_i to be represented as a Fourier series whose coefficients are the equi-spaced values of the autocorrelation function of X_i.

As already discussed, the Fourier transform pair provides a natural connection between the autocorrelation function and the power spectral density of a process $X(t)$ which is continuous in m.s. and at least wide-sense stationary. For the discrete-time case, the Fourier series and the expression for its coefficients provide the same connection.

Consider a real function $f(\omega)$ which is periodic with period 2π. This function may be represented formally in the interval $[-\pi, \pi]$ as the Fourier series

$$f(\omega) = \sum_{n=\infty}^{\infty} c_n\, e^{-jn\omega} \qquad (3.8\text{-}39)$$

where the coefficient c_n is given by

$$c_n = \frac{1}{2\pi} \int_{-\pi}^{\pi} f(\omega) e^{jn\omega} d\omega \qquad (3.8\text{-}40)$$

If $f(\omega)$ is bounded and has a finite number of relative maxima and minima and discontinuities in the interval $[-\pi,\pi]$, then the Fourier series converges to $f(\omega)$ at points of continuity in $[-\pi,\pi]$ and to the mean of the discontinuity at points of discontinuity in that interval [12]. Also, Parseval's theorem yields

$$\frac{1}{2\pi} \int_{-\pi}^{\pi} f^2(\omega) d\omega = \sum_{n=-\infty}^{\infty} |c_n|^2 \qquad (3.8\text{-}41)$$

We return now to the discrete-time process $\{X_i = X(t_i); t_i = 0, \pm 1, \pm 2, \ldots\}$ which is assumed to have zero mean so that

$$E\{X_i\} = 0 \quad , \text{ all } i \qquad (3.8\text{-}42)$$

The process is at least wide-sense stationary with autocorrelation function $R(n)$ given by

$$R(j-i) = E\{X_i X_j\} \qquad (3.8\text{-}43)$$

and constant variance

$$R(0) = E\{X_i^2\} = \sigma^2 \quad , \quad \text{ all } i \qquad (3.8\text{-}44)$$

As in the continuous-parameter case, define the random variable $f_N(\omega)$ by the N-th order Fourier polynomial

$$f_N(\omega) = \sum_{k=-N}^{N} X_k e^{-jk\omega} \qquad (3.8\text{-}45)$$

Form the nonnegative quantity $\phi_N(\omega)$ where

$$\phi_N(\omega) = \frac{1}{2N+1} E\{f_N(\omega) f_N(-\omega)\} = \frac{1}{2N+1} E\{|f_N(\omega)|^2\}$$

$$(3.8\text{-}46)$$

$$= \frac{1}{2N+1} \sum_{l=-N}^{N} \sum_{k=-N}^{N} R(l-k) e^{-jl\omega} e^{jk\omega} \geq 0$$

After a change in the indices of summation, this last expression can be rewritten as

$$\phi_N(\omega) = \sum_{m=-2N}^{0} [1 + \frac{m}{2N+1}] R(m) e^{-jm\omega}$$

$$+ \sum_{m=1}^{2N} [1 - \frac{m}{2N+1}] R(m) e^{-jm\omega}$$

or

$$\phi_N(\omega) = \sum_{m=-2N}^{2N} [1 - \frac{|m|}{2N+1}] R(m) e^{-jm\omega} \qquad (3.8\text{-}47)$$

In the limit as $N \to \infty$, this last expression, when it exists*, is called the *power spectral density* $\phi(\omega)$ of the discrete-time process X_i and is given by

$$\phi(\omega) = \lim_{n \to \infty} \frac{1}{2N+1} E\{f_N(\omega) f_N(-\omega)\}$$

$$= \sum_{m=-\infty}^{\infty} R(m) e^{-jm\omega} \qquad (3.8\text{-}48)$$

As mentioned previously, $\phi(\omega)$ is a Fourier series on the interval $[-\pi, \pi]$ using the values of the autocorrelation function $R(m)$ as the Fourier coefficients c_m.

As before define a function $R_N(m)$ by

$$R_N(m) = \begin{cases} [1 - \dfrac{|m|}{2N+1}] R(m) & , \quad |m| \le 2N \\ \\ 0 & , \quad |m| > 2N \end{cases} \qquad (3.8\text{-}49)$$

Now Eq. (3.8-47) can be written as

$$\phi_N(\omega) = \sum_{m=-\infty}^{\infty} R_N(m) e^{-jm\omega} \qquad (3.8\text{-}50)$$

Let us multiply both sides of this expression by $e^{jn\omega} d\omega$ and integrate from $-\pi$ to π to obtain

$$R_N(n) = \frac{1}{2\pi} \int_{-\pi}^{\pi} \phi_N(\omega) e^{jn\omega} d\omega$$

In the limit as $N \to \infty$, this last expression becomes the integral for the Fourier coefficient

$$R(n) = \frac{1}{2\pi} \int_{-\pi}^{\pi} \phi(\omega) e^{jn\omega} d\omega \qquad (3.8\text{-}51)$$

where $R(m)$ is also given by (3.8-43) and $\phi(\omega)$ by Eq. (3.8-48). Note that $\phi(\omega)$ is periodic with period 2π and that

$$R(0) = E\{X_i^2\} = \sigma^2 = \frac{1}{2\pi} \int_{-\pi}^{\pi} \phi(\omega) d\omega \qquad (3.8\text{-}52)$$

---

*It is clear that a sufficient condition for $\phi(\omega)$ to exist is that $\sum |R(m)| < \infty$.

The power spectral density $\phi(\omega)$ in the discrete-time case has the same set of properties as were developed for a continuous parameter process. Also cross-spectral densities can be defined in the same way.

Example 3.11

Let the autocorrelation function $R(n)$ of a discrete-time process X_i be given by

$$R(n) = \sigma^2 \; \frac{\sin^2(n\,\pi/2)}{(n\,\pi/2)^2} \quad , \quad n = 0, \pm 1, \pm 2,...$$

or, equivalently, by

$$R(n) = \begin{cases} \sigma^2 & , \; n = 0 \\[2em] \dfrac{4\sigma^2}{n^2\pi^2} & , \; n \text{ odd} \\[3em] 0 & , \; n \text{ even}, \; n \neq 0 \end{cases}$$

The power spectral density $\phi(\omega)$ is

$$\phi(\omega) = \sum_{n=-\infty}^{\infty} R(n)\, e^{-jn\omega} \quad , \quad -\pi \le \omega \le \pi$$

or

$$\phi(\omega) = \sigma^2 + \frac{8\sigma^2}{\pi^2} \left[\cos\omega + \frac{1}{9}\cos 3\omega + \frac{1}{25}\cos 5\omega + \cdots \right]$$

However, this last expression is the Fourier series expansion in the interval $[-\pi,\pi]$ for the triangle

$$\phi(\omega) = \begin{cases} 2\sigma^2 \, [1 - \dfrac{|\omega|}{\pi}] & , \; |\omega| \le \pi \\[2em] 0 & , \; |\omega| > \pi \end{cases}$$

The result is easily verified by finding the coefficient $R(n)$ through

$$R(n) = \frac{1}{2\pi} \int_{-\pi}^{\pi} \phi(\omega)\, e^{jn\omega}\, d\omega = \frac{2\sigma^2}{\pi} \int_{0}^{\pi} (1 - \frac{\omega}{\pi}) \cos n\omega \, d\omega$$

or

$$R(n) = \sigma^2 \frac{\sin^2(n\,\pi/2)}{(n\,\pi/2)^2}$$

3.9 *Shot Noise* - In this section we begin a brief study of the *shot noise process*. One of the simplest examples of such a process occurs in the temperature- limited vacuum tube diode. Because of its underlying importance as well as its analytic simplicity, this case will now be treated in some detail.

Consider, for example, a diode connected to a resistive load R_L and with proper plate and filament voltages to be operated temperature limited as shown in Fig. 3.6(a).

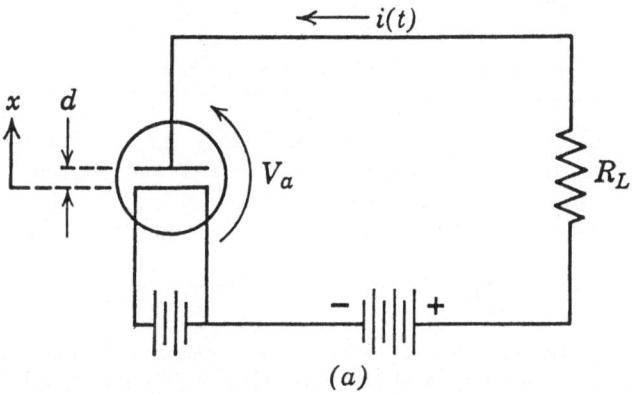

(a)

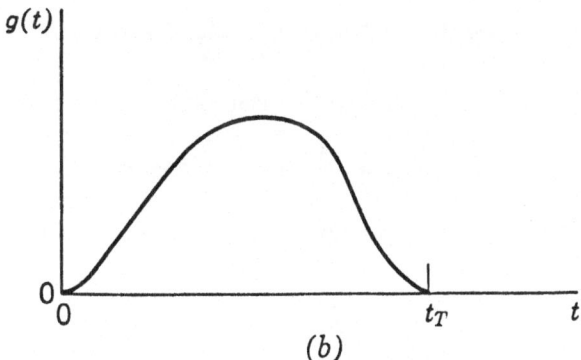

(b)

Fig. 3.6 The temperature limited diode

Since the electrons possess a discrete charge and since they are emitted randomly from the cathode, the plate current $i(t)$ and, hence, the voltage across the load resistor will fluctuate. This is the *shot effect* and the fluctuating component is called the *shot noise*. The effect was first predicted and treated theoretically by Schottky [13] in 1918.

In the temperature-limited case the electron emissions are independent and the current may be considered as the superposition of all of the current pulses due to the individual electrons. A given electron (assumed to be emitted at time $t=0$) will produce a current pulse of some form $g(t)$ as shown in Fig. 3.6(b) where t_T is the transit time and the shape of $g(t)$ depends on the tube geometry, potential distribution, and electron velocity. Note that time t_T is very short (in this respect the picture is very misleading) so that $g(t)$ is almost like a δ-function. From the properties of the electron, we have

$$\int_{-\infty}^{\infty} g(t)dt = \int_0^{t_T} g(t)dt = e \text{ (electronic charge)} \qquad (3.9\text{-}1)$$

Consider the diode current in a long interval $-T \leq t \leq T$ where T is very large compared to t_T. If N electrons are emitted in the time $2T$, then the diode current $i_T(t)$ is just the linear superposition of the N currents due to each electron so that

$$i_T(t) = \sum_{i=1}^{N} g(t-t_i) \quad , \quad -T \leq t \leq T \qquad (3.9\text{-}2)$$

where t_i is the emission time of the i-th electron. Note that N is a function of the length of the interval $(-T,T)$ and could be written $N(2T)$.

The time average value of this current is just

$$A\{i_T(t)\} = <I> = \lim_{T \to \infty} \frac{1}{2T} \int_{-T}^{T} i_T(t)dt \qquad (3.9\text{-}3)$$

On substituting Eq. (3.9-2) into this last expression, we have

$$<I> = \lim_{T \to \infty} \frac{1}{2T} \int_{-T}^{T} \sum_{i=1}^{N} g(t-t_i)dt \qquad (3.9\text{-}4)$$

Interchanging the order of integration and summation, we obtain

$$<I> = \lim_{T \to \infty} \frac{1}{2T} \sum_{i=1}^{N} \int_{-\infty}^{\infty} g(t-t_i)dt \qquad (3.9\text{-}5)$$

or

$$<I> = \lim_{T \to \infty} \frac{1}{2T} \sum_{i=1}^{N} e = \lim_{T \to \infty} \frac{1}{2T} Ne \qquad (3.9\text{-}6)$$

or

$$<I> = <n>e \qquad (3.9\text{-}7)$$

where $<n>$ is the average number of electrons per second and is given by

$$<n> = \lim_{T \to \infty} \frac{N}{2T} \qquad (3.9\text{-}8)$$

Suppose now that the diode circuit of Fig. 3.6(a) is only one of a very large number with identical characteristics. Then the current given by Eq. (3.9-2) can be considered as one of the sample functions or realizations of a random process $i_T(t)$. In this case both N, the number of electrons emitted in the interval $(-T,T)$, and t_i, the emission time of the i-th electron, will be random variables defined on the ensemble of diode currents which are the sample functions. It is apparent that each realization of N is just the number of electrons involved in that particular sample function. Ultimately we are interested in the process $i(t)$ obtained when T becomes infinitely large

$$i(t) = \lim_{T \to \infty} i_T(t) \qquad (3.9\text{-}9)$$

We may write the mean current or ensemble average as

$$E\{i_T(t)\} = I_T =$$

$$(3.9\text{-}10)$$

$$\int\limits_{-\infty}^{\infty} \int\limits_{-\infty}^{\infty} \cdots \int\limits_{-\infty}^{\infty} i_T(t) f_{N+1}(t_1, t_2, \ldots, t_N, N) dt_1 dt_2 \cdots dt_n \, dN$$

where f_{N+1} is the joint distribution (density function) of the N emission times t_i and N itself. In order to proceed further, it is now necessary to determine this joint density f_{N+1}. From the conditional probability relationship

$$P(AB) = P(A)P(B/A) \qquad (3.9\text{-}11)$$

we can rewrite f_{N+1} as

$$f_{N+1}(t_1, \ldots, t_N, N) = f(N) f_C[(t_1, \ldots, t_n)/N] \qquad (3.9\text{-}12)$$

where f_C is a conditional probability density and $f(N)$ is the (univariate) density function of N, the number of electrons emitted in $(-T,T)$. We now proceed to find the two density functions on the right side of Eq. (3.9-12), beginning with $f(N)$.

3.10 *Random Events in Time* - Let the random variable K be the number of events in an interval of length t; for example, the interval $(0,t)$. We assume that

a) the probability of a single event in a very small time interval Δt is proportional to Δt so that

$$P\{\text{one event in } (t,t+\Delta t)\} = \lambda \Delta t \qquad (3.10\text{-}1)$$

b) the probability of more than one event in Δt is negligible so that

$$P\{\text{no event in } (t,t+\Delta t)\} = 1-\lambda \Delta t \qquad (3.10\text{-}2)$$

c) the events are independent.

From these assumptions it is possible to show that K is Poisson distributed with mean λt. The Poisson distribution is discussed in Section 2.17. We proceed as follows:

Let the probability of exactly k events in time t be denoted by $p_k(t)$. In the small time interval Δt, there will be either no events or one event since we have assumed that the probability of more than one event is negligible. Therefore the probability of k events in the time interval $(t+\Delta t)$ is equal to the product of the probabilities of k events in time t and no events in time Δt plus the product of the probabilities of $(k-1)$ events in time t and one event in time Δt. In other words,

$$p_k(t+\Delta t) = p_k(t)(1-\lambda \Delta t)+p_{k-1}(t)\lambda \Delta t \quad , \quad k \geq 1 \quad (3.10\text{-}3)$$

From the definition of derivative, we have

$$\frac{dp_k(t)}{dt} = \lim_{\Delta t \to 0} \frac{p_k(t+\Delta t)-p_k(t)}{\Delta t} \qquad (3.10\text{-}4)$$

Thus Eq. (3.10-3) may be rewritten as

$$\frac{dp_k(t)}{dt} = \lambda[p_{k-1}(t)-p_k(t)] \qquad (3.10\text{-}5)$$

This equation may be solved by induction as follows: Let $k=0$, then

$$\frac{dp_0(t)}{dt} = -\lambda p_0(t) \qquad (3.10\text{-}6)$$

This is an ordinary homogeneous differential equation in the dependent variable $p_0(t)$ with solution

$$p_0(t) = C_1 e^{-\lambda t} \qquad (3.10\text{-}7)$$

where C_1 is an arbitrary constant determined from the boundary condition

$$p_0(0) = 1 = C_1 \qquad (3.10\text{-}8)$$

Thus, for $k=0$, we have

$$p_0(t) = e^{-\lambda t} \qquad (3.10\text{-}9)$$

Now let $k=1$ and use this last result to write Eq. (3.10-5) as

$$\frac{dp_1(t)}{dt} + \lambda p_1(t) = \lambda e^{-\lambda t} \qquad (3.10\text{-}10)$$

This is an ordinary non-homogeneous differential equation with solution

$$p_1(t) = C_2 e^{-\lambda t} + \lambda t e^{-\lambda t} \qquad (3.10\text{-}11)$$

The appropriate boundary condition now is

$$p_1(0) = 0 = C_2 \qquad (3.10\text{-}12)$$

so that

$$p_1(t) = \lambda t e^{-\lambda t} \qquad (3.10\text{-}13)$$

We continue in the same way and find that

$$p_k(t) = e^{-\lambda t} \frac{(\lambda t)^k}{k!} \quad , \quad k = 0,1,\ldots \qquad (3.10\text{-}14)$$

This result may be checked by direct substitution into Eq. (3.10-5). It is the Poisson distribution of Eq. (2.17-7) with the parameter λ replaced by λt. Thus the random variable K with density function given by Eq. (3.10-14) has a mean

$$E\{K\} = \lambda t \qquad (3.10\text{-}15)$$

and a variance

$$E\{(K - \lambda t)^2\} = \lambda t \qquad (3.10\text{-}16)$$

Both of these results follow from Eqs. (2.17-11) and (2.17-14).

Note that $p_k(t)$ is the probability that exactly k events occur in a time interval t. The term $f(N)$ of Eq. (3.9-12) is the probability that exactly N electrons are emitted in the interval $2T$. On setting up the correspondences

$$t \sim 2T \qquad (3.10\text{-}17)$$

$$\lambda \sim \frac{E\{N\}}{2T} \qquad (3.10\text{-}18)$$

and hence

$$\lambda t \sim E\{N\} = \overline{N} \qquad (3.10\text{-}19)$$

we have

$$f(N) = \frac{e^{-\overline{N}}(\overline{N})^N}{N!} \quad , \quad N = 0,1,\ldots \qquad (3.10\text{-}20)$$

where $\overline{N}$ has been written for $E\{N\}$ in order to simplify notation. It should be kept in mind that $f(N)$ is a discrete density defined on the non-negative integers.

The next problem is to determine the conditional density $f_C[(t_1,...,t_N)/N]$ of Eq. (3.9-12). Let us use Eq. (3.9-11) and construct the probability

$$P(C) = \frac{P(D)}{f(N)} \qquad (3.10\text{-}21)$$

where C is the event "one electron is emitted in each of the intervals $(t_1,t_1+dt_1)...(t_N,t_N+dt_N)$ given that N electrons are emitted in $(-T,T)$ where all $t_i \in (-T,T)$", D is the event "no electrons emitted in the intervals $(-T,t_1)$, $(t_1+dt_1,t_2)...(t_N+dt_N,T)$ and one electron emitted in each of the intervals (t_1,t_1+dt_1), $(t_2,t_2+dt_2)...(t_N,t_N+dt_N)$" and $f(N)$ has been defined previously by Eq. (3.10-20) and is the event "exactly N electrons emitted in $(-T,T)$". or

$$P\{C\} =$$

$$\frac{\left[e^{-\lambda(t_1+T)} e^{-\lambda(t_2-t_1-dt_1)}...e^{-\lambda(T-t_N-dt_N)} \right] \left[\lambda dt_1 e^{-\lambda dt_1}...\lambda dt_N e^{-\lambda dt_N} \right]}{\dfrac{(\lambda 2T)^N}{N!} e^{-\lambda 2T}}$$

where λ has been written for $\overline{N}/2T$. This expression is readily reduced to

$$P\{C\} = \frac{N!}{(2T)^N} dt_1...dt_N \qquad (3.10\text{-}22)$$

But the probability $P\{C\}$ can also be written as

$$P\{C\} = \sum_{\substack{\text{all orderings} \\ \text{of the } x_i}} p(x_1,x_2,...,x_N)dx_1 dx_2...dx_N \qquad (3.10\text{-}23)$$

where

$$p(x_1,x_2,...,x_N)dx_1 dx_2...dx_N = P(E) \qquad (3.10\text{-}24)$$

Here E is the event "the emission time t_i falls in the interval (x_i,x_i+dx_i) for each $i=1,2,...,N$". Since there are exactly $N!$ orderings of N objects [see Section 2.7], we easily calculate Eq. (3.10-23) to be

$$P\{C\} = N!p(t_1,t_2,...,t_N)dt_1 dt_2...dt_N \qquad (3.10\text{-}25)$$

By direct comparison of Eqs. (3.10-22) and (3.10-25) we conclude that

$$p(t_1,t_2,...,t_N) = f_1(t_1)f_1(t_2)...f_1(t_N) \qquad (3.10\text{-}26)$$

where each univariate density $f_1(t_i)$ is given by

$$f_1(t_i) = \begin{cases} \dfrac{1}{2T} & , \ -T \leq t_i \leq T \\[2ex] 0 & , \ \text{elsewhere} \end{cases} \quad , \quad i = 1,2,...,N \quad (3.10\text{-}27)$$

The results of the last two equations are intuitively reasonable. The emission times of the electrons are independent and each is as likely to occur in any one small interval as in any other.

3.11 The Mean and Autocorrelation Function of Shot Noise -
We now return to the evaluation of Eq. (3.9-10). It follows from Eqs. (3.10-1) and (3.10-26) that

$$f_{N+1}(t_1,...,t_N,N) = f(N)f_1(t_1)...f_1(t_N) \quad (3.11\text{-}1)$$

where the $f_1(t_i)$ are the uniform distributions of Eq. (3.10-27) and $f(N)$ is the discrete density function given by Eq. (3.10-20). Therefore Eq. (3.9-10) may be written as

$$I_T = \sum_{N=0}^{\infty} \int_{-T}^{T} \cdots \int_{-T}^{T} \sum_{i=1}^{N} g(t-t_i)f(N)f_1(t_1)...f_1(t_N)dt_1...dt_N \quad (3.11\text{-}2)$$

Let us now bring each integral with respect to each t_j inside the summation. It is clear that $(N-1)$ of these (where $i \neq j$) will be of the form

$$\int_{-T}^{T} f_1(t_j)dt_j = \frac{1}{2T} \int_{-T}^{T} dt_j = 1 \quad (3.11\text{-}3)$$

Thus Eq. (3.11-3) may be rewritten as

$$I_T = \sum_{N=0}^{\infty} f(N)\left\{ \sum_{i=1}^{N} \frac{1}{2T} \int_{-T}^{T} g(t-t_i)dt_i \right\} \quad (3.11\text{-}4)$$

Consider now the integral

$$\int_{-T}^{T} g(t-t_i)\,dt_i$$

If the function $g(t)$ has finite support and if the emission time t_i occurs far enough inside the interval $(-T,T)$, then the integral becomes e, the electronic charge, in accordance with Eq. (3.9-1). Shortly, we will let $T \rightarrow \infty$ and "end-point" problems with this integral will disappear. Meanwhile, we will assume that the integral is e and proceed but, until $T \rightarrow \infty$, the result is an approximation which becomes better and better as T gets large compared to t_T, the support of $g(t)$. Now, Eq. (3.11-4) becomes

$$I_T = \frac{e}{2T} \sum_{N=0}^{\infty} Nf(N) = \frac{e\bar{N}}{2T} = \lambda e \qquad (3.11\text{-}5)$$

where $\bar{N}$ is the mean value of N and is given by

$$\bar{N} = E\{N\} = \sum_{N=0}^{\infty} Nf(N) \qquad (3.11\text{-}6)$$

and λ is the mean number of electrons per unit of time and is

$$\lambda = \frac{\bar{N}}{2T} \qquad (3.11\text{-}7)$$

as previously defined by Eq. (3.10-18). Notice that the result of Eq. (3.11-5) is independent of T [taking into account the discussion before Eq. (3.11-5) and letting T become large enough] so that we have

$$\lim_{T \to \infty} E\{i_T(t)\} = E\{i(t)\} = I = \lambda e \qquad (3.11\text{-}8)$$

as the final result for the mean value of the shot current process $i(t)$.

We now proceed to calculate the autocorrelation function of $i(t)$. We begin by forming

$$E\{i_T(t)i_T(t+\tau)\} = \qquad (3.11\text{-}9)$$

$$\sum_{N=0}^{\infty} \int_{-\infty}^{\infty} \cdots \int_{-\infty}^{\infty} \left[\sum_{i=1}^{N} g(t-t_i) \right] \left[\sum_{j=1}^{N} g(t+\tau-t_j) \right] f_{N+1}(t_1,...,t_N,N) dt_1...dt_N$$

We now make the substitution of Eq. (3.11-1) for f_{N+1} and write

$$E\{i_T(t)i_T(t+\tau)\} = \qquad (3.11\text{-}10)$$

$$\sum_{N=0}^{\infty} f(N) \left[\sum_{i=1}^{N}\sum_{j=1}^{N} (\frac{1}{2T})^N \int_{-T}^{T}....\int_{-T}^{T} g(t-t_i)g(t+\tau-t_j) dt_1...dt_N \right]$$

For each of the N terms where $i=j$ in the double summation, we have

$$\frac{1}{2T}\int_{-T}^{T} g(t-t_i)g(t+\tau-t_i) dt_i$$

and for each of the N^2-N terms where $i \neq j$, we find

$$\frac{1}{2T}\int_{-T}^{T} g(t-t_i) dt_i \; \frac{1}{2T}\int_{-T}^{T} g(t+\tau-t_j) dt_j$$

Thus Eq. (3.11-10) becomes

$$E\{i_T(t)i_T(t+\tau)\} = \sum_{N=0}^{\infty} f(N)[\frac{N}{2T}\int_{-T}^{T} g(t-t_i)g(t+\tau-t_i)dt_i$$

(3.11-11)

$$+ \frac{N^2-N}{(2T)^2} \int_{-T}^{T} g(t-t_i)dt_i \int_{-T}^{T} g(t+\tau-t_j)dt_j]$$

In the limit as $T \to \infty$, we obtain the autocorrelation function $R(\tau)$ of $i(t)$. Note that

$$\lim_{T\to\infty} \int_{-T}^{T} g(t-t_i)dt_i = \lim_{T\to\infty} \int_{-T}^{T} g(t+\tau-t_j)\,dt_j = e$$

where e is the electronic charge. Also we have

$$\lim_{T\to\infty} \int_{-T}^{T} g(t-t_i)g(t+\tau-t_i)dt_i = \int_{-\infty}^{\infty} g(t)g(t+\tau)dt$$

We may now write

$$R(\tau) = \lim_{T\to\infty} E\{i_T(t)i_T(t+\tau)\}$$

$$= \int_{-\infty}^{\infty} g(t)g(t+\tau)dt \; \frac{1}{2T} \sum_{N=0}^{\infty} Nf(N)$$

(3.11-12)

$$+ (\frac{e}{2T})^2 \sum_{N=0}^{\infty} (N^2-N)f(N)$$

Note that

$$\sum_{N=0}^{\infty} N \frac{(\bar{N})^N}{N!} e^{-\bar{N}} = \bar{N}$$

(3.11-13)

and

$$\sum_{N=0}^{\infty} N(N-1) \frac{(\bar{N})^N}{N!} e^{-\bar{N}} = (\bar{N})^2$$

(3.11-14)

from the properties of the Poisson distribution as given by Eqs. (2.7-11) and (2.17-13) respectively. Thus we have finally

$$R(\tau) = E\{i(t)i(t+\tau)\} = \lambda \int_{-\infty}^{\infty} g(t)g(t+\tau)dt + \lambda^2 e^2$$

(3.11-15)

It is convenient to define a new shot current $i_a(t)$ with zero mean by

$$i_a(t) = i(t) - \lambda e$$

(3.11-16)

with autocorrelation function

$$R_a(\tau) = E\{i_a(t)i_a(t+\tau)\} = R(\tau) - \lambda^2 e^2$$

(3.11-17)

or

$$R_a(\tau) = \lambda \int_{-\infty}^{\infty} g(t)g(t+\tau)dt \qquad (3.11\text{-}18)$$

Thus the shape of the autocorrelation function is determined only by the shape of the basic current pulses due to the individual electrons.

Let us consider briefly the power spectral density $\phi_a(\omega)$ where

$$\phi_a(\omega) = \int_{-\infty}^{\infty} R_a(\tau)e^{-j\omega\tau}d\tau \qquad (3.11\text{-}19)$$

This expression is easily evaluated as follows:

$$\phi_a(\omega) = \lambda \int_{-\infty}^{\infty} \int_{-\infty}^{\infty} g(t)g(t+\tau)dte^{-j\omega\tau}d\tau \qquad (3.11\text{-}20)$$

We make the linear change in variable $\tau = \sigma - t$ to obtain

$$\phi_a(\omega) = \lambda \int_{-\infty}^{\infty} g(\sigma)e^{-j\omega\sigma}d\sigma \int_{-\infty}^{\infty} g(t)e^{j\omega t}\,dt \qquad (3.11\text{-}21)$$

or

$$\phi_a(\omega) = \lambda G(\omega)G(-\omega) = \lambda \mid G(\omega)\mid^2 \qquad (3.11\text{-}22)$$

where $G(\omega)$ is the Fourier transform of the basic pulse shape.

The low frequency behavior of the power spectral density is given by

$$\lim_{\omega \to 0} \phi_a(\omega) = \lambda \mid \int_{-\infty}^{\infty} g(t)dt \mid^2 = \lambda e^2 \qquad (3.11\text{-}23)$$

from Eq. (3.11-21). Note that this last expression may be written as

$$\lim_{\omega \to 0} \phi_a(\omega) = e(\lambda e) = eI \qquad (3.11\text{-}24)$$

which is often called the *Schottky Formula*.

Example 3.11

We consider a parallel-plane diode as shown in Fig. 3.6(a) operated temperature limited and with cathode-anode spacing d and plate voltage V_a. Since the diode is operated temperature limited, there is no space charge and Laplace's Equation is satisfied in the cathode-anode space

$$\nabla^2 V = 0 = \frac{\partial^2 V}{\partial^2 x}$$

The solution is

$$V = ax + b$$

At $x = 0$, $V = 0$ so that $b = 0$; also at $x = d$, $V = V_a$ so that $a = V_a / d$. Therefore we have

$$V = \frac{V_a}{d} x \quad , \quad 0 \le x \le d$$

for the potential distribution in the cathode-anode space.

The energy ξ gained by an electron in moving through a potential V is

$$\xi = eV$$

and the corresponding power flow is

$$\frac{\partial \xi}{\partial t} = e \frac{dV}{dt} = \frac{eV_a}{d} \frac{dx}{dt} \quad V_a i$$

Thus the current flow due to a single electron is

$$i = g(t) = \frac{e}{d} \frac{dx}{dt} = \frac{e}{d} v_x$$

where v_x is the electron velocity . The electron acceleration a_x is a constant since

$$a_x = \frac{e}{m} \frac{V_a}{d}$$

hence

$$v_x = \frac{e}{m} \frac{V_a}{d} t$$

where the additive constant of integration is zero from our assumption that v_x is zero at $t = 0$; that is, we assume that the electron is emitted with zero initial velocity* from the cathode. Thus the current is

$$i = g(t) = \frac{e^2 V_a}{md^2} t$$

If we call the transit time t_T, then the average value of the velocity v_x is given by

$$\frac{d}{t_T} = \frac{1}{2} \frac{e}{m} \frac{V_a}{d} t_T$$

We may now write the current due to a single electron as

---

*This assumption of zero initial velocity is not correct since the actual velocity is a random variable with Maxwellian distribution. For details see [14].

$$i = g(t) = \begin{cases} \dfrac{2e}{t_T^2}\, t & , \ 0 \le t \le t_T \\[3mm] 0 & , \ \text{elsewhere} \end{cases}$$

as shown in Fig. 3.7(a).

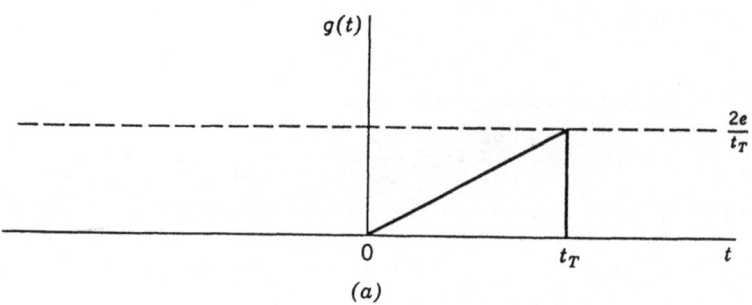

(a)

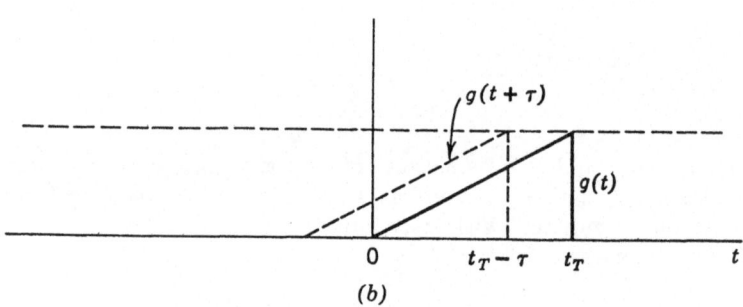

(b)

Fig. 3.7 - The current in a parallel-plane diode

The autocorrelation function is easily found from Eq. (3.11-18) and Fig. 3.7(b). For the region $0 \le \tau \le t_T$ we see that

$$R_a(\tau) = \lambda \int_0^{t_T-\tau} \frac{2e}{t_T^2}\, t\; \frac{2e}{t_T^2}\,(t+\tau)dt = \frac{4e^2\lambda}{t_T^4} \left[\frac{t^3}{3} + \frac{t^2\tau}{2} \right]_{t=0}^{t=t_T-\tau}$$

$$= \frac{4}{3}\, \frac{\lambda e^2}{t_T}\,(1-\frac{\tau}{t_T})^2(1+\frac{1}{2}\,\frac{\tau}{t_T})$$

Since $I = \lambda e$ and since $R_a(\tau)$ is even, we have

$$
R_a(\tau) =
\begin{cases}
\dfrac{4}{3}\dfrac{eI}{t_T}\left(1-\dfrac{|\tau|}{t_T}\right)^2\left(1+\dfrac{1}{2}\dfrac{|\tau|}{t_T}\right) & , \quad 0 \le |\tau| \le t_T \\[2em]
0 & , \quad \text{elsewhere}
\end{cases}
$$

This function is plotted in Fig. 3.8(a).

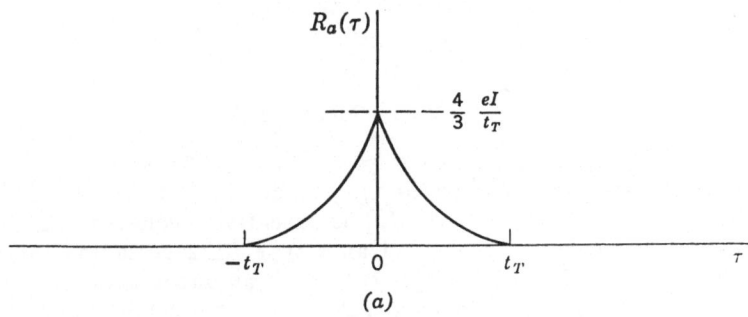

(a)

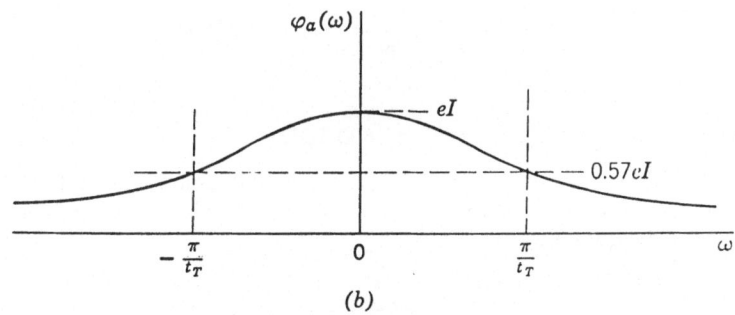

(b)

Fig. 3.8 - The autocorrelation function and power
spectral density of Example 3.11

The power spectral density $\phi_a(\omega)$ is obtained by direct transformation or from Eq. (3.16-22). We have

$$
\phi_a(\omega) = 2 \int_0^{\infty} R(\tau) \cos \omega\tau \, d\tau = \frac{8eI}{t_T^2} \int_0^{t_T} \left(\frac{t_T}{3} - \frac{\tau}{2} + \frac{1}{6}\frac{\tau^3}{t_T^2}\right) \cos \omega\tau \, d\tau
$$

or, finally,

$$\phi_a(\omega) = \frac{4eI}{(\omega t_T)^4} \left\{ (\omega t_T)^2 + 2(1 - \cos \omega t_T - \omega t_T \sin \omega t_T) \right\}$$

This spectrum is plotted in Fig. 3.8(b).

At low frequencies ($\omega \ll \pi/t_T$) the spectrum is nearly constant at a value of eI. Thus, at *low frequencies,* the shot noise power in a bandwidth $\Delta f = f_2 - f_1$ is approximately

$$P_N \approx \frac{1}{2\pi} \int_{-\omega_2}^{\omega_1} \phi(0) d\omega + \frac{1}{2\pi} \int_{\omega_1}^{\omega_2} \phi(0) d\omega$$

or

$$P_N \approx \frac{1}{\pi} \int_{\omega_1}^{\omega_2} eI d\omega = 2eI(f_2 - f_1) \approx 2eI \Delta f$$

Some practical ideal of the magnitude of t_T and hence of the frequency range over which $\phi(\omega)$ is essentially constant can be obtained by considering the Sylvania type 5722 noise generating diode. This is a 7-pin miniature tube with an anode-cathode spacing of 0.0375 inches. When operated as recommended, it has a transit angle of 58° at 500 MHz., or a transit time t_T of approximately 300 $\mu\mu$ seconds. Hence the frequency at which $\phi(\omega)$ has dropped to 0.57 of its value at zero frequency is approximately

$$f = \frac{1}{2t_T} \approx \frac{10^{12}}{600} \approx 1.6 \times 10^9 \text{ Hz.}$$

or

$$f \approx 1600 \text{ MHz.}$$

The 5722 is used as a noise generator at frequencies to 400 or 500 MHz. and has a reasonably flat power spectral density up to this range of frequencies.

3.12 *The Amplitude Distribution of Shot Noise* - As defined in Section 2.15, the characteristic function $M_X(jv)$ of a random variable X is given by

$$M_X(jv) = E\{e^{jvX}\} = \int_{-\infty}^{\infty} f_X(x) e^{jvX} dx \qquad (3.12\text{-}1)$$

where $f_X(x)$ is the density function of X. As in Section 2.19 we consider the case where X is made up of a sum of *independent* random variables; that is,

$$X = X_1 + X_2 + \dots + X_n = \sum_{i=1}^{n} X_i \qquad (3.12\text{-}2)$$

Then the characteristic function becomes

$$M_X(jv) = E\{e^{jvX}\} = E\{e^{jv\sum\limits_{i=1}^{n}X_i}\} = \prod_{i=1}^{n} M_{X_i}(jv) \quad (3.12\text{-}3)$$

where $M_{X_i}(jv)$ is the characteristic function of X_i; that is,

$$M_{X_i}(jv) = E\{e^{jvX_i}\} \quad (3.12\text{-}4)$$

If the X_i are identically distributed so that each has the same characteristic function, say $M_1(jv)$, then Eq. (3.12-3) becomes

$$M_X(jv) = [M_1(jv)]^n \quad (3.12\text{-}5)$$

Let us now consider the shot current $i(t)$ of Eq. (3.9-9) and find

$$M_X(jv) = E\{e^{jvi(t)}\} = \lim_{T \to \infty} E\{e^{jvi_T(t)}\} \quad (3.12\text{-}6)$$

where $i_T(t)$ is given by Eq. (3.9-2). Note that $i_T(t)$ is a sum of independent random variables $g(t-t_i)$ and that the number N of the terms in the sum is also a random variable with density function $f(N)$ given by Eq. (3.10-20). We can write

$$M_X(jv) = \sum_{N=0}^{\infty} f(N)M_X(jv/N) \quad (3.12\text{-}7)$$

where $M_X(jv/N)$ is the characteristic function for a given N. It follows from Eq. (3.12-3) that

$$M_X(jv) = \lim_{T \to \infty} \sum_{N=0}^{\infty} f_N(N)[M_1(jv)]^N \quad (3.12\text{-}8)$$

where $M_1(jv)$ is the characteristic function of one of the $g(t-t_k)$ and can be written as

$$M_1(jv) = E\{e^{jvg(t-t_i)}\} = \frac{1}{2T} \int_{-T}^{T} e^{jvg(t-\sigma)} d\sigma \quad (3.12\text{-}9)$$

from Eq. (3.10-27).

We may use Eq. (3.10-20) to rewrite Eq. (3.12-8) as

$$M_X(jv) = \lim_{T \to \infty} \sum_{N=0}^{\infty} \frac{e^{-\overline{N}}(\overline{N})^N}{N!} [M_1(jv)]^N \quad (3.12\text{-}10)$$

$$= \lim_{T \to \infty} e^{-\overline{N}} \sum_{N=0}^{\infty} \frac{[\overline{N}M_1(jv)]^N}{N!}$$

or

$$M_X(jv) = \lim_{T \to \infty} e^{-\overline{N}} e^{\overline{N}M_1(jv)} = \lim_{T \to \infty} e^{\overline{N}[M_1(jv)-1]} \quad (3.12\text{-}11)$$

We now use Eq. (3.12-9) to obtain finally

$$M_X(jv) = \lim_{T \to \infty} e^{\frac{\bar{N}}{2T} \int_{-T}^{T} [e^{jvg(t-\sigma)}-1]d\sigma} \tag{3.12-12}$$

or

$$M_X(jv) = e^{\lambda \int_{-\infty}^{\infty} [e^{jvg(\sigma)}-1]d\sigma} \tag{3.12-13}$$

In general this last expression is difficult to evaluate. However for the high density case ($\lambda \to \infty$) we can show that $M_X(jv)$ is the characteristic function of a Gaussian random variable. For convenience, we introduce the normalized current

$$Y(t) = \frac{i(t)-\lambda e}{\sqrt{\lambda}} \tag{3.12-14}$$

with zero mean; that is,

$$E\{Y\} = \frac{E\{i\}-\lambda e}{\sqrt{\lambda}} = \frac{\lambda e-\lambda e}{\sqrt{\lambda}} = 0 \tag{3.12-15}$$

and autocorrelation function $R_y(\tau)$

$$R_y(\tau) = E\{Y(t)Y(t+\tau)\} = \frac{R(\tau)-\lambda^2 e^2}{\lambda} \tag{3.12-16}$$

or

$$R_y(\tau) = \int_{-\infty}^{\infty} g(t)g(t+\tau)d\tau \tag{3.12-17}$$

We now note that the characteristic function of Y is

$$M_Y(jv) = E\{e^{jvY}\} = E\{e^{jv\frac{i-\lambda e}{\sqrt{\lambda}}}\} \tag{3.12-18}$$

$$= e^{-j\sqrt{\lambda}ev} E\{e^{\frac{jiv}{\sqrt{\lambda}}}\} = e^{-j\sqrt{\lambda}ev} M_X(j\frac{v}{\sqrt{\lambda}})$$

where $M_X(jv)$ is given by Eq. (3.12-13). Thus

$$M_Y(jv) = e^{\lambda \int_{-\infty}^{\infty} [e^{\frac{jv}{\sqrt{\lambda}}g(\sigma)}-1-j\frac{v}{\sqrt{\lambda}}g(\sigma)]d\sigma} \tag{3.12-19}$$

since

$$e^{j\sqrt{\lambda}ev} = e^{\lambda \int_{-\infty}^{\infty} \frac{jv}{\sqrt{\lambda}}g(\sigma)d\sigma} \tag{3.12-20}$$

On expanding the exponential inside the integral of Eq. (3.12-19) in a power series, we obtain

$$M_Y(jv) = e^{\lambda \int_{-\infty}^{\infty} \sum_{k=2}^{\infty} \frac{j^k}{k!} (\frac{1}{\sqrt{\lambda}})^k [vg(\sigma)]^k d\sigma} \tag{3.12-21}$$

On allowing λ to become large, the leading term predominates and

$$\lim_{\lambda \to \infty} M_Y(jv) = e^{\int_{-\infty}^{\infty} \frac{(j)^2}{2} v^2 g^2(\sigma) d\sigma} = e^{-\frac{v^2}{2} \int_{-\infty}^{\infty} g^2(\sigma) d\sigma} \tag{3.12-22}$$

We see from Eq. (3.12-17) that

$$\lim_{\lambda \to \infty} M_Y(jv) = e^{-\frac{v^2}{2} R_y(0)} \tag{3.12-23}$$

As shown in Section 2.18, this is the characteristic function of a Gaussian random variable with zero mean and variance $R_y(0)$. Thus the shot current has a limiting first order probability amplitude distribution that is Gaussian or normal. Note also that this probability amplitude distribution will be independent of time so that the shot process is stationary to order one.

The second order distribution may be found in the same way. We are interested in the joint density of $i(t)$ and $i(t+\tau)$ with corresponding characteristic function $M_{X_1,X_2}(jv_1, jv_2)$ given by

$$M_{X_1,X_2}(jv_1, jv_2) = \lim_{T \to \infty} E\{e^{j[v_1 i_T(t) + v_2 i_T(t+\tau)]}\} \tag{3.12-24}$$

As in Eq. (3.12-10), this may be written as

$$M_{X_1,X_2}(jv_1, jv_2) = \tag{3.12-25}$$

$$\lim_{T \to \infty} \sum_{N=0}^{\infty} \frac{e^{-\bar{N}}(\bar{N})^N}{N!} \prod_{i=1}^{N} \frac{1}{2T} \int_{-T}^{T} e^{j[v_1 g(t-t_i) + v_2 g(t+\tau-t_i)]} dt_i$$

We now follow the previous development of Eqs. (3.12-10) to (3.12-13) to obtain

$$M_{X_1,X_2}(jv_1, jv_2) = e^{\lambda \int_{-\infty}^{\infty} [e^{j[v_1 g(\sigma) + v_2 g(\sigma+\tau)]} - 1] d\sigma} \tag{3.12-26}$$

Again this expression is difficult to evaluate except in the limit of large λ. As in Eq. (3.12-14) we define a normalized current $Y(t)$ with joint characteristic function $M_{Y_1,Y_2}(jv_1, jv_2)$ where

$$M_{Y_1,Y_2}(jv_1, jv_2) = E\{e^{j[v_1 Y(t) + v_2 Y(t+\tau)]}\}$$

$$= E\{e^{j[v_1 \frac{i_1 - \lambda e}{\sqrt{\lambda}} + v_2 \frac{i_2 - \lambda e}{\sqrt{\lambda}}]}\} \tag{3.12-27}$$

$$= e^{-j\sqrt{\lambda}e(v_1+v_2)} M_{X_1,X_2}(\frac{jv_1}{\sqrt{\lambda}}, \frac{jv_2}{\sqrt{\lambda}})$$

where i_1 has been written for $i(t)$ and i_2 for $i(t+\tau)$ and where $M_{X_1, X_2}(jv_1, jv_2)$ is given by Eq. (3.12-26). Thus we have

$$M_{Y_1, Y_2}(jv_1, jv_2) = e^{\lambda \int\limits_{-\infty}^{\infty} \sum\limits_{k=2}^{\infty} \frac{j^k}{k!}(\frac{1}{\sqrt{\lambda}})^k |v_1 g(\sigma) + v_2 g(\sigma+\tau)|^k d\sigma} \qquad (3.12\text{-}28)$$

On passing to the limit we obtain

$$\lim_{\lambda \to \infty} M_{Y_1, Y_2}(jv_1, jv_2) = e^{-\frac{1}{2}\int\limits_{-\infty}^{\infty} |v_1 g(\sigma)+v_2 g(\sigma+\tau)|^2 d\sigma} \qquad (3.12\text{-}29)$$

We now use Eq. (3.12-17) to rewrite this last expression as

$$\lim_{\lambda \to \infty} M_{Y_1, Y_2}(jv_1, jv_2) = e^{-\frac{1}{2}[v_1^2 R_y(0) + 2v_1 v_2 R_y(\tau)+v_2^2 R_y(0)]} \qquad (3.12\text{-}30)$$

As shown by Eq. (2.22-7), this is the characteristic function for the bivariate normal distribution with mean of zero and covariance function $R_y(\tau)$. Thus the second-order amplitude distribution of the shot current is jointly normal in the limit of high density. Note that the second-order amplitude distribution of shot noise is independent of t_1 and t_2 but depends only on the difference $\tau = t_2 - t_1$. Thus the shot process is stationary to order two.

PROBLEMS

1. Consider the random process $E(t) = A \sin(\omega t + \theta)$ where A and ω are constants and θ is a random variable. What are necessary and sufficient conditions on the random variable θ such that the random process $E(t)$ is wide-sense stationary? Give a specific example of such a random variable.

2. Consider the random process $Y(t) = X \cos t$ where X is a normally distributed random variable with zero mean and unit variance. What is the probability density function of $Y(0)$? What is the joint density function of $Y(0)$ and $Y(\pi)$? Is the random process stationary?

3. Give at least two reasons why the following matrix cannot be a correlation matrix of two jointly wide sense stationary processes.

$$\begin{bmatrix} A^2 \cos \tau & 2A^2 \cos \dfrac{3\tau}{2} \\[3mm] 2A^2 \cos \dfrac{3\tau}{2} & A^2 \sin 2\tau \end{bmatrix}$$

4. Consider two stationary random processes $X(t)$ and $Y(t)$. Find the autocorrelation function of the sum of these two processes assuming, (a) that the processes are correlated, (b) that the processes are uncorrelated, (c) that the processes are uncorrelated and have zero means.

5. Consider the random process $E_1(t) = A \sin(\omega t + \theta)$ where A and ω are constants and θ is a uniformly distributed random variable on $(-\pi, +\pi)$. Let $E_2(t) = |E_1(t)|$. Find the autocorrelation function of $E_2(t)$ and the cross-correlation function of $E_1(t)$ and $E_2(t)$.

6. Consider the constant random process $X(t){=}A$ where A is a random variable uniformly distributed on $(-1,+1)$. Find the autocorrelation function and time autocorrelation function of this random process. Is the process ergodic?

7. If $x(t)$ is a bounded periodic function with fundamental period P, show that

$$\lim_{T \to \infty} \frac{1}{2T} \int_{-T}^{+T} x(t)x(t+\tau)dt = \frac{1}{P} \int_{0}^{P} x(t)x(t+\tau)dt$$

8. Find the time autocorrelation function of the periodic function $x(t)$ shown below.

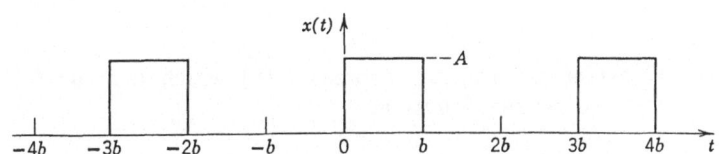

9. Find the time cross-correlation function of the two periodic functions $x(t)$ and $y(t)$ shown below.

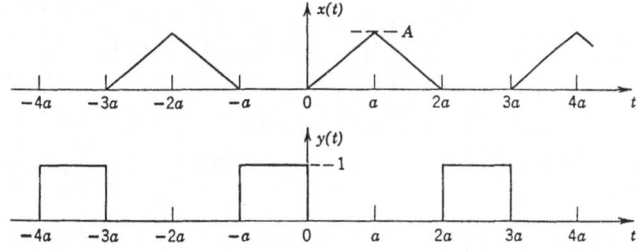

10. Consider two different periodic functions $x(t)$ and $y(t)$ having the same fundamental period. Show that their time cross-correlation function depends only on those harmonics which are present in both $x(t)$ and $y(t)$.

11. Is the following function the power spectral density of a physical process? If so, what is the autocorrelation function of the process? If not, what condition does it violate?

$$\frac{A^2\omega^2}{a^2 + \omega^2}$$

12. Find the Fourier transform of $\dfrac{1}{b^2 + t^2}$ using contour integration.

13. Consider a wide-sense stationary random process having the power spectral density $\phi(\omega)$ as shown. Find the autocorrelation function of the random process.

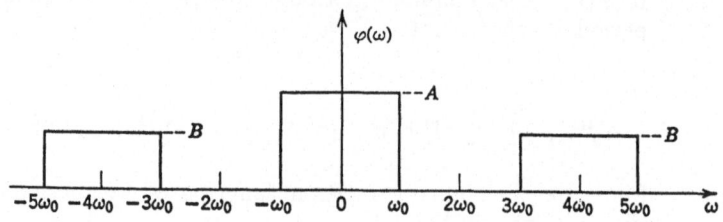

14. Consider the random process $X(t)$ which has the following discrete density function:

$$f(x) = \begin{cases} 1-1/t^2 & , \; x = 0, \; t > 1 \\[2mm] 1/2t^2 & , \; x = t, \; t > 1 \\[2mm] 1/2t^2 & , \; x = -t, \; t > 1 \\[2mm] 1/2 & , \; x = 1, \; t \leq 1 \\[2mm] 1/2 & , \; x = -1, \; t \leq 1 \end{cases}$$

Find the mean and variance of $X(t)$. Is the process stationary?

15. A random process $X(t)$ has the autocorrelation function shown.

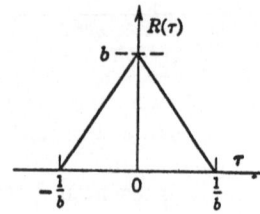

Find the power spectral density of this process. As b becomes large what does the process tend towards?

16.   Consider the following sequences of random variables:

$$(a) \quad f(x_k) = \begin{cases} 2^{-(k+1)} & \text{for } x_k = -2^k \\ 1-2^{-k} & \text{for } x_k = 0 \qquad \text{for } k = 1,2,... \\ 2^{-(k+1)} & \text{for } x_k = 2^k \end{cases}$$

$$(b) \quad f(x_k) = \begin{cases} 2^{-(2k+1)} & \text{for } x_k = -2^k \\ 1-2^{-2k} & \text{for } x_k = 0 \qquad \text{for } k = 1,2,... \\ 2^{-(2k+1)} & \text{for } x_k = 2^k \end{cases}$$

$$(c) \quad f(x_k) = \begin{cases} (1/2)k^{-1/2} & \text{for } x_k = -k \\ 1-k^{-1/2} & \text{for } x_k = 0 \qquad \text{for } k = 2,3,... \\ (1/2)k^{-1/2} & \text{for } x_k = k \end{cases}$$

Which of these sequences converge to zero (a) in probability, (b) in the mean?

17.   Consider the random process:

$$Y(t) = A \sin t + B \cos t$$

where A and B are independent random variables with zero means and equal variances σ^2. Show that the process is wide-sense stationary but not strictly stationary if the density function of A differs from the density function of B.

REFERENCES

1.   E. Parzen, *Stochastic Processes*, Holden-Day, Inc., San Francisco, CA ; 1962.

2.   M Rosenblatt, *Random Processes*, Oxford University Press, New York, N.Y.; 1962.

3. M. S. Bartlett, *An Introduction to Stochastic Processes with Special Reference to Methods and Applications,* Cambridge University Press, Cambridge; 1978.

4. J. L. Doob, *Stochastic Processes,* John Wiley & Sons, Inc., New York, N.Y.; 1953.

5. E. C. Titchmarsh, *Introduction to the Theory of Fourier Integrals,* Oxford University Press; 1950.

6. W. Kaplan, *Operational Methods for Linear Systems,* Addison-Wesley Publishing Company, Inc., Reading, Mass.; 1962.

7. M. E. VanValkenburg, *Network Analysis,* Prentice-Hall, Inc., Englewood Cliffs, N.J.; 1974.

8. R. S. Burington, *Handbook of Mathematical Tables and Formulas,* Handbook Publishers, Inc., Sandusky, Ohio; 1965.

9. D. Middleton, *An Introduction to Statistical Communication Theory,* McGraw-Hill Book Company, Inc., New York, N.Y.; 1960.

10. W. B. Davenport, Jr., and W. L. Root, *An Introduction to the Theory of Random Signals and Noise,* McGraw-Hill Book Company, Inc., New York, N.Y.; 1958.

11. W. Feller, *An Introduction to Probability Theory and Its Applications, Vol. I,* John Wiley & Sons, Inc., New York, N.Y.; 1968.

12. T.M. Apostal, *Mathematical Analysis,* Addison-Wesley Publishing Co., Inc., Reading, Mass.; 1974.

13. W. Schottky, "Theory of Shot Effect", Ann. Phys., Vol. 57, December 1918; pp. 541-568.

14. W. R. Bennett, *Electrical Noise,* McGraw-Hill Book Company, Inc., New York, N.Y., 1960; Section 4.2.

Chapter 4

LINEAR FILTERING OF STATIONARY PROCESSES: STEADY-STATE ANALYSIS

4.1 *Introduction* - In a number of important applications of the theory of random processes, various linear operations are performed on stationary random processes. The generic name "linear filtering" is given to the operation indicated in Fig. 4.1. The box denoted by L is a *linear filter*

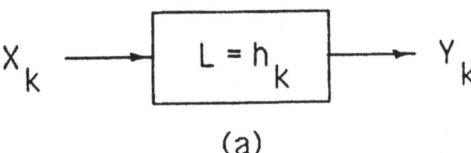

(a)

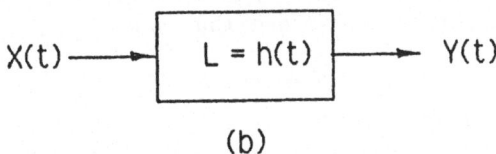

(b)

Fig. 4.1 - Schematic representation of a linear system. (a)Linear filtering of a discrete-parameter random process. (b) Linear filtering of a continuous-parameter random process.

or *linear operator*. The *input* may be, for example, the continuous-parameter random process $\{X(t); t \in T_x\}$ and *output* the random process

$$L\{X(t); t \in T_x\} = \{Y(t); t \in T_y\} \qquad (4.1\text{-}1)$$

as shown in Fig. 4.1(b). The system L is linear if, for any input sample function $x_i(t)$ and corresponding output sample function

$$L\left[x_i(t)\right] = y_i(t) \qquad (4.1\text{-}2)$$

and for any two constants α_i and α_j, it is true that

$$L\left[\alpha_i\, x_i(t) + \alpha_j\, x_j(t)\right] = \alpha_i\, y_i(t) + \alpha_j\, y_j(t) \qquad (4.1\text{-}3)$$

In other words, linearity means that the output resulting from the linear superposition of any two inputs is the linear superposition of the corresponding two outputs. Note that the linear filter or operator L is defined for some class of inputs, in this case the random process $\{X(t);\ t \in T_z\}$. In the physical world, a filter may be linear for a given class of inputs and nonlinear for some other class.

A linear filter or operator L is called *time-invariant* (or *fixed parameter*) for some class of inputs $\{X(t);\ t \in T_z\}$ if

$$y_i(t) = L\left[x_i(t)\right] \to y_i(t+t_0) = L\left[x_i(t+t_0)\right] \qquad (4.1\text{-}4)$$

for all $x_i(t) \in \{X(t);\ t \in T_z\}$ and all t_0 such that $(t+t_0) \in T_z$. A time-invariant linear filter is one where the only effect of a delay in the input is a corresponding delay in the output.

A linear filter or operator L is said to be *casual* (or *non-anticipatory*) if

$$x_i(t) = 0,\ t < t_0 \to y_i(t) = L\left[x_i(t)\right] = 0,\ t < t_0 \qquad (4.1\text{-}5)$$

that is, if there is no output prior to there being an input. A casual filter is sometimes called *physically-realizable* although this latter term is also used with a slightly different meaning.

A linear filter or operator L is said to be *stable* if its response to any bounded input is bounded; that is, if

$$|x| < M \to |y| < MI \qquad (4.1\text{-}6)$$

where M and I are constants.

Although the preceding discussion concerns Fig. 4.1(b) and continuous-parameter random processes, a similar set of remarks and definitions hold in an obvious way for discrete-parameter random processes as illustrated in Fig. 4.1(a).

Two types of (linear time-invariant*) filters will be treated in this chapter. The first will be called a *discrete-time filter* and will be identified by its *impulse response* h_n relating a discrete-time input x_n to a discrete time output y_n through the *convolution sum*

$$y_n = \sum_{k \in T_z} h_{n-k}\, x_k \qquad (4.1\text{-}7)$$

---

*Subsequently, "filter" will be taken to mean "linear, time-invariant filter".

The output parameter set $T_y = \{n\}$ will be determined by the non-zero values of h_{n-k} and by the input parameter set $T_x = \{k\}$. If h_n is a casual filter, then

$$h_{n-k} = 0 \quad , \quad n < k \tag{4.1-8a}$$

or

$$h_m = 0 \quad , \quad m < 0 \tag{4.1-8b}$$

and Eq. (4.1-7) can be written as

$$y_n = \sum_{k \in T_x} h_{n-k} \, x_k \quad , \quad n \geq \lim \inf T_x \tag{4.1-9a}$$

or, equivalently, as

$$y_n = \sum_{k \geq 0} h_k \, x_{n-k} \quad , \quad (n-k) \in T_x \tag{4.1-9b}$$

Suppose the input x_n to the discrete-time filter h_n is the *unit impulse* given by

$$x_n = \begin{cases} 1 \ , \ n = 0 \\ 0 \ , \ n \neq 0 \end{cases} \tag{4.1-10}$$

It is clear from Eq. (4.1-9) that the corresponding output y_n is

$$y_n = h_n \tag{4.1-11}$$

Thus, the impulse response h_n is the filter's response to the unit impulse of Eq. (4.1-10) occurring at time $t = 0$. If the filter were not time invariant, then the impulse response would be written $h_{n,k}$ to indicate the response of the filter at time $t = n$ to a unit impulse occurring at time $t = k$.

The *continuous-time* filter will be identified by its *impulse response* $h(t)$, relating a continuous-time input $x(t)$ to a continuous time output $y(t)$ through the *convolution integral*

$$y(t) = \int_{\sigma \in T_x} h(t - \sigma) x(\sigma) d\sigma$$

As in the discrete-time case, the output parameter set T_y will be an interval or set of intervals determined by the non-zero values of $h(t - \sigma)$ and of the input parameter set T_x. If $h(t)$ is a casual filter, then

$$h(t - \sigma) = 0 \quad , \quad t < \sigma \tag{4.1-13a}$$

or

$$h(\tau) = 0 \quad , \quad \tau < 0 \tag{4.1-13b}$$

In this case Eq. (4.1-12) can be written as

$$y(t) = \int_{\sigma \in T_x} h(t - \sigma) x(\sigma) d\sigma \quad , \quad t \geq \lim \inf T_x \tag{4.1-14a}$$

or, equivalently, as

$$y(t) = \int\limits_{\sigma \geq 0} h(\sigma)x(t-\sigma)d\sigma \quad , \quad (t-\sigma) \in T_x \qquad (4.1\text{-}14b)$$

Suppose the input $x(t)$ to the continuous-time filter $h(t)$ is the Dirac delta-function of Appendix B; that is

$$x(t) = \delta(t) \qquad (4.1\text{-}15)$$

Now Eq. (4.1-14) becomes, from Eq. (B.2-1),

$$y(t) = h(t) \qquad (4.1\text{-}16)$$

and the impulse response $h(t)$ is the filter's response to the Dirac delta-function occurring at $t=0$. If the filter were not time invariant, the impulse response would be written as $h(t,\sigma)$ to indicate the response of the filter at time t to a delta-function occurring at time σ.

It should be apparent that a discrete-time filter can be associated with a continuous parameter input process and that a continuous-time filter can have a discrete-time input. We shall not discuss these cases explicitly, but it should be clear from the subsequent development how to deal with them. We proceed now to consider separately discrete-time and continuous-time filters.

4.2 *Discrete-Time Filters* - Unless specifically stated otherwise, let it be assumed that the discrete-parameter random process $\{X_n = X(t_n); t_n = 0, \pm 1, \pm 2, \cdots \}$ exists over the set of all integers and that the discrete-time filter h_n is not necessarily casual so that Eq. (4.1-9) becomes

$$y_n = \sum_{k=-\infty}^{\infty} h_{n-k}\, x_k = \sum_{k=-\infty}^{\infty} h_k\, x_{n-k} \qquad (4.2\text{-}1)$$

Define the *generating function* of h_k by

$$\mathbf{G}\{h_k\} = H(z) = \sum_{k=-\infty}^{\infty} h_k\, z^{-k} \qquad (4.2\text{-}2)$$

where z is a complex variable. The generating function of the sequence h_k serves the same purpose in the analysis of discrete-time linear filters as the Fourier transform of the function $h(t)$ serves in the analysis of continuous-time linear filters. The function $H(z)$ will be called the *(complex) transfer function* of the discrete-time filter h_n.

Example 4.1

Let h_k be the sequence defined by

$$h_k = \begin{cases} 0 & , k < 0 \\ 1 & , k \geq 0 \end{cases}$$

The generating function $H(z)$ is given by

$$H(z) = G\{h_k\} = \sum_{k=0}^{\infty} z^{-k} = \frac{z}{z-1}$$

This generating function exists iff. $|z| > 1$. On the other hand let h_k be given by

$$h_k = \begin{cases} 1 & , k < 0 \\ 0 & , k \geq 0 \end{cases}$$

so that $H(z)$ is

$$H(z) = \sum_{k=-\infty}^{-1} z^{-k} = \sum_{n=1}^{\infty} z^n = \frac{1}{1-z} - 1 = \frac{z}{1-z}$$

which exists iff. $|z| < 1$.

For the discrete-parameter random process X_n with sample function x_n, we shall denote generating functions by

$$G\{X_n\} = G_X(z) = \sum_{k=-\infty}^{\infty} X_k z^{-k} \tag{4.2-3}$$

and by

$$G\{x_n\} = G_x(z) = \sum_{k=-\infty}^{\infty} x_k z^{-k} \tag{4.2-4}$$

the function $G_X(z)$ is a random variable, of course. It can also be regarded as a new random process with parameter z generated by the operation of Eq. (4.2-3). The function $G_x(z)$ is simply one of the realizations or sample functions of $G_X(z)$.

Consider now the convolution sum given by Eq. (4.2-1) and find the generating function for y_n. We have

$$G\{y_n\} = G_y(z) = \sum_{n=-\infty}^{\infty} y_n z^{-n} = \sum_{n=-\infty}^{\infty} \sum_{k=-\infty}^{\infty} h_{n-k} x_k z^{-n} \tag{4.2-5}$$

On interchanging the order of summation and replacing $n-k$ by j, we obtain

$$G_y(z) = \sum_{k=-\infty}^{\infty} x_k z^{-k} \sum_{j=-\infty}^{\infty} h_j z^{-j} \tag{4.2-6}$$

It follows from Eqs. (4.2-2) and (4.2-4) that this last expression can be rewritten as

$$G_y(z) = G_z(z)\, H(z) \qquad (4.2\text{-}7)$$

Thus the generating function of the filter output is equal to the product of the generating function of the filter input and the transfer function of the filter.

Let us now consider that the input to h_n is the random process X_n. From Eq. (4.2-1), an output random process Y_n is defined by

$$Y_n = \sum_{k=-\infty}^{\infty} h_k\, X_{n-k} \qquad (4.2\text{-}8)$$

The mean of Y_n can be written as

$$E\{Y_n\} = \sum_{k=-\infty}^{\infty} h_k\, E\{X_{n-k}\} \qquad (4.2\text{-}9)$$

If X_n has a constant mean m_z, then

$$E\{Y_n\} = m_y = m_z \sum_{k=-\infty}^{\infty} h_k = m_z\, J \qquad (4.2\text{-}10)$$

where J is some constant. The autocorrelation function of Y_n is

$$R_{yy}(n\,,\,n+m) = E\{Y_n\, Y_{n+m}\} \qquad (4.2\text{-}11)$$

$$= \sum_{k=-\infty}^{\infty} \sum_{j=-\infty}^{\infty} h_k\, h_j\, E\{X_{n-k}\, X_{n+m-j}\}$$

If the process X_n is at least wide-sense stationary so that

$$E\{X_{n-k}\, X_{n+m-j}\} = R_{zz}(m+k-j) \qquad (4.2\text{-}12)$$

then Eq. (4.2-11) becomes

$$R_{yy}(n\,,\,n+m) = R_{yy}(m) = \sum_{k=-\infty}^{\infty} \sum_{j=-\infty}^{\infty} h_k\, h_j\; R_{zz}(m+k-j) \quad (4.2\text{-}13)$$

Thus the output process Y_n is at least wide-sense stationary if the filter is time-invariant and if the input process is at least wide-sense stationary.

In Section 3.8, the power spectral density $\phi(\omega)$ and the autocorrelation function $R(n)$ of a wide-sense stationary discrete-time random process were related through Eqs. (3.8-48) and (3.8-51); that is

$$\phi(\omega) = \sum_{n=-\infty}^{\infty} R(n)\, e^{-jn\omega} \qquad (4.2\text{-}14)$$

and

$$R(n) = \frac{1}{2\pi} \int_{-\pi}^{\pi} \phi(\omega)\, e^{jn\omega}\, d\omega \qquad (4.2\text{-}15)$$

Note that $\phi(\omega)$ is a Fourier series whose coefficients are the sequence $R(n)$. In a formal sense, the generating function $\phi(z)$ of the sequence $R(n)$ is obtained from Eq. (4.2-14) by the substitution

$$z = e^s \mid_{s=j\omega} \qquad (4.2\text{-}16)$$

so that*

$$\phi(z) = \mathbf{G}\{R(n)\} = \sum_{n=-\infty}^{\infty} R(n) z^{-n} \qquad (4.2\text{-}17)$$

Here s and z are both complex variables and the inversion formula corresponding to Eq. (4.2-15) is

$$R(n) = \frac{1}{2\pi j} \int_C \phi(z) z^{n-1} \, dz \qquad (4.2\text{-}18)$$

where C is an appropriate contour in the complex z-plane. We shall not use Eq. (4.2-18) but the interested reader should refer to [1] for further details. The pair defined by Eqs. (4.2-17) and (4.2-18) are often called a two-sided *z-transform pair* in the engineering literature related to sampled-data.

Let us return to Eq. (4.2-13) and find the power spectral density $\phi_{yy}(z)$ by applying Eq. (4.2-17). We have

$$\phi_{yy}(z) = \mathbf{G}\{R_{yy}(m)\}$$

$$= \sum_{m=-\infty}^{\infty} \sum_{k=-\infty}^{\infty} \sum_{j=-\infty}^{\infty} h_k h_j R_{zz}(m+k-j) z^{-m} \qquad (4.2\text{-}19)$$

After the change in index of summation $m = j - k + n$, this last expression becomes

$$\phi_{yy}(z) = \sum_{k=-\infty}^{\infty} h_k z^k \sum_{j=-\infty}^{\infty} h_j z^{-j} \sum_{n=-\infty}^{\infty} R_{zz}(n) z^{-n} \qquad (4.2\text{-}20)$$

With the aid of Eqs. (4.2-2) and (4.2-17), the power spectral density $\phi_{yy}(z)$ may be written as

$$\phi_{yy}(z) = H(1/z) H(z) \phi_{zz}(z) \qquad (4.2\text{-}21)$$

This equation relates the output power spectral density of a linear discrete-time filter to the input power spectral density through the complex transfer function of the filter.

Example 4.2

Consider a discrete-time filter with impulse response given by

---

*To be correct, $\phi(z)$ should be written as $\phi(-j \ln z)$, but we shall not do this. The reader should keep in mind that $\phi(z)$ is not $\phi(\omega)$ with ω replaced by z.

$$
h_k = \begin{cases} 0 & , \ k < 0 \\[2mm] e^{-\alpha k} & , \ k \geq 0 \ , \ \alpha > 0 \end{cases}
$$

Such a filter is sometimes called a *lowpass* filter. The complex transfer function $H(z)$ of this causal time-invariant filter is the generating function of Eq. (4.2-2):

$$
H(z) = \sum_{k=0}^{\infty} e^{-\alpha k} z^{-k} = \sum_{k=0}^{\infty} (e^{\alpha} z)^{-k}
$$

or

$$
H(z) = \frac{1}{1 - (e^{\alpha} z)^{-1}} = \frac{e^{\alpha}}{e^{\alpha} - z^{-1}}
$$

Let the input random process X_n be wide-sense stationary with zero mean and autocorrelation function $R_{xx}(n)$ given by

$$
R_{xx}(n) = \begin{cases} N_0/2 \ , & n = 0 \\[2mm] 0 & , \ n \neq 0 \end{cases}
$$

so that X_n is a sequence of uncorrelated random variables with a common variance $N_o/2$. Such a process is sometimes called *white noise* in the engineering literature. The power spectral density $\phi_{xx}(\omega)$ is given by Eq. (4.2-17) as

$$
\phi_{xx}(z) = \sum_{n=-\infty}^{\infty} R_{xx}(n) z^{-n} = N_0/2
$$

It follows from Eq. (4.2-21) that the output power spectral density $\phi_{yy}(z)$ is

$$
\phi_{yy}(z) = \frac{e^{\alpha}}{e^{\alpha} - z} \ \frac{e^{\alpha}}{e^{\alpha} - z^{-1}} \ \frac{N_0}{2} = \frac{N_0 \, e^{2\alpha}}{2(e^{\alpha} - z)(e^{\alpha} - z^{-1})}
$$

The output autocorrelation function is obtained by inverting $\phi_{yy}(z)$ through Eq. (4.2-18). In practice, it is more convenient to construct a table of those z-transform pairs most commonly encountered [1]. Note that $\phi_{yy}(z)$ may be written in factored form as

$$
\phi_{yy}(z) = \frac{N_0}{2} \left[\frac{e^{2\alpha}(e^{2\alpha}-1)^{-1} z}{e^{\alpha} - z} + \frac{e^{3\alpha}(e^{2\alpha}-1)^{-1}}{e^{\alpha} - z^{-1}} \right]
$$

A comparison of each of these terms with Example 4.1 shows that $\phi_{yy}(z)$ may be expressed as

$$
\phi_{yy}(z) = \frac{N_0}{2} \ \frac{e^{2\alpha}}{e^{2\alpha}-1} \left[\sum_{n=-\infty}^{-1} e^{\alpha n} z^{-n} + \sum_{n=0}^{\infty} e^{-\alpha n} z^{-n} \right]
$$

Thus, as shown by Eq. (4.2-17), the autocorrelation function $R_{yy}(n)$ is

$$R_{yy}(n) = \frac{N_0}{2} \frac{e^{2\alpha}}{e^{2\alpha}-1} e^{-\alpha|n|} \quad , \quad n = 0, \pm 1, \pm 2, \cdots$$

The various quantities are illustrated in Fig. 4.2.

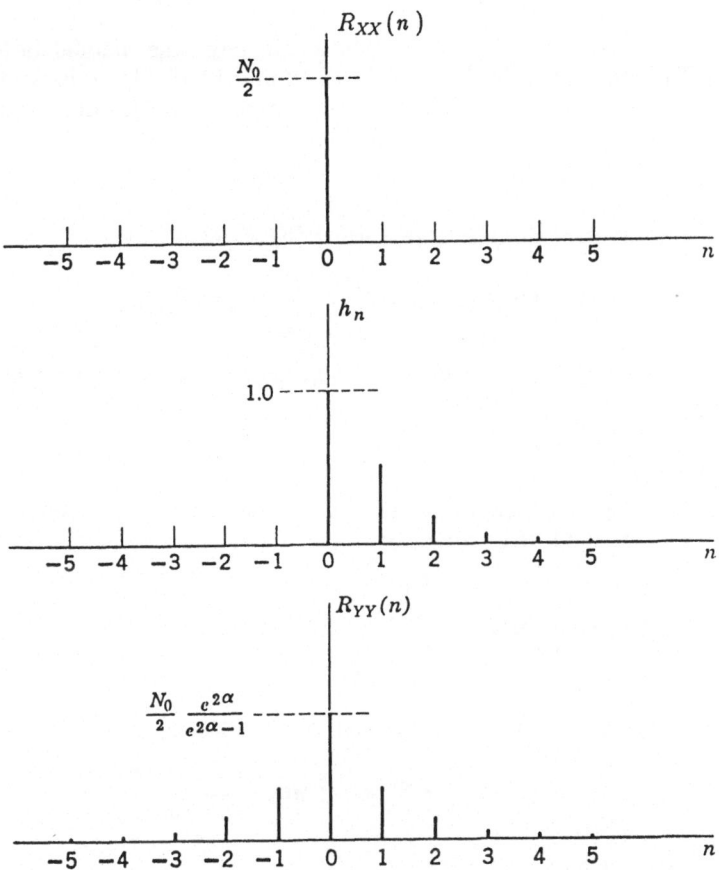

Fig. 4.2 - White noise through a lowpass filter - discrete time

4.3 *Continuous-Time Filters* - In this section it will be assumed, unless explicitly stated otherwise, that $X(t) = \{X(t); -\infty < t < \infty\}$ is a continuous-parameter random process. The output $y(t)$ of the continuous time filter $h(t)$ for an input sample function $x(t)$ will be given by

$$y(t) = \int_{-\infty}^{\infty} h(t-\sigma)x(\sigma)d\sigma = \int_{-\infty}^{\infty} h(\sigma)x(t-\sigma)d\sigma \qquad (4.3\text{-}1)$$

The filter $h(t)$ is not necessarily causal, but if it is, then $h(t)=0$, $t<0$ and Eq. (4.3-1) becomes

$$y(t) = \int_{-\infty}^{t} h(t-\sigma)x(\sigma)d\sigma = \int_{0}^{\infty} h(\sigma)x(t-\sigma)d\sigma \qquad (4.3\text{-}2)$$

The concept of a Fourier transform pair was defined in Section 2.15; for example, by Eqs. (2.15-9) and (2.15-11) or in Section 3.8. Let us denote the (direct) Fourier transform $H(\omega)$ of $h(t)$ by

$$\mathbf{F}\{h(t)\} = H(\omega) = \int_{-\infty}^{\infty} h(t)\, e^{-j\omega t}\, dt \qquad (4.3\text{-}3)$$

and the (inverse) Fourier transform $h(t)$ of $H(\omega)$ by

$$\mathbf{F}^{-1}\{H(\omega)\} = h(t) = \frac{1}{2\pi} \int_{-\infty}^{\infty} H(\omega)\, e^{j\omega t}\, d\omega \qquad (4.3\text{-}4)$$

The function $H(\omega)$ is called the *(complex) transfer function $H(\omega)$* of the continuous-time filter $h(t)$.

Example 4.3

Let the impulse response $h(t)$ of a continuous-time linear time-invariant filter be given by

$$h(t) = \begin{cases} 0 & , t < 0 \\ e^{-\alpha t} & , t \geq 0,\ \alpha > 0 \end{cases}$$

The complex transfer function $H(\omega)$ of the filter is

$$H(\omega) = \int_{0}^{\infty} e^{-\alpha t}\, e^{-j\omega t}\, dt = \frac{1}{\alpha + j\omega}$$

On the other hand, let $h(t)$ be

$$h(t) = \begin{cases} e^{\alpha t} & , t \leq 0,\ \alpha > 0 \\ 0 & , t > 0 \end{cases}$$

so that

$$H(\omega) = \int_{-\infty}^{0} e^{\alpha t}\, e^{-j\omega t}\, dt = \frac{1}{\alpha - j\omega}$$

For the continuous-parameter random process $X(t)$ with sample function $x(t)$, we shall denote Fourier transforms by

$$\mathbf{F}\{X(t)\} = F_X(\omega) = \int\limits_{-\infty}^{\infty} X(t)\, e^{-j\omega t}\; dt \qquad (4.3\text{-}5)$$

and

$$\mathbf{F}\{x(t)\} = F_x(\omega) = \int\limits_{-\infty}^{\infty} x(t)\, e^{-j\omega t}\; dt \qquad (4.3\text{-}6)$$

As in the discrete-time case, $F_X(\omega)$ is a random variable (or a random process with parameter ω) and $F_x(\omega)$ is one of the realizations (or sample functions).

Consider now the convolution integral given by Eq. (4.3-1). The Fourier transform of $y(t)$ is

$$\mathbf{F}\{y(t)\} = F_y(\omega) = \int\limits_{-\infty}^{\infty} e^{-j\omega t} \int\limits_{-\infty}^{\infty} h(t-\sigma)x(\sigma)d\sigma dt \qquad (4.3\text{-}7)$$

On interchanging the order of integration and making the change of variable $\tau = t - \sigma$, we obtain

$$F_y(\omega) = \int\limits_{-\infty}^{\infty} x(\sigma)e^{-j\omega\sigma}\,d\sigma \int\limits_{-\infty}^{\infty} h(\tau)e^{-j\omega\tau}\,d\tau \qquad (4.3\text{-}8)$$

With the use of Eqs. (4.3-3) and (4.3-6), this last expression becomes

$$F_y(\omega) = F_x(\omega)H(\omega) \qquad (4.3\text{-}9)$$

Thus the Fourier transform of the filter output is equal to the product of the Fourier transform of the filter input and the transfer function of the filter.

Let us now consider that the input to $h(t)$ is the random process $X(t)$. From Eq. (4.3-1), an output random process $Y(t)$ is defined by

$$Y(t) = \int\limits_{-\infty}^{\infty} h(\sigma)X(t-\sigma)d\sigma \qquad (4.3\text{-}10)$$

The mean of $Y(t)$ can be written as

$$E\{Y(t)\} = \int\limits_{-\infty}^{\infty} h(\sigma)E\{X(t-\sigma)\}d\sigma \qquad (4.3\text{-}11)$$

If $X(t)$ has a constant mean m_x, then

$$E\{Y(t)\} = m_y = m_x \int\limits_{-\infty}^{\infty} h(\sigma)d\sigma = m_x K \qquad (4.3\text{-}12)$$

where K is a constant. The autocorrelation function of $Y(t)$ is

$$R_{yy}(t,t+\tau) = E\{Y(t)Y(t+\tau)\} =$$

$$\int\limits_{-\infty}^{\infty} \int\limits_{-\infty}^{\infty} h(\sigma)h(\mu)E\{X(t-\sigma)X(t+\tau-\mu)\}d\sigma d\mu \qquad (4.3\text{-}13)$$

If the process $X(t)$ is at least wide-sense stationary so that

$$E\{X(t-\sigma)X(t+\tau-\mu)\} = R_{xx}(\tau+\sigma-\mu) \qquad (4.3\text{-}14)$$

then Eq. (4.3-13) becomes

$$R_{yy}(t,t+\tau) = R_{yy}(\tau) =$$

$$\int\limits_{-\infty}^{\infty}\int\limits_{-\infty}^{\infty} h(\sigma)h(\mu)R_{xx}(\tau+\sigma-\mu)d\sigma d\mu \qquad (4.3\text{-}15)$$

Thus the output process $Y(t)$ is at least wide-sense stationary if the filter is time-invariant and if the input process is at least wide-sense stationary.

Let us now take the Fourier transform of $R_{yy}(\tau)$ to obtain the power spectral density $\phi_{yy}(\omega)$ given by Eq. (4.3-31). We have

$$\phi_{yy}(\omega) = \int\limits_{-\infty}^{\infty} R_{yy}(\tau)\, e^{-j\omega\tau}\, d\tau \qquad (4.3\text{-}16)$$

It follows from Eq. (4.3-15) that

$$\phi_{yy}(\omega) = \int\limits_{-\infty}^{\infty}\int\limits_{-\infty}^{\infty}\int\limits_{-\infty}^{\infty} h(\sigma)h(\mu)R_{xx}(\tau+\sigma-\mu)e^{-j\omega\tau}\, d\sigma d\mu d\tau \qquad (4.3\text{-}17)$$

The power spectral density $\phi_{xx}(\omega)$ of the input is

$$\phi_{xx}(\omega) = \int\limits_{-\infty}^{\infty} R_{xx}(\tau)\, e^{-j\omega\tau}\, d\tau \qquad (4.3\text{-}18)$$

After the change in variable $\gamma = \tau + \sigma - \mu$, Eq. (4.3-17) may be rearranged to yield

$$\phi_{yy}(\omega) = \int\limits_{-\infty}^{\infty} h(\sigma)e^{j\omega\sigma}\, d\sigma \int\limits_{-\infty}^{\infty} h(\mu)e^{-j\omega\mu}\, d\mu \int\limits_{-\infty}^{\infty} R_{xx}(\gamma)e^{-j\omega\gamma}\, d\gamma \qquad (4.3\text{-}19)$$

With the aid of Eqs. (4.3-3) and (4.3-18), this last expression becomes

$$\phi_{yy}(\omega) = H(-\omega)H(\omega)\phi_{xx}(\omega) = |H(\omega)|^2\,\phi_{xx}(\omega) \qquad (4.3\text{-}20)$$

This last equation relates the output power spectral density of a linear continuous-time filter to the input power spectral density through the complex transfer function of the filter.

Example 4.4

Consider a continuous-time filter with impulse response

$$h(t) = \begin{cases} 0 & ,\ t < 0 \\[2mm] e^{-\alpha t} & ,\ t \geq 0,\ \alpha > 0 \end{cases}$$

Such a filter is sometimes called a *lowpass* filter. The complex transfer function $H(\omega)$ of this causal filter has been given in Example 4.3 as

$$H(\omega) = \frac{1}{\alpha + j\,\omega}$$

Let the random process $X(t)$ be wide-sense stationary with zero mean and autocorrelation function $R_{zz}(\tau)$ given by

$$R_{zz}(\tau) = \frac{N_0}{2}\,\delta(\tau)$$

where $\delta(\tau)$ is the Dirac delta-function of Appendix B. The power spectral density $\phi_{zz}(\omega)$ is given by Eq. (4.3-18) as

$$\phi_{zz}(\omega) = \frac{N_0}{2} \int\limits_{-\infty}^{\infty} \delta(\tau)\, e^{-j\omega t}\, d\tau = \frac{N_0}{2}$$

The process $X(t)$ with constant power spectral density is sometimes called *white noise*.

It follows from Eq. (4.3-20) that the output power spectral density $\phi_{yy}(\omega)$ is

$$\phi_{yy}(\omega) = \frac{1}{\alpha - j\,\omega}\ \frac{1}{\alpha + j\,\omega}\ \frac{N_0}{2} = \frac{N_0}{2}\ \frac{1}{\alpha^2 + \omega^2}$$

The output autocorrelation function $R_{yy}(\tau)$ is found by using Eq. (3.8-4) to obtain

$$R_{yy}(\tau) = \frac{N_0}{4\pi} \int\limits_{-\infty}^{\infty} \frac{e^{j\omega t}}{\alpha^2 + \omega^2}\, d\omega$$

or

$$R_{yy}(\tau) = \frac{N_0}{4\alpha}\, e^{-\alpha|\tau|}$$

The various quantities are illustrated in Fig. 4.3.

4.4 *Complete Statistical Description of the Output of a Linear System* - In the previous two sections, the first and second moments of the output of a linear system subject to a random input has been given in terms of the impulse response of the system and the corresponding moment of the (random)input. These techniques can be extended in an obvious way to yield expressions for the higher order moments of the output in terms of the corresponding input moments.

As a matter of notational convenience, let the convolution integral of Eq. (4.3-1) be expressed as

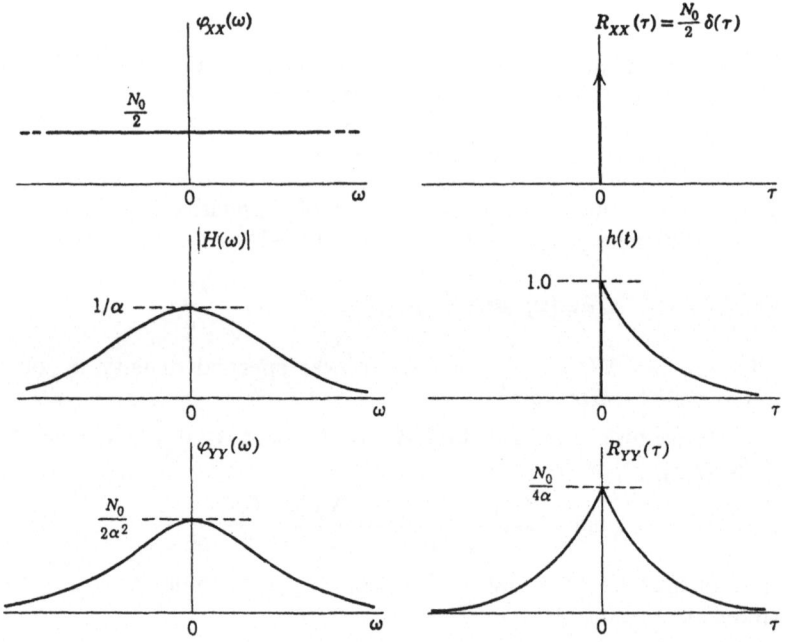

Fig. 4.3 - White noise through a low-pass filter - continuous time

$$y(t) = h * x(t) = x * h(t) \qquad (4.4\text{-}1)$$

In the same way, for the discrete-time filter, let Eq. (4.2-1) be written as

$$y_n = h * x_n = x * h_n \qquad (4.4\text{-}2)$$

When no confusion will arise or when it is desired to represent both the continuous-time and the discrete-time case, both of these last two expressions will be written as

$$y = h * x = x * h \qquad (4.4\text{-}3)$$

In this latter case, however, it will be easy to lose track of the time dependence. Now both Eq. (4.2-9) and (4.3-11) can be written as

$$E\{Y\} = E\{h * X\} = h * E\{X\} \qquad (4.4\text{-}4)$$

With the notation of Eq. (4.4-1), the output autocorrelation function of Eq. (4.3-13) becomes

$$E\{Y(t)Y(t+\tau)\} = E\{[h \ast X(t)][h \ast X(t+\tau)]\}$$
$$= h \ast h \ast E\{X(t)X(t+\tau)\} \qquad (4.4\text{-}5)$$

If the input process is at least wide-sense stationary with autocorrelation function $R_{xx}(\tau)$, this last expression reduces to

$$E\{Y(t)Y(t+\tau)\} = R_{yy}(\tau) = h \ast h \ast R_{xx}(\tau) \qquad (4.4\text{-}6)$$

in agreement with Eq. (4.3-15). If the notation of Eq. (4.4-3) is used, then both Eqs. (4.2-11) and (4.3-13) can be written as

$$E\{YY\} = E\{[h \ast X][h \ast X]\} = h \ast h \ast E\{XX\} \qquad (4.4\text{-}7)$$

However, the time dependence has been suppressed and it must be kept in mind that $E\{YY\} \neq E\{Y^2\}$ and that $E\{XX\} \neq E\{X^2\}$.

It is clear that explicit expressions for higher order output moments are easily written. For example, the third order output moment is

$$E\{YYY\} = h \ast h \ast h \ast E\{XXX\} \qquad (4.4\text{-}8)$$

In the continuous-time case, this becomes

$$E\{Y(t_1)Y(t_2)Y(t_3)\} = h \ast h \ast h \ast E\{X(t_1)X(t_2)X(t_3)\} \qquad (4.4\text{-}9)$$

or, more explicitly,

$$E\{Y(t_1)Y(t_2)Y(t_3)\} \qquad (4.4\text{-}10)$$

$$= \int_{-\infty}^{\infty} \int_{-\infty}^{\infty} \int_{-\infty}^{\infty} h(\alpha)h(\beta)h(\gamma)E\{X(t_1-\alpha)X(t_2-\beta)X(t_3-\gamma)\}\,d\alpha\,d\beta\,d\gamma$$

In the same fashion, all higher order moments can be found, in principle. It is apparent, however, that the amount of work involved will become prohibitively large even for the first few moments. Thus there would appear to be no practical way to characterize completely the output of a linear system subjected to an arbitrary random input.

It was shown in Section 2.23 that the result of a linear transformation on a normally distributed random variable was another normally distributed random variable. In other words, *if a linear system has an input which is a Gaussian random process, the output will also be a Gaussian random process.* In addition, if the input is wide-sense stationary (and, hence, strictly stationary) and if the linear system is time-invariant, then the output will be stationary also. In the Gaussian case, only the first two moments are required to characterize the process. Thus, the output of a linear system with a Gaussian input is easily described and only second moment theory, as developed in Sections 3.8, 4.2, and 4.3, is required.

4.5 *The Orthogonal Decomposition of Random Processes; Fourier Series* - In a large number of applications involving a continuous parameter random process $X(t)$, it is useful and convenient to represent the process in some interval $[a,b]$ by an orthogonal decomposition of the form

$$X(t) = E\{X(t)\} + \sum_{n=1}^{\infty} V_n \Psi_n(t) \ , \ a \leq t \leq b \qquad (4.5\text{-}1)$$

Here the coefficients V_n are random variables and the $\Psi_n(t)$ are non-random functions forming a complete orthonormal set in $[a,b]$ so that

$$\int_a^b \Psi_m(t)\Psi_n^*(t)dt = \delta_{mn} \qquad (4.5\text{-}2)$$

where $\Psi_n^*(t)$ is the complex conjugate of $\Psi_n(t)$ and δ_{mn} is the Kronecker delta given by

$$\delta_{mn} = \begin{cases} 1 \ , \ m = n \\ \\ 0 \ , \ m \neq n \end{cases}$$

For convenience, let it be assumed that only real random processes will be considered and that the mean $E\{X(t)\}$ is zero in $[a,b]$. Now Eq. (4.5-1) becomes

$$X(t) = \sum_{n=1}^{\infty} V_n \Psi_n(t) \qquad (4.5\text{-}3)$$

The random coefficient V_n is found from Eq. (4.5-2) to be

$$V_n = \int_a^b X(t)\Psi_n*(t)dt \qquad (4.5\text{-}4)$$

and has zero mean since

$$E\{V_n\} = \int_a^b E\{X(t)\} \ \Psi_n*(t)dt = 0 \qquad (4.5\text{-}5)$$

A question of great importance at this point is the following: in what sense does the right side of Eq. (4.5-3) represent the random process $X(t)$; that is, how does the series converge to $X(t)$? Define the truncated sum $X_N(t)$ by

$$X_N(t) = \sum_{n=1}^{N} V_n \ \Psi_n(t) \ , \ a \leq t \leq b \qquad (4.5\text{-}6)$$

If the relationship

$$\lim_{N \to \infty} E\{[X(t)\text{-}X_N(t)]^2\} = 0 \ , \ a \leq t \leq b \qquad (4.5\text{-}7)$$

holds, then the right side of Eq. (4.5-3) converges mean square to the process $X(t)$. We will investigate this type of convergence for each orthogonal decomposition that we consider.

The principle reason for investigating such representations as Eq. (4.5-3) is to allow us to use the joint density of the denumerable set $\{V_n\}$ as a complete statistical description of the random process $\{X(t); a \leq t \leq b\}$ which contains a non-denumerable set of random variables.

We proceed now to consider several different types of series - each suitable for a particular class of random processes. We begin with the Fourier series which, in addition to its historical significance [2,3,4,5], will be useful when we treat narrowband processes in Sections 4.9 and 4.10.

THE FOURIER SERIES

One of the simplest cases to consider is that where the random process $X(t)$ is *periodic*. A process $X(t)$ which is at least wide sense stationary is said to be periodic with period T if its autocorrelation function

$$R_{xx}(\tau) = E\{X(t)X(t+\tau)\} \qquad (4.5\text{-}8)$$

is periodic with period T so that

$$R_{xx}(\tau) = R_{xx}(\tau+T) \quad , \quad \text{all } \tau \qquad (4.5\text{-}9)$$

In this case, the autocorrelation function obviously possesses a Fourier series expansion of the form

$$R_{xx}(\tau) = R_{xx}(\tau+T) = \sum_{n=-\infty}^{\infty} r_n \; e^{jn\omega_1\tau} \; , \quad \frac{-T}{2} \leq \tau \leq \frac{T}{2} \qquad (4.5\text{-}10)$$

where

$$\omega_1 = \frac{2\pi}{T} \qquad (4.5\text{-}11)$$

and r_n is the (complex)Fourier coefficient given by

$$r_n = \frac{1}{T} \int_{-T/2}^{T/2} R_{xx}(\tau) \; e^{-jn\omega_1\tau} \; d\tau \qquad (4.5\text{-}12)$$

Note that $R_{xx}(\tau) = R_{xx}(-\tau)$ for the (real) random process $X(t)$; consequently, in this case, the coefficient r_m is real.

If the process $X(t)$ is at least mean-square continuous, then $R_{xx}(\tau)$ is a continuous function of τ and the right side of Eq. (4.5-10) converges everywhere to $R(\tau)$. The process itself can be represented in a mean square sense (as will be shown later) as

$$X(t) = \sum_{n=-\infty}^{\infty} c_n \; e^{jn\omega_1 t} \qquad (4.5\text{-}13)$$

or as

$$X(t) = \frac{a_0}{2} + \sum_{n=1}^{\infty} \left[a_n \cos n \omega_1 t + b_n \sin n \omega_1 t \right] \quad (4.5\text{-}14)$$

where

$$a_n = c_n + c_{-n} \quad (4.5\text{-}15)$$

$$b_n = j(c_n - c_{-n}) \quad (4.5\text{-}16)$$

These Fourier coefficients are random variables given by

$$a_n = \frac{2}{T} \int_{-T/2}^{T/2} X(t) \cos n \omega_1 t \ dt \quad (4.5\text{-}17)$$

$$b_n = \frac{2}{T} \int_{-T/2}^{T/2} X(t) \sin n \omega_1 t \ dt \quad (4.5\text{-}18)$$

and

$$c_n = \frac{1}{T} \int_{-T/2}^{T/2} X(t) \, e^{-jn \omega_1 t} \ dt \quad (4.5\text{-}19)$$

The Fourier coefficients c_n of a periodic random process possess an interesting property. Using Eq. (4.5-19), let us form

$$E\{c_m \, c_n{}^*\} = \frac{1}{T^2} \int_{-T/2}^{T/2} \int_{-T/2}^{T/2} E\{X(s)X(t)\} \, e^{j \omega_1 (ns - mt)} dt ds \quad (4.5\text{-}20)$$

Since $E\{X(s)X(t)\} = R_{xx}(t-s)$ can be expanded in the Fourier series of Eq. (4.5-10), this last expression can be written as

$$E\{c_m \, c_n{}^*\} = \frac{1}{T^2} \int_{-T/2}^{T/2} \int_{-T/2}^{T/2} \sum_{k=-\infty}^{\infty} r_k \, e^{jk \omega_1 (t-s)} \, e^{j \omega_1 (ns - mt)} dt ds$$

or as

$$E\{c_m \, c_n{}^*\} = \frac{1}{T^2} \sum_{k=-\infty}^{\infty} r_k \int_{-T/2}^{T/2} e^{j \omega_1 (k-m)t} \ dt \int_{-T/2}^{T/2} e^{j \omega_1 (n-k)s} \ ds$$

This last expression becomes

$$E\{c_m \, c_n{}^*\} = \sum_{k=-\infty}^{\infty} r_k \, \delta_{km} \, \delta_{nk} = r_m \, \delta_{mn} \quad (4.5\text{-}21)$$

Thus the complex Fourier coefficients c_n are *uncorrelated* random variables for a periodic random process. This is an important property since it can simplify greatly certain problems. Suppose that $X(t)$ is a Gaussian random process. Then the c_n are normal random variables since they result [see Eq. (4.5-19)] from linear operations on a Gaussian process. Normal random variables which are uncorrelated are independent. Thus, for example, the joint density of a set of these random variables will be the product of the univariate densities of each of the members of the set.

The coefficients c_k are related to a_k and b_k by

$$c_k = \frac{1}{2}(a_k - jb_k), \quad c_{-k} = \frac{1}{2}(a_k + jb_k) = c_k^* \qquad (4.5\text{-}22)$$

Since $R(\tau)$ is a real function (for a real random process), it is clear from Equations (4.5-12) and (4.5-21) that the imaginary part of $E\{c_m\,c_n^*\}$ is zero:

$$\operatorname{Im} E\{c_m\,c_n^*\} = 0 = \frac{1}{4}[E\{a_m\,b_n\} - E\{a_n\,b_m\}] \qquad (4.5\text{-}23)$$

$$\operatorname{Im} E\{c_m\,c_{-n}^*\} = 0 = \frac{1}{4}[E\{a_m\,b_n\} + E\{a_n\,b_m\}] \qquad (4.5\text{-}24)$$

or

$$E\{a_m\,b_n\} = E\{a_n\,b_m\} = E\{a_k\,b_k\} = 0 \qquad (4.5\text{-}25)$$

Of course most processes in which we are interested will not be periodic and, strictly speaking, Fourier-series representation cannot be applied. We can avoid this problem by assuming that the processes are periodic with period $T \to \infty$. Actually it can be shown that the coefficients are approximately uncorrelated for any process whose spectral density is approximately constant over intervals large compared to $1/T$. Thus, for the white-noise case, where the spectral density is some constant K for all frequencies ω, the coefficients are always uncorrelated. This is easily shown by considering Eq. (4.5-20) with

$$E\{X(s)X(t)\} = K\,\delta(t - s) \qquad (4.5\text{-}26)$$

Now Eq. (4.5-20) becomes

$$E\{c_m\,c_n^*\} = \frac{K}{T^2} \int_{-T/2}^{T/2} e^{\,j\omega_1(m - n)t}\,dt = \frac{K}{T}\,\delta_{mn} \qquad (4.5\text{-}27)$$

as was to be proved.

We digress briefly at this point to obtain some further statistical properties of the Fourier coefficients which will be useful later. It is apparent from Eqs. (4.5-12) and (4.5-21) that

$$\lim_{T \to \infty} Tr_n = \lim_{T \to \infty} TE\{c_n\,c_n^*\} = \phi_{xx}(n\,\omega_1) \qquad (4.5\text{-}28)$$

where $\phi_{xx}(\omega)$ is the power spectral density of the wide-sense stationary process $X(t)$. We also have, from Eq. (4.5-22) that

$$c_n\,c_n^* = c_n\,c_{-n} = |c_n|^2 = \frac{1}{4}(a_n^2 + b_n^2) \qquad (4.5\text{-}29)$$

Thus,

$$\lim_{T \to \infty} TE\{a_n^2 + b_n^2\} = 4\phi_{xx}(n\,\omega_1) \qquad (4.5\text{-}30)$$

Now let us use Eq. (4.5-19) to form

$$\lim_{T \to \infty} T^2 E\{c_n^2\} = \lim_{T \to \infty} \int_{-T/2}^{T/2} \int_{-\infty}^{\infty} R_{xx}(t-s) \, e^{-jk\,\omega_1(t+s)} \, dt \; ds \quad (4.5\text{-}31)$$

We make the linear change of variable $y = t - s$ to obtain

$$\lim_{T \to \infty} T^2 E\{c_n^2\} = \phi_{xx}(k\,\omega_1) \lim_{T \to \infty} \int_{-T/2}^{T/2} e^{-2jk\,\omega_1 s} \, ds = 0 \quad (4.5\text{-}32)$$

since $\omega_1 = 2\pi/T$. Now, from Eq. (4.5-22) we have

$$4E\{c_n^2\} = E\{a_n^2 - b_n^2\} - 2jE\{a_n\,b_n\} = 0 \quad (4.5\text{-}33)$$

or, in the limit, as $T \to \infty$

$$4E\{c_n^2\} = E\{a_n^2\} - E\{b_n^2\} = 0 \quad (4.5\text{-}34)$$

and

$$\lim_{T \to \infty} TE\{a_n^2\} = \lim_{T \to \infty} TE\{b_n^2\} = 2\phi_{xx}(n\,\omega_1) \quad (4.5\text{-}35)$$

The mean values of the coefficients are given by

$$E\{a_n\} = \frac{2}{T} \int_{-T/2}^{T/2} E\{X(t)\} \cos n\,\omega_1 t \; dt \quad (4.5\text{-}36)$$

$$E\{b_n\} = \frac{2}{T} \int_{-T/2}^{T/2} E\{X(t)\} \sin n\,\omega_1 t \; dt \quad (4.5\text{-}37)$$

and

$$E\{c_n\} = \frac{1}{T} \int_{-T/2}^{T/2} E\{X(t)\} \, e^{-jn\,\omega_1 t} \; dt \quad (4.5\text{-}38)$$

If the process $X(t)$ has a zero mean, then

$$E\{a_n\} = E\{b_n\} = E\{c_n\} = 0$$

We have already shown [see Eq. (4.5-25)] that the a_k's are uncorrelated with the b_k's. In general, however, the a_k's are not uncorrelated with each other and the b_k's are not uncorrelated with each other. However, in the limit as $T \to \infty$, we can show [see Problem #13] that, for $k \neq m$,

$$\lim_{T \to \infty} E\{a_k\,a_m\} = \lim_{T \to \infty} E\{b_k\,b_m\} = 0 \quad (4.5\text{-}39)$$

An important question is the meaning of the equality in Eq. (4.5-13) or (4.5-14); that is, in what sense do the right sides of these equations converge to the random process $X(t)$? We proceed now to answer this question.

Let us form the partial sum

$$X_N(t) = \sum_{n=-N}^{N} c_n \, e^{jn\omega_1 t} \qquad (4.5\text{-}40)$$

Then we have (dropping the subscripts on R)

$$E\{\,|\,X(t) - X_N(t)\,|^2\} \qquad (4.5\text{-}41)$$

$$= R(0) - 2E\{X(t)X_N(t)\} + E\{\,|\,X_N(t)\,|^2\}$$

The last two terms in this expression may be identified as follows:

$$E\{X(t)X_N(t)\} = \sum_{n=-N}^{N} e^{jn\omega_1 t} \, E\{c_n\,X(t)\}$$

which becomes, from Eq. (4.5-19),

$$E\{X(t)X_N(t)\} = \sum_{n=-N}^{N} e^{jn\omega_1 t} \, \frac{1}{T} \int_{-T/2}^{T/2} R_{zz}(\tau - t) \, e^{-jn\omega_1 \tau} \, d\tau$$

$$= \sum_{n=-N}^{N} \frac{1}{T} \int_{-T/2}^{T/2} R_{zz}(\tau - t) \, e^{-jn\omega_1(\tau - t)} \, d\tau$$

$$= \sum_{n=-N}^{N} r_n \qquad (4.5\text{-}42)$$

In the same way, we form

$$E\{\,|\,X_N(t)\,|^2\} = \sum_{m=-N}^{N} \sum_{n=-N}^{N} E\{c_m\,c_n^*\} \, e^{j\omega_1(m-n)t} \quad (4.5\text{-}43)$$

We substitute Eq. (4.5-21) into (4.5-43) to obtain

$$E\{\,|\,X_N(t)\,|^2\} = \sum_{n=-N}^{N} r_n \qquad (4.5\text{-}44)$$

Thus, in the limit, Eq. (4.5-41) becomes

$$\lim_{N \to \infty} E\{\,|\,X(t) - X_N(t)\,|^2\} = R(0) - 2R(0) + R(0) = 0 \qquad (4.5\text{-}45)$$

and the representations of Eq. (4.5-13) or (4.5-14) converge mean-square to $X(t)$.

4.6 *The Karhunen-Loeve Expansion* - Most random processes of interest will not be periodic and hence their autocorrelation functions cannot be expanded in a Fourier series. However, as pointed out in Section 3.7, the autocorrelation function (if it exists) of any random process is not only symmetric but non-negative definite and satisfies Eq. (3.7-15) or Eq. (3.7-16). There is a theorem in analysis [6,7] called *Mercer's Theorem* which states that any real function $R_{zz}(t,s)$ which is symmetric, continuous, and non-negative definite in the square $a \le t \le b$ and $a \le s \le b$ may be expanded in the absolutely and uniformly convergent series

$$R_{xx}(t,s) = \sum_{k=1}^{\infty} \frac{\Psi_k(t)\Psi_k(s)}{\lambda_k} \quad , \quad \begin{array}{c} a \le t \le b \\ a \le s \le b \end{array} \qquad (4.6\text{-}1)$$

Here the set of functions $\{\Psi_k(t)\}$ and the set of constants $\{\lambda_k\}$ are called the *eigenfunctions* and the *eigenvalues*, respectively, of the homogeneous integral equation

$$\Psi(s) = \lambda \int_a^b R_{xx}(t,s)\,\Psi(t)dt \quad , \quad a \le s \le b \qquad (4.6\text{-}2)$$

The known function $R_{xx}(t,s)$ is called the *kernel* of the integral equation. In the most general form of Mercer's Theorem, the ψ_k may be complex. We will show later that, if R_{xx} is real (and symmetric), then the ψ_k are real also.

It is clear that

$$\Psi(t) \equiv 0 \qquad (4.6\text{-}3)$$

is a general solution to this integral equation. It can be shown [6] that there is a set of functions $\{\Psi_k(t)\}$ and a corresponding set of values $\{\lambda_k\}$ of λ which are also solutions. In addition, if $R_{xx}(t,s)$ is positive definite, the $\{\Psi_k(t)\}$ comprise a complete orthonormal set in $[a,b]$. Thus, any mean-square continuous process $X(t)$ with autocorrelation function $R_{xx}(t,s)$ may be represented in the mean-square sense in $[a,b]$ by the orthonormal *Karhunen-Loeve* expansion of Eq. (4.5-3); that is, by

$$X(t) = \sum_{k=1}^{\infty} V_k\,\Psi_k(t) \quad , \quad a \le t \le b \qquad (4.6\text{-}4)$$

where

$$V_n = \int_a^b X(t)\,\Psi_n(t)dt \qquad (4.6\text{-}5)$$

The $\{\Psi_n(t)\}$ are an orthonormal set satisfying Eq. (4.6-2), and, if $R_{xx}(t,s)$ is positive definite, the set is complete.

As in the Fourier series expansion of a periodic random process, the coefficients V_n are uncorrelated random variables. From Eq. (4.6-5), form the expression (dropping the subscripts on R)

$$E\{V_m V_n\} = \int_a^b \int_a^b R(t,s)\,\Psi_m(s)\Psi_n(t)dtds \qquad (4.6\text{-}6)$$

Now Eq. (4.6-2) may be used to yield

$$E\{V_m V_n\} = \frac{1}{\lambda_n} \int_a^b \Psi_n(s)\Psi_m(s)ds \qquad (4.6\text{-}7)$$

or, from the orthonormality of the $\{\Psi_n(t)\}$,

$$E\{V_m V_n\} = \frac{\delta_{mn}}{\lambda_n} \qquad (4.6\text{-}8)$$

Thus the coefficients are uncorrelated random variables with second moments given by

$$E\{[V_n]^2\} = \frac{1}{\lambda_n} \qquad (4.6\text{-}9)$$

It is easily shown that these eigenfunctions form an *orthonormal* set as follows: Let $\Psi_0(\sigma)$ and $\Psi_1(\sigma)$ be the eigenfunctions corresponding to the different eigenvalues λ_0 and λ_1. We may write

$$\Psi_0(\sigma)\Psi_1(\sigma) = \lambda_1 \int_a^b R(\sigma - \tau)\Psi_0(\sigma)\Psi_1(\tau)d\tau \qquad (4.6\text{-}10)$$

On integrating both sides of this equation with respect to σ, we have

$$\int_a^b \Psi_0(\sigma)\Psi_1(\sigma)d\sigma = \lambda_1 \int_a^b \int_a^b R(\sigma - \tau)\Psi_0(\sigma)\Psi_1(\tau)d\tau d\sigma \qquad (4.6\text{-}11)$$

In the same way

$$\Psi_0(\sigma)\Psi_1(\sigma) = \lambda_0 \int_a^b R(\sigma - \tau)\Psi_0(\tau)\Psi_1(\sigma)d\tau \qquad (4.6\text{-}12)$$

or

$$\int_a^b \Psi_0(\sigma)\Psi_1(\sigma)d\sigma = \lambda_0 \int_a^b \int_a^b R(\tau - \sigma)\Psi_0(\sigma)\Psi_1(\tau)d\sigma d\tau \qquad (4.6\text{-}13)$$

by interchanging τ and σ in the right side of this equality. Now, since $R(\tau - \sigma) = R(\sigma - \tau)$ and if $\lambda_0 \neq \lambda_1$, we must conclude from Eq. (4.6-11) and (4.6-13) that

$$\int_a^b \Psi_0(\sigma)\Psi_1(\sigma)d\sigma = 0 \qquad (4.6\text{-}14)$$

Since the subscripts 0 and 1 were chosen arbitrarily, the eigenfunctions $\Psi_j(\tau)$ are orthogonal and, in fact, are normalized so that

$$\int_a^b \Psi_m(\sigma)\Psi_n(\sigma)d\sigma = \delta_{mn} \qquad (4.6\text{-}15)$$

A set of orthonormal functions $\{\Psi_k(t)\}_{k=1}^{\infty}$ is said to be *complete* in the interval $a \leq t \leq b$ if an arbitrary function $f(t)$, defined on the same interval, can be expanded as

$$f(t) = \sum_{n=1}^{\infty} a_n \Psi_n(t) \qquad (4.6\text{-}16)$$

where the $a_n's$ are constant coefficients. This is equivalent to requiring that there exists a function for which the coefficient a_n given by

$$a_n = \int_a^b f(t)\Psi_n(t)dt \qquad (4.6\text{-}17)$$

is not zero, all n.

Suppose an autocorrelation function $R_1(t,s)$ can be represented by a finite sum of the form of Eq. (4.6-1); that is,

$$R_1(t,s) = \sum_{k=1}^{N} \frac{\Psi_k(t)\Psi_k(s)}{\lambda_k} \qquad (4.6\text{-}18)$$

Since $R_1(t,s)$ is non-negative definite, we have

$$\int_a^b\int_a^b R_1(t,s)f(t)f(s)dtds \geq 0 \qquad (4.6\text{-}19)$$

for *arbitrary* $f(t)$. Using Eq. (4.6-18) in this expression, we obtain

$$\sum_{k=1}^{N} \frac{1}{\lambda_k} \int_a^b f(t)\Psi_k(t)dt\int_a^b f(s)\Psi_k(s)ds = \sum_{k=1}^{N} \frac{a_k^2}{\lambda_k} \geq 0 \qquad (4.6\text{-}20)$$

In this case, the $\{\Psi_k(t)\}_{k=1}^{N}$ is not a complete set. Suppose that $f(t) = \Psi_{N+1}(t)$. Then $a_k = 0$, $k=1,2,...,N$ and

$$\sum_{k=1}^{N} \frac{a_k^2}{\lambda_k} = 0 \qquad (4.6\text{-}21)$$

Thus, an autocorrelation function which is non-negative definite, but *not* positive definite, will have the finite series representation of Eq. (4.6-18). Such an autocorrelation function will be called *degenerate*.

Conversely, suppose $R_2(t,s)$ is an autocorrelation function which is positive definite so that, for *arbitrary* $f(t)$,

$$\int_a^b\int_a^b R_2(t,s)f(t)f(s)dt\,ds > 0 \qquad (4.6\text{-}22)$$

It is clear from the previous discussion that $R_2(t,s)$ has the infinite series representation of Eq. (4.6-1) and that the $\{\Psi_k(t)\}$ are complete; if not, then there is some Ψ_i for which

$$\int_a^b \Psi_i(t)[\int_a^b R_2(t,s)\Psi_i(s)ds]dt = \frac{1}{\lambda_i} \int_a^b \Psi_i(t)\Psi_i(t)dt = 0 \qquad (4.6\text{-}23)$$

which is a contradiction. Furthermore, we have

$$\frac{1}{\lambda_k} > 0 \quad , \quad \text{all } k \tag{4.6-24}$$

where the λ_k are generated, of course, by $R_2(t,s)$ through Eq. (4.6-2).

Since the kernel $R(\tau - \sigma)$ of Eq. (4.6-2) is real and symmetric, all of the eigenvalues λ_i are real. Suppose that λ_0 is a complex eigenvalue. Then it is clear that its complex conjugate λ_1 is also an eigenvalue. The corresponding eigenfunctions are

$$\Psi_0(x) = \mu(x) + j \, \upsilon(x) \tag{4.6-25}$$

$$\Psi_1(x) = \mu(x) - j \, \upsilon(x) \tag{4.6-26}$$

Then, from Eq. (4.6-15), we have

$$\int\limits_a^b [\mu(x)]^2 + [\upsilon(x)]^2 dx = 0 \tag{4.6-27}$$

which implies that

$$\mu(x) = \upsilon(x) = 0 \tag{4.6-28}$$

Thus, the integral equation has no non-zero solutions corresponding to complex eigenvalues and complex eigenfunctions.

The mean-square convergence of Eq. (4.6-4) is apparent when Eq. (4.5-7) is expanded to give

$$\lim_{N \to \infty} \left\{ R(t,t) - 2 \sum_{n=1}^{N} E[X(t)V_n]\Psi_n(t) + \sum_{m=1}^{N} \sum_{n=1}^{N} E[V_m V_n]\Psi_m(t)\Psi_n(t) \right\} \tag{4.6-29}$$

We have that $E[V_m V_n] = \delta_{mn}/\lambda_n$ and

$$E[X(t)V_n] = \int\limits_a^b R(t-s)\Psi_n(s)ds = \frac{1}{\lambda_n} \Psi_n(t) \tag{4.6-30}$$

so that Eq. (4.6-29) becomes

$$\lim_{N \to \infty} \left\{ R(t,t) - \sum_{n=1}^{N} \frac{\Psi_n(t)\Psi_n(t)}{\lambda_n} \right\} \tag{4.6-31}$$

which is zero by Mercer's Theorem [Eq. (4.6-1)].

A heuristic derivation of Mercer's theorem proceeds as follows: Suppose we expand $R(t,s)$ as

$$R(t,s) = \sum_{i=1}^{\infty} \alpha_i(t)\Psi_i(s) \tag{4.6-32}$$

Then the coefficient $\alpha_i(t)$ is given by

$$\alpha_i(t) = \int_a^b R(t,s)\Psi_i(s)ds \tag{4.6-33}$$

or

$$\alpha_i(t) = \Psi_i(t)/\lambda_i \tag{4.6-34}$$

if Eq. (4.6-2) generates the $\{\Psi_i\}$.

The Karhunen-Loeve expansion of Eq. (4.6-4) has been used extensively in problems of statistical design and optimization. It is most useful as a formal representation since, from a practical point of view, it has some disadvantages. The orthogonal functions $\{\Psi_n(t)\}$ depend on the autocorrelation function $R_{xx}(t,s)$ and on the interval (a,b). Furthermore, the $\{\Psi_n(t)\}$ are relatively difficult to determine since the solution of Eq. (4.6-2) may not be easy [4,6]. We now consider two examples.

Example 4.5

Consider the sinusoid of random phase given by $X(t) = A\cos[\omega_0 t + \theta]$ where A, ω_0 are constants and θ is a random variable uniformly distributed on $(-\pi,\pi)$. This process if w.s.s. and periodic with period $T = 2\pi/\omega_0$. Its autocorrelation function has already been shown to be [see Section 3.3]

$$R(\tau) = \frac{A^2}{2}\cos\omega_0\tau$$

which is a periodic function with period $T = 2\pi/\omega_0$.

The integral equation to be solved is

$$\Psi(t) = \lambda \int_{-T/2}^{T/2} \frac{A^2}{2}\cos\omega_0(t-s)\Psi(s)ds \quad, \quad -T/2 \le t \le T/2 \quad (A)$$

The solution to this equation is easily found by using Mercer's Theorem to write

$$R(t-s) = \sum_{i=1}^{\infty} \frac{\Psi_i(t)\Psi_i(s)}{\lambda_i} = \frac{A^2}{2}\cos\omega_0(t-s)$$

Since $\cos(x-y) = \cos x \cos y + \sin x \sin y$, we have

$$R(t-s) = \frac{A^2}{2} \cos \omega_0 t \, \cos \omega_0 s + \frac{A^2}{2} \sin \omega_0 t \, \sin \omega_0 s$$

Hence we see, after normalization, that the integral equation possesses two eigenfunctions and two eigenvalues given by

$$\Psi_1(t) = \sqrt{2/T} \, \cos \omega_0 t \quad , \quad \lambda_1 = 4/TA^2$$

$$\Psi_2(t) = \sqrt{2/T} \, \sin \Psi_0 t \quad , \quad \lambda_2 = 4/TA^2$$

Note that the eigenvalues are not distinct; although $\sqrt{2/T} \cos \omega_0 t$ and $\sqrt{2/T} \sin \omega_0 t$ are orthonormal, they obviously do not form a complete set. The $K-L$ expansion becomes

$$X(t) = A \, \cos[\omega_0 t + \theta] = V_1 \sqrt{2/T} \cos \omega_0 t + V_2 \sqrt{2/T} \, \sin \omega_0 t$$

where the random coefficients V_1 and V_2 are given by

$$V_1 = A \sqrt{T/2} \cos \theta$$

and

$$V_2 = -A \sqrt{T/2} \sin \theta$$

Also, we have

$$E\{V_1 V_2\} = 0 = E\{\frac{A^2 T}{2} \sin \theta \cos \theta\} = \frac{A^2 T}{4} E\{\sin 2\theta\}$$

$$= \frac{A^2 T}{4} \frac{1}{2\pi} \int_{-\pi}^{\pi} \sin 2\theta \, d\theta = 0$$

as expected, and

$$E\{V_1^2\} = \frac{A^2 T}{2} E\{\cos^2 \theta\} = \frac{A^2 T}{4}$$

$$E\{V_2^2\} = \frac{A^2 T}{2} E\{\sin^2 \theta\} = \frac{A^2 T}{4}$$

This example was particularly easy to solve since the Mercer Theorem expansion of the autocorrelation function was degenerate and obvious.

Example 4.6

Let an autocorrelation function $R(\tau)$ be given by

$$R(\tau) = A \, e^{-\alpha|\tau|} \tag{A}$$

where $A > 0$ and $\alpha > 0$ and take the interval of expansion to be $-a \leq t \leq a$. We require the solution to the integral equation

$$\Psi(t) = \lambda \int_{-a}^{a} A \ e^{-\alpha|t-s|} \ \Psi(s)ds \ , \quad -a \leq t \leq a \tag{B}$$

This last expression may be written as

$$\Psi(t) = \lambda \int_{-a}^{t} A \ e^{-\alpha(t-s)}\Psi(s)ds \ + \lambda \int_{t}^{a} A \ e^{-\alpha(s-t)}\Psi(s)ds$$

Let us differentiate twice with respect to t to obtain

$$\Psi'(t) = -\lambda\alpha \int_{-a}^{t} A \ e^{-\alpha(t-s)}\Psi(s)ds \ + \lambda A \ \Psi(t)$$

$$+ \lambda\alpha \int_{t}^{a} A \ e^{-\alpha(s-t)}\Psi(s)ds \ - \lambda A \ \Psi(t)$$

and

$$\Psi''(t) = \lambda\alpha^2 \int_{-a}^{a} A e^{-\alpha|t-s|} \Psi(s)ds \ - 2\lambda\alpha A \ \Psi(t)$$

or

$$\Psi''(t) = \alpha^2\Psi(t) - 2\lambda\alpha A \ \Psi(t) = (\alpha^2 - 2\lambda\alpha A)\Psi(t)$$

where the quantity $(\alpha^2 - 2\lambda\alpha A)$ is real since λ is real. This last expression is a linear homogeneous differential equation

$$\Psi''(t) - (\alpha^2 - 2\lambda\alpha A) \ \Psi(t) = 0 \tag{C}$$

which $\Psi(t)$ must satisfy in order to be a solution of the integral equation given by Eq. (B).

In general, the homogeneous linear differential equation

$$\Psi''(t) - c^2\Psi(t) = 0 \ , \quad c^2 \text{ real}$$

has the solutions whose form depends on c^2 so that

$$\Psi(t) = B_1 e^{ct} + B_2 e^{-ct} \ , \quad 0 < c^2 < \infty$$

or

$$\Psi(t) = C_1 e^{jct} + C_2 e^{-jct} \ , \quad -\infty < c^2 < 0$$

or

$$\Psi(t) = D_1 t + D_2 \ , \quad c^2 = 0$$

The proper procedure to follow now is to consider separately each of these three possible general solutions for $\Psi(t)$. In each case subsititute the general solution into Eq. (A), perform the prerequisite integration, and equate like terms on each side of the resulting expression - giving solutions for the unknown constants B_1 and B_2 (or C_1 and C_2 or D_1 and D_2).

This example is typical of a class of problems where the solution of the homogeneous integral equation reduces to the solution of an associated homogeneous differential equation [4].

A complete solution to this example is given in [4] and is suggested as Problem #14.

4.7 Optimal Truncation Properties of the Karhunen-Loeve Expansion - The problem of truncation error is likely to arise in any practical application of an infinite series representation. Specifically we are interested in the error that arises when the summation of Eq. (4.6-4) is taken over a finite number of terms. Let this truncated sum be denoted by

$$X_N(t) = \sum_{n=1}^{N} V_n \Psi_n(t) \quad , \quad a \le t \le b \qquad (4.7\text{-}1)$$

and let us take as a measure of the error the quantity

$$E\left\{ \epsilon_N^2 \right\} = E\left\{ \int_a^b [X(t) - X_N(t)]^2 dt \right\} \qquad (4.7\text{-}2)$$

Note that this error involves a time averaging. We shall now prove that the mean square truncation error given by Eq. (4.7-2) is a minimum if the $\Psi(t)$ of Eq. (4.7-1) are a particular subset of the eigenfunctions of Eq. (4.6-2). In other words, the Karhunen-Loeve expansion is the optimal truncated expansion in the sense that it yields a minimum mean square error [8,9].

Eq. (4.7-2) may be expanded as in Eq. (4.5-45) to give

$$E\left\{ \epsilon_N^2 \right\} = \int_a^b E\left\{ X^2(t) \right\} dt - \sum_{n=1}^{N} E\left\{ V_N^2 \right\} \qquad (4.7\text{-}3)$$

It is apparent that the first term in Eq. (4.7-3) is independent of the choice of the $\Psi_n(t)$; consequently the error is minimized when the last term

$$\sum_{i=1}^{N} E\left\{ V_n^2 \right\}$$

is maximized. Since each term in this sum is non-negative, the sum is a maximum when each term is a maximum. From Eq. (4.6-5), we write one of the terms as

$$E\left\{ V_n^2 \right\} = \int_a^b \int_a^b R(s-t) \Psi_n(s) \Psi_n(t) ds dt \qquad (4.7\text{-}4)$$

and maximize this quantity with respect to Ψ_n subject to the orthonormality constraint of Eq. (4.5-2); that is, we consider

$$\int_a^b \int_a^b R(s,t)\Psi_n(s)\Psi_n(t)dsdt - \sigma\int_a^b \Psi_n^2(t)dt \qquad (4.7\text{-}5)$$

where σ is a Lagrangian multiplier. The minimization procedure is as follows:

1. In Eq.(7-5), replace Ψ_n by $\Psi_n + \epsilon\delta\Psi_n$, where ϵ is a real variable and $\delta\Psi_n$ is an arbitrary variation in Ψ_n.

2. Differentiate the resulting expression with respect to ϵ; set the result equal to zero; and allow ϵ to approach zero. The result is

$$\int_a^b \int_a^b R(s,t)[\delta\Psi_n(s)\Psi_n(t) + \delta\Psi_n(t)\Psi_n(s)]dsdt$$

$$(4.7\text{-}6)$$

$$-\sigma\int_a^b 2\delta\Psi_n(t)\Psi_n(t)dt = 0$$

Since $R(s,t)=R(t,s)$, Eq. (4.7-6) may be rearranged to give

$$\int_a^b \delta\Psi_n(t)\left[\int_a^b R(s,t)\Psi_n(s)ds - \sigma\Psi_n(t)\right]dt = 0 \qquad (4.7\text{-}7)$$

If this expression is to be zero for arbitrary $\delta\Psi_n$, it is necessary that the expression in brackets be zero for all $t \in [a,b]$, or

$$\frac{1}{\sigma}\int_a^b R(s,t)\Psi_n(s)ds = \Psi_n(t), \quad a \leq t \leq b \qquad (4.7\text{-}8)$$

which is Eq. (4.6-2), as was to be proved. Thus, a necessary condition for Eq. (4.7-2) to be a minimum is that the Ψ_n be the orthonormal functions of the Karhunen-Loeve expansion.

Equation (4.7-3) may be rewritten with the aid of Eqs. (4.7-4) and (4.6-2) as

$$E\left\{\epsilon_N^2\right\} = \int_a^b R(t,t)dt - \sum_{n=1}^N \frac{1}{\lambda_n} \qquad (4.7\text{-}9)$$

We have already pointed out in Section 4.6 that $R(s,t)$ is nonnegative definite if the process $X(t)$ is mean-square continuous. Equivalently, it is clear that the terms in Eq. (4.7-4) are nonnegative or

$$E\left\{V_n^2\right\} = \frac{1}{\lambda_n} \geq 0 \qquad (4.7\text{-}10)$$

that is, the eigenvalues are non-negative. An examination of Eq. (4.7-9) shows that if $\int_a^b R(t,t)dt$ exists, then the λ_n must be positive.

Thus Eq. (4.7-10) has a maximum value when the eigenvalue is chosen which is a minimum (this particular λ will be called λ_1 and we have $\lambda_1 > 0$). Now suppose we form a new symmetric kernel

$$R_{(1)}(s,t) = R(s,t) - \frac{\Psi_1(s)\Psi_1(t)}{\lambda_1} \tag{4.7-11}$$

and maximize the quantity

$$\int_a^b \int_a^b R_{(1)}(s,t)\xi(s)\xi(t)dsdt - \sigma\int_a^b \xi^2(t)dt \tag{4.7-12}$$

By the same procedure as before, the ξ must be a solution of the homogeneous integral equation

$$\int_a^b R_{(1)}(s,t)\xi(s)ds = \sigma\xi(t) \tag{4.7-13}$$

With the aid of Eq. (4.7-11) we write

$$\sigma\xi(t) = \int_a^b R(s,t)\xi(s)ds - \frac{\Psi_1(t)}{\lambda_1}\int_a^b \xi(s)\Psi_1(s)ds \tag{4.7-14}$$

Now multiply by $\Psi_1(t)$, integrate with respect to t, and interchange the order of integration. We have

$$\sigma\int_a^b \xi(t)\Psi_1(t)dt = 0 \tag{4.7-15}$$

since the two terms on the right side are equal. It follows immediately from Eq. (4.7-11) that

$$\int_a^b R_{(1)}(s,t)\xi(s)ds = \int_a^b R(s,t)\xi(s)ds \tag{4.7-16}$$

and hence, that ξ is an eigenfunction of Eq. (4.7-8). Let us call this eigenfunction Ψ_2 and its corresponding eigenvalue λ_2. It is apparent that $0 < \lambda_1 \leq \lambda_2$.

In the same way, the procedure can be continued by constructing $R_2(s,t)$, etc.. Thus the eigenvalues should be numbered in nondecreasing order

$$0 < \lambda_1 \leq \lambda_2 \leq \lambda_3 \ldots\ldots \tag{4.7-17}$$

and the first N of them used in Eq. (4.7-9). The least mean square error representation of Eq. (4.7-1) is obtained when the eigenfunctions corresponding to the first N eigenvalues from Eq. (4.7-17) are used. The error expression of Eq. (4.7-9) can be written in an alternative form. From Mercer's theorem [Eq. (4.6-1)] we have

$$\int_a^b R(t,t)dt = \sum_{n=1}^{\infty} \frac{1}{\lambda_n} \int_a^b \Psi_n{}^2(t)dt = \sum_{n=1}^{\infty} \frac{1}{\lambda_n} \qquad (4.7\text{-}18)$$

Consequently Eq. (4.7-9) becomes

$$E\left\{\epsilon_N^2\right\} = \sum_{n=N+1}^{\infty} \frac{1}{\lambda_n} \qquad (4.7\text{-}19)$$

4.8 *The Sampling Theorem* - Suppose we consider a given arbitrary signal $f(t)$ available for all time as shown in Fig. 4.4(a).

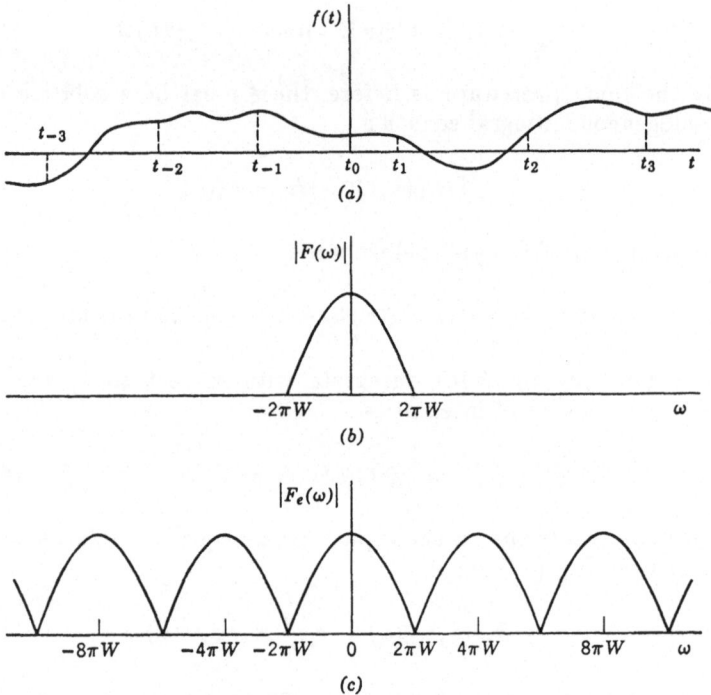

Fig. 4.4 - Sampling and bandlimited signals

We ask ourselves if it is necessary to know the amplitude of the signal for every value of time in order to characterize it uniquely? In other words, can $f(t)$ be represented (and reconstructed) from some set of *sample values* or *samples* $...,f(t_{-1}),f(t_0),f(t_1),...$? Surprisingly enough, it turns out that, under certain fairly reasonable conditions, a signal may be represented exactly by samples spaced relatively far apart. The reasonable conditions are that the signal be *strictly bandlimited*.

As defined previously, a (real) signal $f(t)$ will be called *strictly bandlimited* $(-2\pi W, 2\pi W)$ if its Fourier transform $F(\omega)$ has the property

$$F(\omega) = 0 \quad , \quad |\omega| > 2\pi W \qquad (4.8\text{-}1)$$

The magnitude of each a spectrum is shown in Fig. 4.4(b) where the shape is arbitrary. It is clear that this spectrum could be extended into a periodic frequency function with period $4\pi W$ as shown in Fig. 4.4(c). In other words, we define a new function $F_e(\omega)$ by

$$F_e(\omega) = \sum_{n=-\infty}^{\infty} F(\omega + n\,4\pi W) \qquad (4.8\text{-}2)$$

This function is periodic with period $4\pi W$ since

$$F_e(\omega + k\,4\pi W) = \sum_{n=-\infty}^{\infty} F[\omega + (n+k)4\pi W]$$

$$= \sum_{m=-\infty}^{\infty} F(\omega + m\,4\pi W) = F_e(\omega) \qquad (4.8\text{-}3)$$

where $m = n + k$.

For reasonably well behaved* $F(\omega)$, the periodic function $F_e(\omega)$ can be expanded in a Fourier series with period $4\pi W$, and, in the interval $-2\pi W < \omega < 2\pi W$, this Fourier Series will converge to $F(\omega)$; that is,

$$F(\omega) = \sum_{k=-\infty}^{\infty} F_k\, e^{-jk\,2\pi\omega/4\pi W} = \sum_{k=-\infty}^{\infty} F_k\, e^{-jk\,\omega/2W}, \quad |\omega| \le 2\pi W \quad (4.8\text{-}4)$$

where $j = \sqrt{-1}$ and F_k is the Fourier coefficient given by

$$F_k = \frac{1}{4\pi W} \int_{-2\pi W}^{2\pi W} F(\omega) e^{jk\,\omega/2W}\, d\omega \qquad (4.8\text{-}5)$$

Since $F(\omega)$ is band-limited as described by Eq. (4.8-1), its inverse Fourier transform is

$$f(t) = \frac{1}{2\pi} \int_{-2\pi W}^{2\pi W} F(\omega) e^{j\omega t}\, d\omega \qquad (4.8\text{-}6)$$

If we define the *Nyquist instants* as the set of times

$$\{t_n\} = \{t_n \mid t_n = \frac{n}{2W}, \; n = ...,-1,0,1,...\} \qquad (4.8\text{-}7)$$

---

*The conditions that a function must satisfy in order to be expanded in a Fourier series are discussed in Appendix D, particularly Section D.4.

then it is clear that $f(t_n)$ is given from Eq. (4.8-6) as

$$f(t_n) = f(\frac{n}{2W}) = \frac{1}{2\pi} \int_{-2\pi W}^{2\pi W} F(\omega) e^{jn\omega/2W} d\omega \qquad (4.8\text{-}8)$$

A comparison of this last equation with Eq. (4.8-5) shows that the Fourier coefficient F_k is related to the sample value $f(\frac{k}{2W})$ by

$$F_k = \frac{1}{2W} f(\frac{k}{2W}) \qquad (4.8\text{-}9)$$

If the sample values $f(\frac{n}{2W})$ are given for all time, then the Fourier series

$$F(\omega) = \frac{1}{2W} \sum_{k=-\infty}^{\infty} f(\frac{k}{2W}) e^{-jk\omega/2W} \, , \, |\omega| \leq 2\pi W \qquad (4.8\text{-}10)$$

determines $F(\omega)$ exactly and hence $f(t)$ through the inverse Fourier transform given by Eq. (4.8-6). This completes the proof of the existence of a sampling theorem: *a function f(t), strictly bandlimited $(-2\pi W, 2\pi W)$ radians per second, is uniquely and exactly determined by its sample values spaced $1/2W$ seconds apart throughout the time domain.* Of course, there are an infinite number of such samples.

We now consider the reconstruction of $f(t)$ from its sample values. On substituting Eq. (4.8-10) into Eq. (4.8-6) we have

$$f(t) = \frac{1}{2\pi} \int_{-2\pi W}^{2\pi W} \frac{1}{2W} \sum_{k=-\infty}^{\infty} f(\frac{k}{2W}) e^{-jk\omega/2W} e^{j\omega t} d\omega \qquad (4.8\text{-}11)$$

We may interchange the order of summation and integration to obtain

$$f(t) = \sum_{k=-\infty}^{\infty} f(\frac{k}{2W}) \frac{1}{4\pi W} \int_{-2\pi W}^{2\pi W} e^{j\omega(t - \frac{k}{2W})} d\omega \qquad (4.8\text{-}12)$$

The integral is the Fourier transform of the symmetrical pulse centered on $t = \frac{k}{2W}$. It may be evaluated easily as

$$\frac{1}{4\pi W} \int_{-2\pi W}^{2\pi W} e^{j\omega(t - \frac{k}{2W})} d\omega = \frac{1}{2\pi Wt - k\pi} \left[\frac{e^{j(2\pi Wt - k\pi)} - e^{-j(2\pi Wt - k\pi)}}{2j} \right]$$

$$= \frac{\sin(2\pi Wt - k\pi)}{2\pi Wt - k\pi} \qquad (4.8\text{-}13)$$

This last result may be used in Eq. (4.8-12) to give the *sampling representation*

$$f(t) = \sum_{k=-\infty}^{\infty} f\left(\frac{k}{2W}\right) \frac{sin\left(2\pi Wt - k\pi\right)}{\left(2\pi Wt - k\pi\right)} \qquad (4.8\text{-}14)$$

This expression is sometimes called the *Cardinal Series* or *Shannon's Sampling Theorem*. Although it has been known to mathematicians at least since 1915 [10], its use in engineering dates from some of the pioneer work of Shannon in 1949 in the field of information theory [11,12]. This expression, together with Eq. (4.8-8), related the discrete time domain $\{\frac{k}{2W}\}$ with sample values $f\left(\frac{k}{2W}\right)$ to the continuous time domain $\{t\}$ of the function $f(t)$.

The interpolation function

$$k(t) = \frac{\sin 2\pi Wt}{2\pi Wt} \qquad (4.8\text{-}15)$$

has a Fourier transform $K(\omega)$ given by

$$K(\omega) = \begin{cases} \dfrac{1}{2W} & , \ |\omega| < 2\pi W \\[3mm] 0 & , \ |\omega| > 2\pi W \end{cases} \qquad (4.8\text{-}16)$$

Also the shifted function $k\left(t - \frac{k}{2W}\right)$ has the transform

$$\mathbf{F}\left\{k\left(t - \frac{k}{2W}\right)\right\} = K(\omega)e^{-j\omega\frac{k}{2W}} \qquad (4.8\text{-}17)$$

Therefore, each term on the right side of Eq. (4.8-14) is a time function which is strictly bandlimited $(-2\pi W, 2\pi W)$. Note also that

$$k\left(t - \frac{k}{2W}\right) = \frac{\sin\left(2\pi Wt - k\pi\right)}{2\pi Wt - k\pi} = \begin{cases} 1 & , \ t = t_k = \dfrac{k}{2W} \\[3mm] 0 & , \ t = t_n, \ n \neq k \end{cases} \qquad (4.8\text{-}18)$$

Thus this sampling function $k\left(t - \frac{k}{2W}\right)$ is zero at all Nyquist instants except t_k, where it equals unity.

The expansion of Eq. (4.8-14) might be called a sampling theorem in the *time domain*. It is apparent from the symmetrical properties of the Fourier transform that a similar relationship could be written in the *frequency domain* for strictly time-limited signals.

For the strictly bandlimited signal $f(t)$, the spectrum $F(\omega)$ is nonzero only for $|\omega| \leq 2\pi W$, and the Fourier series of Eq. (4.8-10) is considered only for this frequency interval. On the other hand, the Fourier transform of the sampling expression of Eq. (4.8-14) is $F(\omega)$ only for $|\omega| \leq 2\pi W$ and, for all frequencies, is the function $F_e(\omega)$ given by Eq. (4.8-2)

$$F_e(\omega) = \sum_{n=-\infty}^{\infty} F(\omega - n\,4\pi W) \qquad (4.8\text{-}19)$$

We have already pointed out that $F_e(\omega)$ is periodic with period $4\pi W$. Let us call the set of frequencies ω_k, defined by

$$\omega_k = \omega - k\,4\pi W, \quad k = \ldots,-1,0.1,\ldots, \qquad (4.8\text{-}20)$$

the *aliases* of ω. It is clear that

$$F(\omega) = F_e(\omega_k) \quad , \quad |\omega| < 2\pi W \qquad (4.8\text{-}21)$$

Suppose that a function $h(t)$ is not strictly band-limited to at least $(-2\pi W, 2\pi W)$ radians per second and an attempt is made to reconstruct the function using Eq. (4.8-14) with sample values space $1/2W$ seconds apart. It is apparent that the reconstructed signal [which is strictly bandlimited $(-2\pi W, 2\pi W)$ as already mentioned] will differ from the original. Moreover a given set of sample values $\{f(\frac{k}{2W})\}$ could have been obtained from a whole class of different signals. Thus we emphasize that the reconstruction of Eq. (4.8-14) is unambiguous only for signals strictly bandlimited to at least $(-2\pi W, 2\pi W)$ radians per second. The set of different possible signals with the same set of sample values $\{f(\frac{k}{2W})\}$ are called the *aliases* of the bandlimited signal $f(t)$.

This theorem has also played an important role in the application of random theory to engineering problems. Shannon's purpose in reviving the theorem was to permit the treatment of continuous noisy information channels in the framework of discrete mathematics. As a matter of fact, much of the pioneer work on the statistical design of optimum systems used this form of representation [11,13,14]. It is true that this early work, as was the case with Fourier series use, was characterized by a certain lack of rigor, but this deficiency has been corrected subsequently [15].

Suppose the wide-sense stationary random process $X(t)$ has a power spectral density $\phi(\omega)$ which is zero outside the frequency range $(-W, W)$ cps so that

$$\phi(\omega) = 0 \quad , \quad |\omega| > 2\pi W$$

Such a process will be called *strictly bandlimited* $(-2\pi W, 2\pi W)$. Then $X(t)$ can be represented as

$$X(t) = \sum_{n=-\infty}^{\infty} X(\frac{n}{2W}) \ sinc \ (2Wt - n) \qquad (4.8\text{-}22)$$

where the function *sinc* x is defined as

$$sinc \ x = \frac{\sin\pi x}{\pi x} \qquad (4.8\text{-}23)$$

and where the right side of Eq. (4.8-22) converges mean-square [16] to the process. The proof of this convergence proceeds exactly as in Section 4.5; that is, it is necessary to express the correlation function $R(\tau)$ in a series similar to that of Eq. (4.8-22).

Since we have presumed that $X(t)$ is bandlimited to $(-W, W)$ Hz, we can write the correlation function $R(\tau)$ as

$$R(\tau) = \frac{1}{2\pi} \int_{-2\pi W}^{2\pi W} \phi(\omega) \ e^{j\omega\tau} d\omega \qquad (4.8\text{-}24)$$

where $\phi(\omega)$ is the power spectral density of $X(t)$. It follows from the sampling theorem just developed that

$$R(\tau) = \sum_{n=-\infty}^{\infty} R(\frac{n}{2W}) \ sinc \ (2W\tau - n) \qquad (4.8\text{-}25)$$

or, with a change in index $k = 2W\tau + n$,

$$R(t-\tau) = \sum_{k=-\infty}^{\infty} R(\frac{k}{2W} - \tau) \ sinc \ (2Wt - k) \qquad (4.8\text{-}26)$$

where the series of Eqs. (4.8-25) and (4.8-26) converge as previously discussed to $R(\tau)$ or $R(t-\tau)$.

Suppose we define the partial sum

$$X_M(t) = \sum_{n=-M}^{M} X(\frac{n}{2W}) \ sinc \ (2Wt - n) \qquad (4.8\text{-}27)$$

and form

$$E\left\{ |X(t) - X_M(t)|^2 \right\} = \qquad (4.8\text{-}28)$$

$$R(t-t) - 2E\left\{ X(t)X_M(t) \right\} + E\left\{ |X_M(t)|^2 \right\}$$

as in Section 4.5. Then, as before, we have

$$E\left\{X(t)X_M(t)\right\} = \sum_{n=-M}^{M} E\left\{X(t)X(\frac{n}{2W})\right\} sinc\,(2Wt-n)$$

$$= \sum_{n=-M}^{M} 2W \int_{-\infty}^{\infty} R(s-t)sinc\,(2Ws-n)ds \; sinc\,(2Wt-n)$$

$$= \sum_{n=-M}^{M} R(\frac{n}{2W}-t)\,sinc\,(2Wt-n) \qquad (4.8\text{-}29)$$

where use has been made of the orthogonality relationship

$$\int_{-\infty}^{\infty} sinc\,(x-n)sinc\,(x-m)dx = \delta_{mn} \qquad (4.8\text{-}30)$$

to give

$$X(\frac{n}{2W}) = 2W \int_{-\infty}^{\infty} X(t)\,sinc\,(2Wt-n)dt \qquad (4.8\text{-}31)$$

and

$$R(\frac{n}{2W}-t) = 2W \int_{-\infty}^{\infty} R(s-t)sinc\,(2Ws-n)ds \qquad (4.8\text{-}32)$$

from Eqs. (4.8-22) and (4.8-26) respectively. In the same way

$$E\{|X_M(t)|^2\} = \sum_{n=-M}^{M} \sum_{m=-M}^{M} E\{X(\frac{n}{2W})\,X(\frac{n}{2W})\}sinc\,(2Wt-n)sinc\,(2Wt-m)$$

$$= \sum_{n=-M}^{M} \sum_{m=-M}^{M} R(\frac{m-n}{2W})sinc\,(2Wt-m)sinc\,(2Wt-n)$$

$$= \sum_{n=-m}^{M} R(\frac{n}{2W}-t)\,sinc\,(2Wt-n) \qquad (4.8\text{-}33)$$

Combining Eqs. (4.8-28), (4.8-29), and (4.8-33) we have

$$\lim_{M\to\infty} E\{|X(t)-X_M(t)|^2\}=R(t-t)-\lim_{M\to\infty} \sum_{n=-M}^{M} R(\frac{n}{2W}-t)sinc\,(2Wt-n)$$

$$= 0 \qquad (4.8\text{-}34)$$

by Eq. (4.8-26). The series of Eq. (4.8-22) converges mean-square to $X(t)$ as was to be proved.

In addition to the process being strictly band-limited to $(-2\pi W, 2\pi W)$, assume that its spectral density is the constant K in that interval. Then the correlation function of the process is given by

$$R(\tau) = \frac{K}{2\pi} \int_{-2\pi W}^{2\pi W} e^{j\omega\tau}\,d\omega = 2WK \; sinc\,2W\tau \qquad (4.8\text{-}35)$$

This correlation function is zero at all of the Nyquist instants $\tau = n/2W$ except for $n = 0$. Consequently the samples $X(n/2W)$ are uncorrelated; that is,

$$E[X(m/2W)X(n/2W)] = \delta_{mn}/\gamma_n \qquad (4.8\text{-}36)$$

where $1/\gamma_n$ are the variances of the samples. For this particular choice of spectrum, the desirable condition that all of the coefficients $\{\ X(m/2W)\ \}$ be uncorrelated has been met. However, this condition does not hold in general for arbitrary power spectral densities.

Other forms of the sampling theorem than that given by Eq. (4.8-22) can be used. As previously discussed, a considerable literature has developed on sampling theorem generalizations in recent years. As pointed out by Balakrishnan [16] and Beutler [17], these more general forms have their stochastic equivalent.

Both the Fourier series and the various sampling expansions have considerable appeal. The coefficients in both cases correspond to physical concepts with which most engineers feel very much at home. However, as has been shown, both forms of representation have certain inherent weaknesses. In the case of the Fourier series, the coefficients are not uncorrelated for a finite interval T except for the special case of a periodic random process. For the sample values used in the sampling expansion to be uncorrelated, it is necessary that the power spectral density of the process not only be strictly bandlimited, but also belong to a small class of such processes.

The more general orthogonal expansion of Section 4.6 does not have the problems connected with it that have just been enumerated. A proper choice of coefficients insures that they are uncorrelated for arbitrary spectral density and interval of expansion. Unfortunately neither the coefficients nor the orthogonal functions of this expansion have much intuitive appeal, or, indeed, as clear a physical meaning. In addition the form of the orthogonal functions depends on both the spectral density of the process and the interval of expansion and these functions are generally difficult to determine explicitly.

4.9 *Narrow-Band Systems* - In many communication systems, the situation will arise where broad-band noise is passed through a narrow-band linear filter to produce a narrow-band random process. By a *narrow-band process* we mean one whose spectral density is concentrated in a region $\Delta\omega_0$ which is narrow compared to a center frequency ω_0 as shown in Fig. 4.5. In radio receivers, for example, the *RF* stage is a tunable filter-amplifier with a bandwidth which may be very small compared to its center frequency. In such situations, the input process will frequently be Gaussian or nearly Gaussian and,

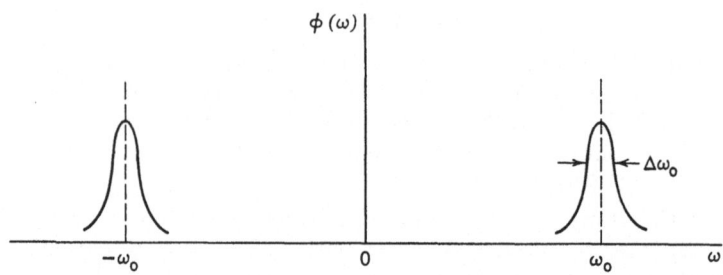

Fig. 4.5 - The spectral density of a narrow-band process

thus, the output will be Gaussian also. Furthermore, the narrow-band filters will frequently be followed by such non-linear operations as rectification, squaring, and clipping. It is pertinent, therefore to examine the properties of narrow-band noise, particularly those properties which will be useful in determining the effect of certain nonlinear operations.

Suppose a broad-band noise is passed through a narrow-band filter with an adjustable bandwidth and the output is displayed on an oscilloscope. As the filter bandwidth is decreased the observed waveform takes on increasingly the appearance of a sinusoid of random amplitude and phase. This suggests that the output might be represented in the form [5]

$$X(t) = E(t) \; cos \, [\omega_0 t + \xi(t)] \qquad (4.9\text{-}1)$$

where $\omega_0/2\pi = f_0$ is the center frequency of the filter and $E(t)$ and $\xi(t)$ are random processes which will be called the *envelope function* and the *phase function*, respectively.

We now make use of the Fourier series as discussed in Section 4.5 to represent the narrow-band process $X(t)$ in the interval $(-T/2, T/2)$:

$$X(t) = \sum_{k=1}^{\infty} [a_k \cos k \, \omega_1 t + b_k \sin k \, \omega_1 t] \qquad (4.9\text{-}2)$$

where $\omega_1 = 2\pi/T$ and where the d-c term has been dropped for convenience. After some algebraic manipulation of Eq. (4.9-2), we will relate the envelope function $E(t)$ and the phase function $\xi(t)$ to the Fourier coefficients a_k and b_k. From a knowledge of the statistical properties of these coefficients, we will determine the statistical properties of $E(t)$ and $\xi(t)$.

We can rewrite Eq. (4.9-1) in the form

$$X(t) = E(t) \cos \xi(t) \cos \omega_0 t - E(t) \sin \xi(t) \sin \omega_0 t \quad (4.9\text{-}3)$$

since

$$\cos(x + y) = \cos x \cos y - \sin x \sin y$$

We also write $\cos k \omega_1 t$ and $\sin k \omega_1 t$ as

$$\begin{aligned}
\cos k \omega_1 t &= \cos[(k\omega_1 - \omega_0)t + \omega_0 t] \\
&= \cos(k\omega_1 - \omega_0)t \cos \omega_0 t - \sin(k\omega_1 - \omega_0)t \sin \omega_0 t \quad (4.9\text{-}4)
\end{aligned}$$

and

$$\begin{aligned}
\sin k \omega_1 t &= \sin[(k\omega_1 - \omega_0)t + \omega_0 t] \\
&= \sin(k\omega_1 - \omega_0)t \cos \omega_0 t + \cos(k\omega_1 - \omega_0)t \sin \omega_0 t \quad (4.9\text{-}5)
\end{aligned}$$

since

$$\sin(x + y) = \sin x \cos y + \cos x \sin y.$$

We may now use Eqs. (4.9-4) and (4.9-5) to connect the alternate expressions for $X(t)$ as given by Eqs. (4.9-2) and (4.9-3). Thus $X(t)$ can be represented as

$$X(t) = e_c(t) \cos \omega_0 t + e_s(t) \sin \omega_0 t \quad (4.9\text{-}6)$$

where the two random processes $e_c(t)$ and $e_s(t)$ are

$$e_c(t) = E(t) \cos \xi(t) \quad (4.9\text{-}7)$$

or

$$e_c(t) = \sum_{k=1}^{\infty} a_k \cos \omega_k^* t + b_k \sin \omega_k^* t \quad (4.9\text{-}8)$$

and

$$e_s(t) = -E(t) \sin \xi(t) \quad (4.9\text{-}9)$$

or

$$e_s(t) = \sum_{k=1}^{\infty} [b_k \cos \omega_k^* t - a_k \sin \omega_k^* t] \quad (4.9\text{-}10)$$

and where, for convenience, we have introduced the notation

$$\omega_k^* = k\omega_1 - \omega_0 \quad (4.9\text{-}11)$$

The random processes $e_c(t)$ and $e_s(t)$ are called the *quadrature components* of $X(t)$.

This procedure has related the envelope and phase functions to the Fourier coefficients since, from Eqs. (4.9-7) and (4.9-8),

$$E(t) = [e_c^2(t) + e_s^2(t)]^{1/2} \quad (4.9\text{-}12)$$

and

$$\xi(t) = \text{arc tan} \frac{-e_s(t)}{e_c(t)} \qquad (4.9\text{-}13)$$

The random processes $e_c(t)$ and $e_s(t)$ are linear combinations (Fourier series) of a_k' s and b_k' s given by Eqs. (4.9-8) and (4.9-10). These coefficients are found from the integrals

$$a_k = \frac{2}{T} \int_{-T/2}^{T/2} X(t) \cos k\,\omega_1 t \ dt \qquad (4.9\text{-}14)$$

$$b_k = \frac{2}{T} \int_{-T/2}^{T/2} X(t) \sin k\,\omega_1 t \ dt \qquad (4.9\text{-}15)$$

If $X(t)$ is a Gaussian process, then a_k and b_k are normal random variables (result of a linear operation on a Gaussian process). Then, from Eqs. (4.9-8) and (4.9-10), it follows that $e_c(t)$ and $e_s(t)$ are Gaussian processes (result of a linear operation on normal random variables). It follows also from the discussion of Section 4.5 that the coefficients a_k, b_k are uncorrelated if $X(t)$ is a periodic random process or if the interval of expansion T is allowed to become infinite.

Let us check to see if $e_c(t)$ and $e_s(t)$ are correlated. We have

$$E\{e_c(t)e_s(t)\} = \sum_{k=1}^{\infty} \sum_{j=1}^{\infty} [E\{a_k b_j\}\cos \omega_k^* t \ \cos \omega_j^* t$$

$$+ E\{b_k b_j\}\sin \omega_k^* t \ \cos \omega_j^* t - E\{a_k a_j\}\cos \omega_k^* t \ \sin \omega_j^* t$$

$$- E\{b_k a_j\}\sin \omega_k^* t \ \sin \omega_j^* t]$$

We will assume that the a_k' s and b_k' s are all mutually uncorrelated as discussed in Section 4.5. Thus, this last expression can be written as

$$E\{e_c(t)e_s(t)\} = 0 + \sum_{k=1}^{\infty} E\{b_k^2\}\sin \omega_k^* t \ \cos \omega_k^* t$$

$$- \sum_{k=1}^{\infty} E\{a_k^2\} \sin \omega_k^* t \ \cos \omega_k^* t - 0$$

We have already shown in Eq. (4.5-36) that $E\{b_k^2\} = E\{a_k^2\} = \frac{2}{T}\phi(k\,\omega_1)$. Hence the *random variables* e_c and e_s generated at the *same* time t are uncorrelated. Since e_c and e_s are also Gaussian random variables, they are independent and their joint density $f_2(e_c, e_s)$ can be written as the product of the univariate densities of e_c and e_s, respectively

$$f_2(e_c, e_s) = f_{e_c}(e_c) f_{e_s}(e_s) \tag{4.9-16}$$

If the process $X(t)$ has zero mean, it follows from Eqs. (4.9-8) and (4.9-10) and from Eq. (4.6-37) that

$$E\{e_c(t)\} = E\{e_s(t)\} = 0 \tag{4.9-17}$$

The variance of e_c and e_s can be found directly from Eqs. (4.9-6), (4.9-8) and (4.9-10) to be

$$E\{e_c^2(t)\} = E\{e_s^2(t)\} = \frac{1}{2} \sum_{k=1}^{\infty} E\{a_k^2 + b_k^2\}$$

$$= \sum_{k=1}^{\infty} E\{a_k^2\} = \sum_{k=1}^{\infty} E\{b_k^2\} \tag{4.9-18}$$

$$= E\{X^2(t)\} = \sigma^2$$

Thus, if $X(t)$ is a Gaussian process, the joint density of e_c and e_s is

$$f_2(e_c, e_s) = \frac{1}{2\pi\sigma^2} e^{-(e_c^2 + e_s^2)/2\sigma^2} \tag{4.9-19}$$

We will now find the joint density $g_2(E, \xi)$ of E and ξ, the envelope and phase functions, by using the functional relationships of Eqs. (4.9-7) and (4.9-9) and the transformation procedure of Section 2.20. We write

$$g_2(E, \xi) = f_2(e_c, e_s) \left| J\left(\frac{e_c, e_s}{E, \xi}\right) \right| \tag{4.9-20}$$

where the determinant of the Jacobian of the transformation is given by

$$\left| J\left(\frac{e_c, e_s}{E, \xi}\right) \right| = \left| \begin{vmatrix} \cos\xi & -E\sin\xi \\ -\sin\xi & -E\cos\xi \end{vmatrix} \right| = E \tag{4.9-21}$$

since

$$e_c = E\cos\xi$$

and

$$e_s = E\sin\xi$$

Now Eq. (4.9-20) can be written as

$$g_2(E, \xi) = \begin{cases} \dfrac{E}{2\pi\sigma^2} e^{-E^2/2\sigma^2} & , \ 0 \le E < \infty, 0 \le \xi < 2\pi \\ \\ 0 & , \ \text{elsewhere} \end{cases}$$

This is the joint density of the envelope and phase functions. Note that E is the square root of the sum of the squares of two Gaussian random variables (or random processes).

The univariate densities $g_E(E)$ and $g_\xi(\xi)$ can be obtained by integrating the joint density $g_2(E,\xi)$ with respect to the proper variable. We have

$$g_E(E) = \int_0^{2\pi} g(E,\xi)d\xi \qquad (4.9\text{-}23)$$

or

$$g_E(E) = \begin{cases} \dfrac{E}{\sigma^2} e^{-E^2/2\sigma^2} , & 0 \le E < \infty \\ 0 , & -\infty < E < 0 \end{cases} \qquad (4.9\text{-}24)$$

This distribution is called the Rayleigh distribution and is plotted in Fig. 4.6 in the normalized form:

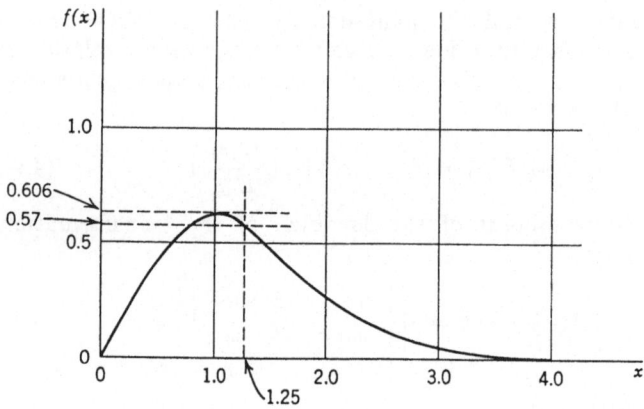

Fig. 4.6 - The normalized Rayleigh distribution

$$f(x) = xe^{-x^2/2}u(x) \qquad (4.9\text{-}25)$$

where E/σ is written as x, $\sigma g_1(E)$ is written as $f(x)$, and $u(x)$ is the unit step function. The value of x for which $f(x)$ is a maximum is found from

$$\frac{d}{dx}f(x) = 0 = (1 - x^2)e^{-x^2/2}$$

or

$$x = 1 \qquad (4.9\text{-}26)$$

This maximum is

$$f_{max} = f(1) = e^{-1/2} \approx 0.606 \qquad (4.9\text{-}27)$$

The mean of $f(x)$ is

$$E(X) = \int_0^\infty x f(x)dx = \int_0^\infty x^2 e^{-x^2/2}\, dx$$

This can be written as

$$E(X) = \sqrt{\frac{\pi}{2}} \left[\frac{1}{\sqrt{2\pi}} \int_{-\infty}^\infty x^2 e^{-x^2/2} dx \right]$$

The quantity in brackets has a value of unity since it is the variance of the unit normal distribution; therefore

$$E(X) = \sqrt{\pi/2} \approx 1.25 \qquad (4.9\text{-}28)$$

These various quantities are indicated in Fig. 4.6.

The distribution of ξ is found from

$$g_\xi(\xi) = \int_0^\infty g(E,\xi)dE$$

$$= \frac{1}{2\pi\sigma^2} \int_0^\infty E e^{-E^2/2\sigma^2} dE \qquad (4.9\text{-}29)$$

$$= -\frac{1}{2\pi} e^{-E^2/2\sigma^2} \Big|_0^\infty = \frac{1}{2\pi}$$

or

$$g_1(\xi) = \begin{cases} \dfrac{1}{2\pi} & , \ 0 \le \xi \le 2\pi \\[2mm] 0 & , \ \text{elsewhere} \end{cases} \qquad (4.9\text{-}30)$$

The phase is uniformly distributed in the interval $(0,2\pi)$ as might be expected since there is no physical basis to prefer one phase angle over another.

It is apparent from Eqs. (4.9-22), (4.9-24), and (4.9-30), that

$$g_2(E,\xi) = g_E(E)g_\xi(\xi) \qquad (4.9\text{-}31)$$

and hence that the random variables E and ξ at any time t are statistically independent. It does not follow, however, that the random processes $E(t)$ and $\xi(t)$ are statistically independent. Joint densities involving two times t_1 and t_2 may not factor. It can be shown [4,5] that the joint densities

$$f_4(E_1,E_2,\xi_1,\xi_2) \neq f_2(E_1,E_2)f_2(\xi_1,\xi_2) \qquad (4.9\text{-}32)$$

where $E_i = E(t_i)$ and $\xi_i = \xi(t_i)$.

4.10 *Narrow-Band Systems with Added Sinusoids* - It may be more realistic in many cases to consider a sinusoidal signal added to the input of the narrow-band filter discussed in the Section 4.9. Then the output would be

$$Y(t) = X(t) + A \cos(\omega_0 t + \Psi) \qquad (4.10\text{-}1)$$

where $X(t)$ is the narrow-band process given by Eqs. (4.5-1) and (4.5-2), $f_0 = \omega_0/2\pi$ is the "center frequency" of the filter, Ψ is an an arbitrary angle in $(0,2\pi)$, and A is a constant. The output $Y(t)$ may be represented in the same form as Eq. (4.5-1) and written as

$$Y(t) = R(t) \cos[\omega_0 t + \theta(t)] \qquad (4.10\text{-}2)$$

where $R(t)$ and $\theta(t)$ are new envelope and phase functions.

In Eq. (4.5-6) the noise process $X(t)$ was written in a form which defined the quadrature components $e_c(t)$ and $e_s(t)$:

$$X(t) = e_c(t) \cos \omega_0 t + e_s(t) \sin \omega_0 t \qquad (4.10\text{-}3)$$

It is clear that the new output $Y(t)$ can be written as

$$Y(t) = e_c'(t) \cos \omega_0 t + e_s'(t) \sin \omega_0 t \qquad (4.10\text{-}4)$$

where

$$e_c'(t) = e_c(t) + A \cos \Psi \qquad (4.10\text{-}5a)$$

$$e_s'(t) = e_s(t) - A \sin \Psi \qquad (4.10\text{-}5b)$$

and $e_c(t)$ and $e_s(t)$ have been defined by the Fourier series of Eqs. (4.5-8) and (4.5-10), respectively. As before the envelope and phase functions are given in terms of these quadrature components by

$$R(t) = \{[e_c'(t)]^2 + [e_s'(t)]^2\}^{1/2} \qquad (4.10\text{-}6)$$

and

$$\theta(t) = \text{arc tan} \frac{-e_s'(t)}{e_c'(t)} \qquad (4.10\text{-}7)$$

It is clear that $e_c{}'(t)$ and $e_s{}'(t)$ are Gaussian if the filter input is Gaussian and that their joint density is

$$f_2(e_c{}', e_s{}') = \frac{1}{2\pi\sigma^2} e^{-\frac{(e_c{}' - A\ cos\ \Psi)^2 + (e_s{}' + A\ sin\ \Psi)^2}{2\sigma^2}} \qquad (4.10\text{-}8)$$

since $e_c{}'(t)$ and $e_s{}'(t)$ have means equal to $A\cos\Psi$ and $-A\sin\Psi$ respectively. As before, the joint density of R and θ is

$$g_2(R,\theta) = f_2(e_c{}', e_s{}') \mid J(\frac{e_c{}', e_s{}'}{R,\theta}) \mid \qquad (4.10\text{-}9)$$

Since

$$e_c{}' = R\ \cos\theta \qquad (4.10\text{-}10)$$

$$e_s{}' = -R\ \sin\theta \qquad (4.10\text{-}11)$$

the Jacobian of the transformation is R. Also the exponent of (4.10-8) can be rewritten in terms of R and θ as follows:

$$(e_c{}' - A\ \cos\ \Psi)^2 + (e_s{}' + A\ \sin\ \Psi)^2 \qquad (4.10\text{-}12)$$

$$= R^2 + A^2 - 2AR\ \cos(\theta - \Psi)$$

Now Eq. (4.10-9) becomes

$$g_2(R,\theta) = \begin{cases} \dfrac{R}{2\pi\sigma^2} e^{-\frac{R^2 + A^2 - 2AR\cos(\theta-\Psi)}{2\sigma^2}} & , \ 0 \leq R < \infty \\ & , \ 0 \leq \theta \leq 2\pi \\ 0 & , \ \text{elsewhere} \end{cases} \qquad (4.10\text{-}13)$$

where σ^2 has been previously defined by Eq. (4.5-18) and is the variance of the Gaussian random process $X(t)$.

This joint density $g_2(R,\theta)$ is not so simply integrated as was $g_2(E,\theta)$ treated in the previous section. The univariate density for R is

$$g_R(R) = \int_0^{2\pi} g_2(R,\theta) d\theta$$

$$= \frac{R}{2\pi\sigma^2} e^{-\frac{R^2 + A^2}{2\sigma^2}} \int_0^{2\pi} e^{\frac{AR\cos(\theta-\Psi)}{\sigma^2}} d\theta$$

or

$$g_R(R) = \frac{R}{\sigma^2} e^{-\frac{R^2 + A^2}{2\sigma^2}} I_o(\frac{AR}{\sigma^2}) \qquad (4.10\text{-}14)$$

where $I_o(x)$ is the modified Bessel function of the first kind and order zero. This last relationship follows from

$$J_n(z) = \frac{(j)^{-n}}{2\pi} \int\limits_{\Psi}^{\Psi+2\pi} e^{jz\cos(\theta-\Psi)} e^{jn(\theta-\Psi)} d\theta \qquad (4.10\text{-}15)$$

where $J_n(z)$ is the Bessel function of the first kind and order n. The modified Bessel function of the first kind and order n (also called a hyperbolic Bessel function) is denoted by $I_n(x)$ and is given by

$$I_n(x) = (-j)^n J_n(jx) \qquad (4.10\text{-}16)$$

Thus $I_o(x)$ is

$$I_o(x) = J_o(jx) \qquad (4.10\text{-}17)$$

This function is plotted in Fig. 4.7 for positive argument and is seen to be monotonically increasing. Further discussion of its properties and tabulations of its values will be found in the literature [6,7,18].

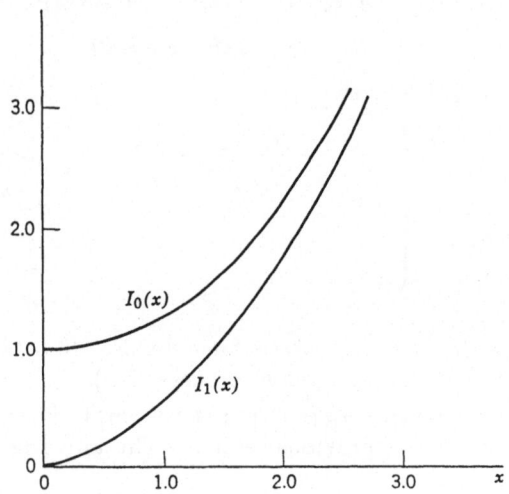

Fig. 4.7 -Modified Bessel functions of the first kind

The density function $g_R(R)$ has been plotted in Fig. 4.8. Again a normalized form has been used:

$$f(x) = xe^{-\frac{1}{2}(x^2+k^2)} I_o(kx) \qquad (4.10\text{-}18)$$

where R/σ is written as x, $\sigma g_1(R)$ as $f(x)$, and A/σ as the parameter k. The modified Bessel function $I_o(x)$ has a series expansion, valid for large x, of the form [19]

$$I_o(x) = \frac{e^x}{\sqrt{2\pi x}} \left(1 + \frac{1}{8x} + \frac{9}{128x^2} + \ldots\ldots\right) \qquad (4.10\text{-}19)$$

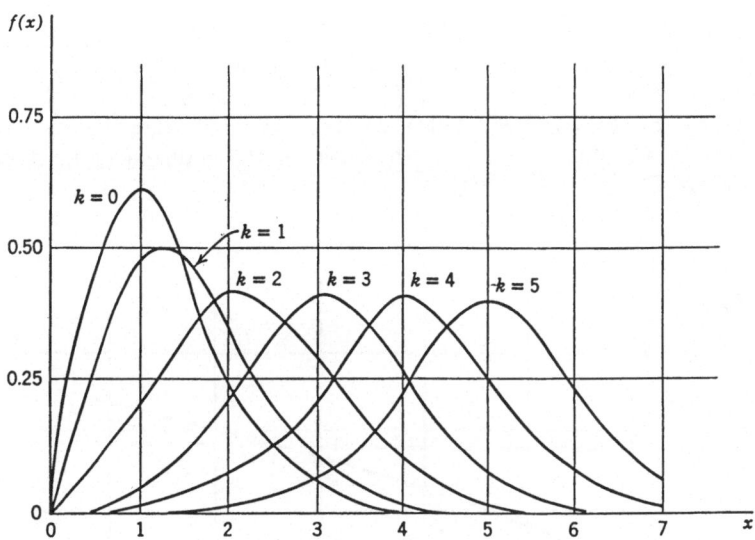

Fig. 4.8 - The probability amplitude distribution of a sinusoid in narrowband Gaussian Noise

Therefore, for the case where $kx = AR/\sigma^2 >> 1$, Eqs. (4.10-14) and (4.10-18) can be written as

$$g_R(R) \to \frac{1}{\sigma} \sqrt{\frac{R}{2\pi A}} e^{-\frac{1}{2}\left(\frac{R-A}{\sigma}\right)^2} \qquad (4.10\text{-}20)$$

and

$$f(x) \to \sqrt{\frac{x}{2\pi k}} e^{-\frac{1}{2}(x-k)^2} \qquad (4.10\text{-}21)$$

It is apparent from these equations and from the curves given in Fig. 4.8 that the envelope distribution is approximately Gaussian in the vicinity of $R = A$ when the sine-wave magnitude A is large compared to the square root of the noise power σ^2.

It is clear from Fig. 4.8 that the mean value of the envelope is a function now of the parameter $k = A/\sigma$ which is, in turn, a measure of the signal-to-noise ratio. This mean value is given by

$$E(X) = E(R/\sigma) = \int_0^\infty x f(x)\,dx$$

$$= e^{-k^2/2} \int_0^\infty x^2 e^{-x^2/2} I_o(kx)\,dx \qquad (4.10\text{-}22)$$

or

$$E(X) = \sqrt{\frac{\pi}{2}}\, e^{-k^2/4}[(1 + \frac{k^2}{2})I_0\,(\frac{k^2}{4}) + \frac{k^2}{2}\, I_1(\frac{k^2}{4})] \qquad (4.10\text{-}23)$$

where $I_1(x)$ is the modified Bessel function of the first kind and order one and is shown in Fig. 4.7 for positive arguments. The value of $E(X)$ as given by Eq. (4.10-23) is plotted on Fig. 4.9 as a function of $k^2 = A^2/\sigma^2$. Also plotted on the same figure and shown by the dashed curve is the case for the sinusoid alone; that is, where $R = A$.

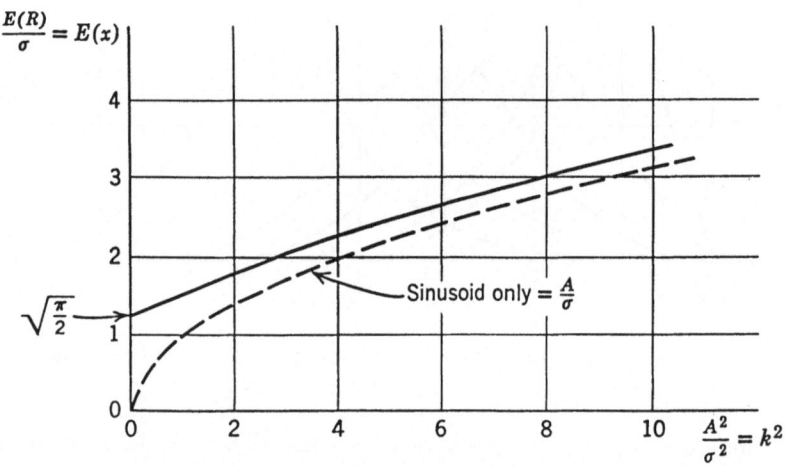

Fig. 4.9 - The expected value of the envelope of a sinusoid in narrowband Gaussian noise

The results of the preceding paragraph are a special case of a more general development [5, Chap. 9]. The r-th moment of the envelope R can be written from Eq. (4.10-14) as

$$E\{R^r\} = \int_0^\infty \frac{R^{r+1}}{\sigma^2}\, e^{-(R^2+A^2)/2\sigma^2}\, I_0\,(\frac{AR}{\sigma^2})dR \qquad (4.10\text{-}24)$$

or

$$E\{R^r\} = (2\sigma^2)^{r/2}\Gamma(r/2 + 1)_1 F_1[-r/2;1;-A^2/2\sigma^2] \qquad (4.10\text{-}25)$$

where $\Gamma(x)$ is the gamma function [6, Chapter 12] defined by

$$\Gamma(x) = \int_0^\infty y^{x-1}e^{-y}\,dy \qquad (4.10\text{-}26)$$

and $_1F_1[\alpha;\beta;\pm y]$ is the confluent hypergeometric function [5,7] given by

$$_1F_1[\alpha;\beta;\pm y] = 1 + \frac{\alpha}{\beta}\frac{(\pm y)}{1!} + \frac{\alpha(\alpha+1)}{\beta(\beta+1)}\frac{(\pm y^2)}{2!} + \dots$$
$$(4.10\text{-}27)$$

$$+ \frac{\alpha(\alpha+1)(\alpha+2)\dots(\alpha+n-1)}{\beta(\beta+1)(\beta+2)\dots(\beta+n-1)}\frac{(\pm y^n)}{n!} + \dots$$

It is the solution to the differential equation

$$y\,\frac{d^2F}{dy^2} + (\beta-y)\,\frac{dF}{dy} - \alpha F = 0 \qquad (4.10\text{-}28)$$

For the case where $r=1$, Eq. (4.10-25) becomes

$$E\{R\} = (2\sigma^2)^{1/2}\Gamma(3/2)_1F_1[-1/2;1;-A^2/2\sigma^2] \qquad (4.10\text{-}29)$$

Since [6, Chapter 12]

$$\Gamma(3/2) = \sqrt{\pi}/2$$

and [5, Appendix A]

$$_1F_1[-1/2;1;-y] = e^{-y/2}[(1+y)I_o(y/2) + yI_1(y/2)] \qquad (4.10\text{-}30)$$

then Eq. (4.10-29) reduces to Eq. (4.10-23) as expected. For the case where $r=2$, we have

$$E\{R^2\} = 2\sigma^2\Gamma(2)_1F_1[-1;1;-A^2/2\sigma^2] \qquad (4.10\text{-}31)$$

But [6,Chapter 12]

$$\Gamma(2) = 1 \qquad (4.10\text{-}32)$$

and, from Eq. (4.10-27),

$$_1F_1[-1;1;-y) = 1 + y \qquad (4.10\text{-}33)$$

so that

$$E\{R^2\} = 2\sigma^2 + A^2 = \sigma^2(2 + k^2) \qquad (4.10\text{-}34)$$

where, as before $k = A/\sigma$. The variance of R is

$$\sigma_R^2 = E\{R^2\} - [E\{R\}]^2 \qquad (4.10\text{-}35)$$

or

$$\frac{\sigma_R^2}{\sigma^2} = 2 + k^2 - \frac{\pi}{2}\,e^{-k^2/2}[(1 + \frac{k^2}{2})I_o(\frac{k^2}{4}) + \frac{k^2}{2}\,I_1(\frac{k^2}{4})]^2 \qquad (4.10\text{-}36)$$

from Eqs. (4.10-23) and (4.10-34). This last relationship is plotted in Fig. 4.10.

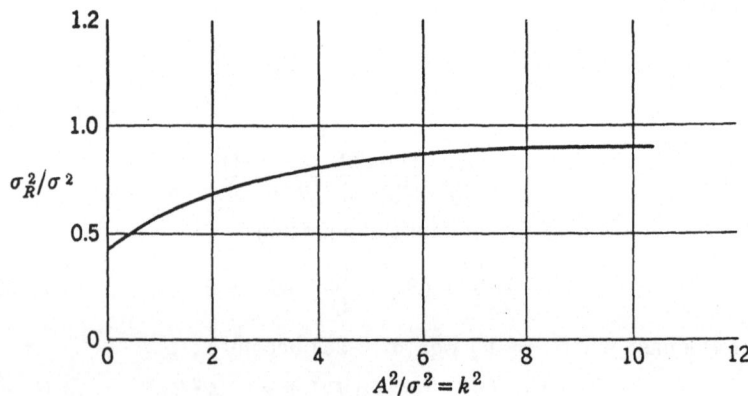

Fig. 4.10 - The normalized power of a sinusoid in
narrowband Gaussian noise

The distribution of the phase function θ is found from

$$g_\theta(\theta) = \int_0^\infty g_2(R,\theta)dR$$

$$= \frac{1}{2\pi\sigma^2} \int_0^\infty R\ e^{-\frac{R^2+A^2-2AR\cos(\theta-\Psi)}{2\sigma^2}}\ dR \qquad (4.10\text{-}37)$$

Let us complete the square in R in the exponent to obtain

$$g_\theta(\theta) = \frac{1}{2\pi\sigma^2}\ e^{-\frac{A^2}{2\sigma^2}[1-\cos^2(\theta-\Psi)]} \int_0^\infty R\ e^{-\frac{1}{2}[\frac{R-A\cos(\theta-\Psi)}{\sigma}]^2}\,dR \qquad (4.10\text{-}38)$$

At this point it is convenient to introduce the notation

$$b = \frac{A\ \cos(\theta-\Psi)}{\sigma} \qquad (4.10\text{-}39)$$

and make a linear change in variable

$$\frac{R}{\sigma} - b = y \qquad (4.10\text{-}40)$$

in Eq. (4.10-38) to obtain

$$g_\theta(\theta) = \frac{1}{2\pi}\ e^{-\frac{A^2\sin^2(\theta-\Psi)}{2\sigma^2}} \int_{-b}^\infty (y+b)e^{-y^2/2}dy \qquad (4.10\text{-}41)$$

The first term in this integral may be evaluated directly while the second may rearranged to yield

$$g_{\phi}(\theta) = e^{-\frac{A^2\sin^2(\theta-\Psi)}{2\sigma^2}} \left[\frac{e^{-b^2/2}}{2\pi} + \frac{b}{\sqrt{2\pi}} \frac{1}{\sqrt{2\pi}} \int\limits_{-b}^{\infty} e^{-y^2/2} dy \right] \quad (4.10\text{-}42)$$

or

$$g_{\phi}(\theta) = \frac{e^{-A^2/2\sigma^2}}{2\pi} + \frac{A \cos(\theta-\Psi)}{\sqrt{2\pi}\,\sigma} e^{-\frac{A^2\sin^2(\theta-\Psi)}{2\sigma^2}} \Phi(b) \quad (4.10\text{-}43)$$

where $\Phi(b)$ is the cumulative distribution function for the unit normal distribution and is given by

$$\Phi(b) = \frac{1}{\sqrt{2\pi}} \int\limits_{-\infty}^{b} e^{-y^2/2} dy \quad (4.10\text{-}44)$$

If, as in Eq. (4.10-18), we let $A/\sigma = k$ then Eq. (4.10-43) becomes

$$f(\theta) = \frac{e^{-k^2/2}}{2\pi} + \frac{k \cos(\theta-\Psi)}{\sqrt{2\pi}} e^{-\frac{k^2\sin^2(\theta-\Psi)}{2}} \Phi[k \cos(\theta-\Psi)]$$

This equation is plotted in Fig. 4.11 with k as the parameter and

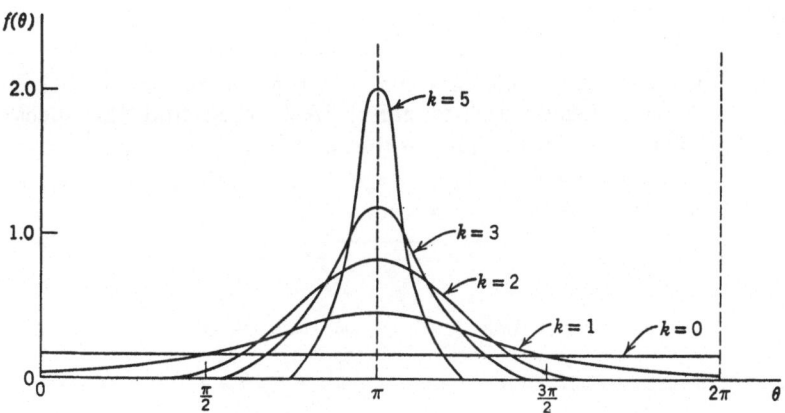

Fig. 4.11 - The density function of the phase of a sinusoid in narrowband Gaussian noise

with $\Psi = \pi$ for convenience. As $k \to 0$ the effect of the sinusoid becomes negligible and the distribution becomes uniform as expected. For strong signal conditions, $k \gg 1$, the distribution approaches a delta-function at $\theta = \pi$.

PROBLEMS

1. Consider the system described by the equation

$$y(t) = a_1 x(t + a_2) + a_3 x\left(\frac{t}{a_4}\right) + a_5 |x(t)| + \frac{a_6}{t} x(t) + a_7$$

where $x(t)$ is the input, $y(t)$ the output, and the a_i are constants. What are the necessary and sufficient conditions on the a_i for the system to be (a) linear, (b) time-invariant, (c) causal, (d) stable?

2. Let $x(t)$ be an input with finite energy to a stable linear time-invariant system h(t); that is,

$$\int_{-\infty}^{\infty} |h(\tau)|^2 d\tau < +\infty \quad \text{and} \quad \int_{-\infty}^{\infty} |x(t)|^2 dt < +\infty$$

Show that the output $y(t)$ has finite energy.

3. Consider the following low-pass filter.

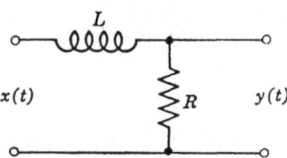

If $x(t)$ is a sample function of a random process with a flat spectral density (white noise) $No/2$ then find the spectral density and mean square value of $Y(t)$.

4. Consider the following bridge circuit with white noise with spectral density N_o as the input.

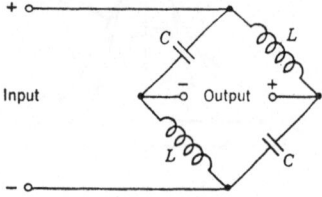

Find the spectral density of the output.

5. Show that if the impulse response of a linear system is not absolutely integrable; that is, if $\int_{-\infty}^{\infty} |h(\tau)| d\tau$ is not finite, then the system is unstable. Hint, use the convolution integral with a properly chosen input.

6. Consider the following linear system composed of a delay line, subtractor, and ideal integrator.

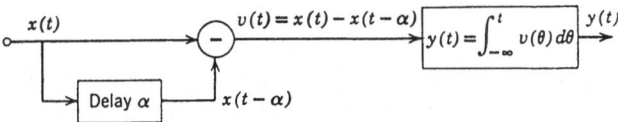

Find the impulse response and transfer function of the system. Prove that the system is stable.

7. Consider the following two linear time-invariant systems:

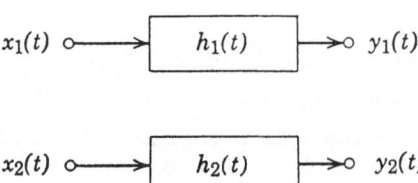

Express the crosscorrelation function $R_{y_1 y_2}(\tau)$ of $Y_1(t)$ and $Y_2(t)$ in terms of $h_1(t)$, $h_2(t)$, and the crosscorrelation function $R_{x_1 x_2}(\tau)$ of the inputs $X_1(t)$ and $X_2(t)$.

8. Consider the following cascade of two linear time-invariant systems.

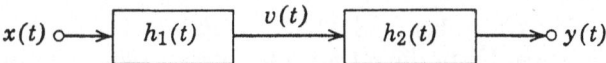

Find the crosscorrelation function $R_{vy}(\tau)$ of $V(t)$ and $Y(t)$ in terms of $h_1(t)$, $h_2(t)$, and the autocorrelation function $R_{xx}(\tau)$ of the input $X(t)$.

9. Consider a resistor R at temperature T in parallel with a capacitor C. find the power spectral density and the mean-square value of the noise voltage developed across C.

10. A linear time-invariant system has the impulse response

$$h(t) = \begin{cases} 2e^{-4t} & , \text{ for } t > 0 \\ 0 & , \text{ for } t < 0 \end{cases}$$

Find the response of the system to the input $x(t) = 10 \cos 4t$.

11. If the sample functions of a stationary random process are periodic; that is, $y(t) = y(t+nT), n = \ldots-1,0,1,2,\ldots$, show that the process is periodic. Conversely if a random process $X(t)$ is periodic show that $X(t)$ converges in the mean to $X(t+nT)$.

12. Let $E(t)$ be the envelope function of a narrow band Gaussian process with variance σ^2. Find the mean and variance of $E(t)$ in terms of σ.

13 Prove that Eq.(4.5-39) is valid.

14 Complete Example 4.6.

REFERENCES

1. E.I. Jury, *Theory and Application of the z-Transform Method,* John Wiley and Sons, Inc., New York, N.Y. 1964.

2. S.O. Rice, *"Mathematical Analysis of Random Noise,"* Bell System Tech. J., vols, 23 and 24, 1944-45.

3. W.L. Root and T.S. Pitcher, *"On the Fourier-Series Expansion of Random Functions,"* Ann. Math. Stat., vol. 26, no. 2, June 1955, pp. 313-318.

4. W. B. Davenport, Jr. and W. L. Root, *An Introduction to the Theory of Random Signals and Noise,* McGraw-Hill Book Company, Inc., New York, N.Y.; 1958.

5. D. Middleton, *An Introduction to Statistical Communication Theory,* McGraw-Hill book Company, Inc., New York, N.Y.; 1960.

6. E. T. Whittaker and G. N. Watson, *A Course in Modern Analysis,* Cambridge University Press, 1963, Chapter XI.

7. P. M. Morse and H. Feshbach, *Methods of Theoretical Physics,* McGraw-Hill Book Company, Inc., New York, N.Y.; 1978.

8. J.L. Brown, *"Mean Square Truncation Error in Series Expansions of Random Functions,"* J. Soc. Indust. Appl. Math., vol. 8, no. 1, March 1960, pp. 28-32.

9. H.P. Kramer, *"On Best Approximations of Random Processes et al,"* IRE Transactions on Information Theory, vol. IT-6, no. 1, March 1960, pp. 52-53.

10. E.T. Whittaker, *"On the Functions which are Represented by Expansions of the Interpolation Theory,"* Proc. Roy. Soc. (Edinburgh), vol. 35, 1914-1915, pp. 181-194.

11. C.E. Shannon, *"A Mathematical Theory of Communication,"* Bell Sys. Tech. J., August 1948.

12. C.E. Shannon, *"Communication in the Presence of Noise,"* Proc. IRE, vol. 37, no. 1, January 1949, pp. 10-21.

13. D. Gabor, *"Theory of Communication,"* J. IEE, vol. 93, part III, November 1946, pp. 429-457.

14. P.M. Woodward, *"Probability and Information Theory, with Applications to Radar,"* Pergamon Press, Ltd., London 1953.

15. S.P. Lloyd, *"A Sampling Theorem for Stationary (Wide Sense) Stochastic Processes,"* Trans. of the Am. Math. Soc., vol. 92, 1959, pp. 1-12.

16. A.V. Balakrishnan, *"A Note on the Sampling Principles of Continous Signals,"* IRE Transactions on Information Theory, vol. IT-3, no. 2, June 1957, pp. 143-146.

17. F.J. Beutler, *"Sampling Theorems and Bases in a Hilbert Space,"* Information and Control, vol. 4, 1961, pp. 97-117.

18. E. Jahnke and F. Emde, *Tables of Functions,* Dover Publications, New York, N.Y.; 1947.

19. H. B. Dwight, *Tables of Integrals,* Macmillan Publishing Company, Inc., New York, N.Y.; 1969.

Chapter 5

OPTIMUM LINEAR SYSTEMS: STEADY-STATE SYNTHESIS

5.1 *Introduction* - The previous chapter was concerned with the *analysis* of signals and systems. We study now a much more difficult problem, the *synthesis* or *design* of systems which are *optimum* in some sense. Since it will turn out that we are interested principally in linear systems with random inputs and with criteria of optimization which are statistical in nature, it might be appropriate to call this general area by the term *statistical design* or *statistical optimization theory for linear systems*.

To proceed with the design of an optimum system at least three factors must be available:

1) *specification of the input* - The input may be random or a combination of random and nonrandom signals and noises. At least some minimum knowledge of the characteristics of the input must be available.

2) *restrictions on system design* - In general, it is necessary that some restricted class of systems be considered, e.g. linear, time-varying systems. In addition, if the systems are to be built, they must be physically realizable. [However an unrealizable system may be useful as a criterion against which to measure other systems.] In practical situations, cost and complexity must be taken into account.

3) *a criterion of optimization* - A measure of the "goodness" of the system must be adopted. This measure must lead to non-trivial and unique systems and must be meaningful in the sense that it is a reasonable criterion to apply to the particular problem. Most important of all, the criterion must yield equations which are mathematically tractable.

It should be apparent that all three of these factors are interrelated and that useful results will be obtained only if a certain compatibility exists among them.

This chapter will be restricted to the study of optimum *linear* systems in the steady state since a large class of optimization problems of this type have been solved and since such problems are relatively simple. We begin with the *matched filter* based on a signal-to-noise ratio criterion.

Part I - The Matched Filter For Continuous-Time Inputs

5.2 *Derivation* - It will turn out that the concept of signal-to-noise ratio is useful in the evaluation of system performance, particularly in linear systems. In fact, the linearity of such systems make it possible to identify uniquely an output signal component and an output noise component whenever the input consists of a uniquely defined sum of a signal component and a noise component.

We now consider a problem of statistical design which is illustrated in Fig. 5.1. This situation will arise in a number of applications where it is desired to decide whether or not a signal is present in additive noise. The following conditions and restrictions are assumed to prevail:

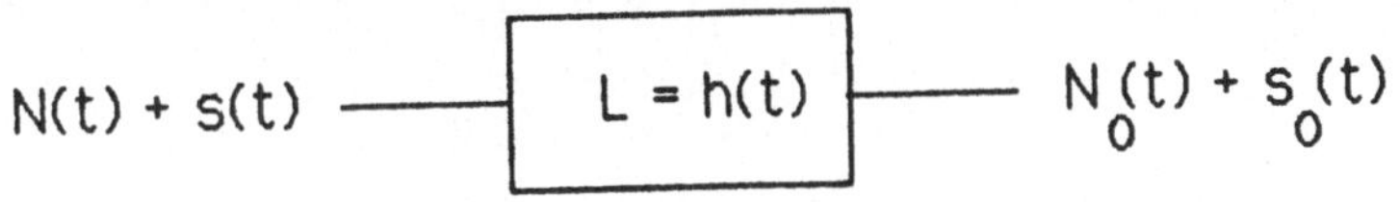

$$N(t) + s(t) \longrightarrow \boxed{L = h(t)} \longrightarrow N_o(t) + s_o(t)$$

Fig. 5.1 - The matched filter in continuous time

1) The input consists of a known signal $s(t)$ and an *additive random* noise process $N(t)$ with continuous parameter t. Additional information about these inputs will be supplied as needed. The corresponding output signal and noise are $s_o(t)$ and $N_o(t)$ respectively.

2) The system is linear and time-invariant with impulse response $h(t)$ as discussed in Section 4.3.

3) The criterion of optimization will be that the output signal-to-noise power ratio R_o be a maximum. Since $N_o(t)$ is random, its mean squared value $E\{N_o^2(t)\}$ will be used as the output noise power. Thus the criterion is that

$$R_o = \frac{s_o^2(t_1)}{E\{N_o^2(t_1)\}} = \text{a maximum} \tag{5.2-1}$$

at some time t_1.

The problem is to find the linear time-invariant system $h(t)$ which accomplishes the maximization of Eq. (5.2-1). This optimum system will be called a *matched filter* for reasons that will become clear as the derivation proceeds.

Since the filter is linear with impulse response $h(t)$, the output signal $s_o(t_1)$ is given by Eq. (4.3-2) as

$$s_o(t_1) = \int_{-\infty}^{t_1} s(\tau)\, h(t_1 - \tau)\, d\tau \tag{5.2-2}$$

and the output noise $N_o(t_1)$ by

$$N_o(t_1) = \int_{-\infty}^{t_1} N(\tau)\, h(t_1 - \tau)\, d\tau \tag{5.2-3}$$

The upper limit in the last two integrals is taken to be t_1 rather than ∞ as would be the case if $h(t)$ were physically realizable; that is, if

$$h(t) = 0 \quad , \quad t < 0$$

It should be made clear that we desire a physically realizable filter, but it may very well turn out (and will do so in general) that the optimum filter is unrealizable. We will deal with this problem later.

The signal-to-noise output ratio of Eq. (5.2-1) now becomes

$$R_o = \frac{\left|\int_{-\infty}^{t_1} s(\tau)\, h(t_1 - \tau)\, d\tau\right|^2}{\int_{-\infty}^{t_1}\int_{-\infty}^{t_1} R_n(\tau,\sigma)\, h(t_1 - \tau)\, h(t_1 - \sigma)\, d\tau\, d\sigma} \tag{5.2-4}$$

where $R_n(\tau,\sigma)$ is the autocorrelation function of the input noise $N(t)$ and is given by

$$R_n(\tau,\sigma) = E\{N(\tau)\, N(\sigma)\} \tag{5.2-5}$$

This function must be known in order to proceed further. Now only the filter $h(t)$ is unknown in Eq. (5.2-4), and the signal-to-noise output power ratio R_o, given by Eq. (5.2-4), must be maximized by the proper choice of $h(t)$. It is apparent that the optimum $h(t)$ (that is, the matched filter) will depend on the noise covariance $R_n(\tau,\sigma)$. We have already required that $h(t)$ be time invariant [that is, $h(t-\tau)$ instead of $h(t,\tau)$]. This last requirement is not in general compatible with nonstationary noise covariance $R_n(\tau,\sigma)$

and it will turn out that at least wide-sense stationarity [that is, $R_n(\tau,\sigma) = R_n(\tau-\sigma)$] must be imposed on the noise for the matched filter to be realizable as a time-invariant linear system. However, we will meet this problem when it arises.

For a given signal $s(t)$ and given noise covariance $R_n(\tau,\sigma)$, Eq. (5.2-4) will yield a maximum value $R_{o\ max} = 1/\alpha$ if the optimum filter $h^\dagger(t-\tau)$ is used. Thus Eq. (5.2-4) can be written as

$$E\{N_o^2(t_1)\} - \alpha\, s_o^2(t_1) \geq 0 \tag{5.2-6}$$

or as

$$\int_{-\infty}^{t_1}\int_{-\infty}^{t_1} R_n(\tau,\sigma)h(t_1-\tau)h(t_1-\sigma)d\tau d\sigma - \alpha\,|\int_{-\infty}^{t_1} s(\tau)h(t_1-\tau)d\tau|^2 =$$

$$K \geq 0 \tag{5.2-7}$$

where the equality $K = 0$ holds only if the optimum filter $h^\dagger$ is used. From this point of view, we wish to minimize the first term (the output noise power) subject to a constraint on the second term (the output signal power). This is a conventional problem in the calculus of variations [1,2,3] with α a Lagrangian multiplier. A brief discussion of the techniques involved is given in Appendix F; however the procedures to be used may be outlined and justified as follows.

Suppose $h^\dagger(\bullet)$, the optimum value of $h(\bullet)$, is replaced by $h^\dagger(\bullet) + \gamma\delta h(\bullet)$ where γ is a real variable and $\delta h(\bullet)$ is an *arbitrary* variation in $h^\dagger(\bullet)$. Then Eq. (5.2-7) becomes a function of γ and can be written as $K(\gamma)$ where

$$K(\gamma) = \int_{-\infty}^{t_1}\int_{-\infty}^{t_1} R_n(\tau,\sigma)[h^\dagger(t_1-\tau)+\gamma\delta h(t_1-\tau)][h^\dagger(t_1-\sigma)+\gamma\delta h(t_1-\sigma)]d\tau d\sigma$$

$$- \alpha\,|\int_{-\infty}^{t_1} s(\tau)[h^\dagger(t_1-\tau) + \gamma\delta h(t_1-\tau)]\,d\tau|^2$$

or

$$K(\gamma) = K(0) + 2\gamma A + \gamma^2 B \geq 0 \tag{5.2-8}$$

Here $K(0)$ has been defined by Eq. (5.2-7), the quantity γ is a real variable, and the quantities A and B are given by

$$A = \frac{1}{2} \int_{-\infty}^{t_1} \delta h\,(t_1 - \tau) \int_{-\infty}^{t_1} R_n\,(\tau,\sigma)\,h^{\dagger}(t_1 - \sigma)\,d\sigma d\tau$$

$$+ \frac{1}{2} \int_{-\infty}^{t_1} \delta h\,(t_1 - \sigma) \int_{-\infty}^{t_1} R_n\,(\tau,\sigma)\,h^{\dagger}(t_1 - \tau)\,d\tau d\sigma$$

$$\text{(5.2-9)}$$

$$- \frac{\alpha}{2} \int_{-\infty}^{t_1} s\,(\sigma)\,h^{\dagger}(t_1 - \sigma)d\sigma \int_{-\infty}^{t_1} s\,(\tau)\,\delta h\,(t_1 - \tau)d\tau$$

$$- \frac{\alpha}{2} \int_{-\infty}^{t_1} s\,(\tau)\,h^{\dagger}(t_1 - \tau)d\tau \int_{-\infty}^{t_1} s\,(\sigma)\,\delta h\,(t_1 - \sigma)\,d\sigma$$

and

$$B = \int_{-\infty}^{t_1} \int_{-\infty}^{t_1} R_n\,(\tau,\sigma)\,\delta h\,(t_1 - \tau)\,\delta h\,(t_1 - \sigma)\,d\tau d\sigma$$

$$- \alpha\,|\int_{-\infty}^{t_1} s\,(\tau)\,\delta h\,(t_1 - \tau)\,d\tau\,|^{\,2} \qquad \text{(5.2-10)}$$

If the function $K\,(\gamma)$ has a minimum, it will occur when $h = h^{\dagger}$. A *necessary* condition for this minimum to occur is

$$\frac{\partial}{\partial \gamma}\,K\,(\gamma)\,|_{\gamma=0} = 0 \quad \text{for all possible } \delta h \qquad \text{(5.2-11)}$$

or, from Eq. (5.2-8),

$$A = 0 \qquad \text{(5.2-12)}$$

The expression for A given by Eq. (5.2-8) can be simplified by noting that

$$R_n\,(\tau,\sigma) = R_n\,(\sigma,\tau) \qquad \text{(5.2-13)}$$

and that

$$\int_{-\infty}^{t_1} s\,(\sigma)\,h^{\dagger}(t_1 - \sigma)\,d\sigma = \int_{-\infty}^{t_1} s\,(\tau)\,h^{\dagger}(t_1 - \tau)\,d\tau = s_o^{\dagger}(t_1) \quad \text{(5.2-14)}$$

is just the optimum output signal at time t_1. Now it is apparent from Eq. (5.2-13) that the first two terms in Eq. (5.2-9) are equal and from Eq. (5.2-14) that the last two terms are equal. The condition $A = 0$ becomes

$$\int_{-\infty}^{t_1} \delta h\,(t_1 - \tau)\left\{\int_{-\infty}^{t_1} R_n\,(\tau,\sigma)h^{\dagger}(t_1 - \sigma)d\sigma - \alpha s_o^{\dagger}(t_1)s\,(\tau)\right\}d\tau = 0 \qquad \text{(5.2-15)}$$

The function $\delta h\,(t_1{-}\tau)$ in this expression was taken to be an *arbitrary* variation in $h^\dagger(t_1{-}\tau)$; therefore, it might have been taken, for example, to be positive when the term in curly brackets was positive and negative or zero when the term was negative. From this type of argument, it is clear that Eq. (5.2-15) can be satisfied for arbitrary $\delta h\,(\bullet)$ for all σ in $(-\infty,t_1)$ only if the term in curly brackets is zero:

$$\int_{-\infty}^{t_1} R_n\,(\tau,\sigma)\ h^\dagger(t_1{-}\sigma)\ d\sigma = \beta\,s\,(\tau)\ ,\ \ -\infty < \tau \le t_1 \qquad (5.2\text{-}16)$$

where β has been written for $\alpha\,s_0^\dagger\,(t_1)$. This *integral equation* must be solved to find $h^\dagger(\bullet)$, the matched filter. Note that $R_n\,(\tau,\sigma)$, the noise correlation function, and $s\,(t)$, the input signal, are presumed known. The unknown function $h^\dagger(\bullet)$ occurs in the integrand (hence the term *integral equation*).

We have so far ignored the quantity $\beta = \alpha\,s_0^\dagger\,(t_1)$: Suppose that $h^\dagger(t)$ is replaced by $b\ h^\dagger(t)$ where b is an arbitrary constant. Then the output signal-to-noise power ratio will be unchanged since b^2 will occur in both numerator and denominator of Eq. (5.2-4) and will cancel. In other words, the signal-to-noise power ratio is unaffected by the system scale factor. Thus the system can be normalized in so far as absolute gain is concerned so that we can let

$$\beta = \alpha\,s_o^\dagger(t_1) = 1 \qquad (5.2\text{-}17)$$

or any other constant for that matter.

As mentioned previously, the restriction that the matched filter be time invariant will require that the input noise be at least wide sense stationary or that

$$R_n\,(\tau,\sigma) = R_n\,(\tau{-}\sigma) = R_n\,(\sigma{-}\tau) = R_n\,(\,|\,\tau{-}\sigma\,|\,) \qquad (5.2\text{-}18)$$

The derivation could have been carried out in terms of a time varying filter $h\,(t\,,\tau)$. In that case, Eq. (5.2-16) would have been obtained with $h^\dagger(t_1{-}\sigma)$ replaced by $h^\dagger(t_1,\sigma)$. However, the solution of Eq. (5.2-16) would then be prohibitively difficult for this treatment. Consequently, we are interested in solving Eq. (5.2-16) with Eqs. (5.2-17) and (5.2-18) applying so that we have

$$\int_{-\infty}^{t_1} R_n\,(\tau{-}\sigma)\ h^\dagger(t_1{-}\sigma)\ d\sigma = s\,(\tau)\ ,\ \ -\infty < \tau \le t_1 \qquad (5.2\text{-}19)$$

as the equation specifying the matched filter.

For the case where the input noise process $N\,(t)$ is white with power spectral density $N_o/2$, the input noise correlation function is

$$R_n\,(\tau{-}\sigma) = \frac{N_o}{2}\ \delta(\tau{-}\sigma) \qquad (5.2\text{-}20)$$

and Eq. (5.2-19) becomes

$$s\left(\tau\right) = \frac{N_o}{2} h^\dagger(t_1 - \tau) \tag{5.2-21}$$

or

$$h^\dagger(t) = \frac{2}{N_o} s\left(t_1 - t\right)$$

This last result is one of the reasons why $h^\dagger(t)$ is called a *matched filter* since the impulse response, is "matched" to the input signal in the white noise case. Note that the constant multiplier is irrelevant and we could write this last equation as

$$h^\dagger(t) = s\left(t_1 - t\right) \tag{5.2-22}$$

The output signal-to-noise power ratio $R_{o \ max}$ of Eq. (5.2-4) can be found for the matched filter by replacing the impulse response $h(t)$ by $h^\dagger(t)$ obtained from a solution of Eq. (5.2-19). Note that the expression for R_o can be simplified by rewriting the denominator with the help of Eq. (5.2-19). The result is

$$R_{o \ max} = \left| \int_{-\infty}^{t_1} s\left(\tau\right) h\left(t_1 - \tau\right) d\tau \right| = \left| s_o(t_1) \right| = \frac{1}{\alpha} \tag{5.2-23}$$

It is worth pointing out that Eq. (5.2-19) is a sufficient, as well as necessary, condition for the function $K(\gamma)$ given by Eq. (5.2-8) to be a minimum. Notice that, if Eq. (5.2-19) is satisfied, then it follows from Eq. (5.2-23) that $K(0)$ given by Eq. (5.2-7) is zero,

$$K(0) = 0 \quad ; \quad \text{for } h(t) = h^\dagger(t) \tag{5.2-24}$$

and that

$$A = 0 \quad ; \quad \text{for } h(t) = h^\dagger(t) \tag{5.2-25}$$

from Eq. (5.2-12). Then Eq. (5.2-3) becomes

$$K(\gamma) = \gamma^2 B \geq 0 \quad ; \quad \gamma \neq 0 \tag{5.2-26}$$

and the filter $h^\dagger(t)$ yields a maximum output signal-to-noise power ratio. In other words, replacing $h^\dagger(t)$ by $h^\dagger(t) + \gamma \delta h(t)$ can only decrease this ratio (or, equivalently, increase K). That $B \geq 0$ in Eq. (5.2-26) follows from Eq. (5.2-10) on noting that the first term in that equation can be written as

$$\int_{-\infty}^{t_1} \int_{-\infty}^{t_1} R_n\left(\tau, \sigma\right) \delta h\left(t_1 - \tau\right) \delta h\left(t_1 - \sigma\right) d\tau d\sigma$$

$$\tag{5.2-27}$$

$$= E\left\{ \left[\int_{-\infty}^{t_1} N(\tau) \delta h\left(t_1 - \tau\right) d\tau \right]^2 \right\} \geq 0$$

and that the second term must be zero

$$\int_{-\infty}^{t_1} s(\tau)\delta h(t_1-\tau)d\tau = 0 \tag{5.2-28}$$

since the constraint in Eq. (5.2-7) was that the output signal power be held constant.

We now look at the white noise case in some detail since there are many applications of matched filters in signal detection and data communications [4,5] where it is reasonable to make this assumption.

On transforming Eq. (5.2-22), we have

$$H^{\dagger}(\omega) = S(-\omega)e^{-j\omega t_1} = S^*(\omega)e^{-j\omega t_1} \tag{5.2-29}$$

where $S(-\omega)=S^*(\omega)$ is the complex conjugate of the Fourier transform of the input signal $s(t)$.

As previously mentioned, early investigators in this field [6,7,8] considered only the case of a white noise background. The term "matched filter" arose from the form of Eq. (5.3-18) which shows that the impulse response of the optimum linear system must be "matched" to the input signal.

Thus far we have ignored the problem of the physical realizability of the filter $h^{\dagger}(t)$. The white noise case is particularly simple in this regard and will be treated now. The general case is deferred until Sections 5.3 and 5.5. The filter is physically realizable if its impulse response vanishes for negative time; that is if

$$h^{\dagger}(t) = 0 \quad , \quad t < 0 \tag{5.2-30}$$

In terms of Eq. (5.2-22), this condition becomes

$$h^{\dagger}(t) = \begin{cases} 0 & , \quad t < 0 \\ s(t_1-t) & , \quad t \geq 0 \end{cases} \tag{5.2-31}$$

If all of the input signal $s(t)$ is to contribute to the output signal component $s_0(t_1)$, then it follows that

$$s(t) = 0 \quad , \quad t > t_1 \tag{5.2-32}$$

This last relationship states that all of the signal $s(t)$ must have entered the filter $h(t)$ by the time $t=t_1$ at which it is desired to have a maximum signal-to-noise power ratio at the output.

The signal output $s_0(t)$ is given by Eq. (5.2-2) as

$$s_0(t) = \int_{-\infty}^{t} s(x)h^{\dagger}(t-x)dx \tag{5.2-33}$$

or, from Eq. (5.2-22) as

$$s_0(t) = \int_{-\infty}^{t} s(x)s(x-t+t_1)dx \tag{5.2-34}$$

This last expression looks very similar to the time autocorrelation function $P(\tau)$ defined in Eq. (3.4-2) as

$$P(\tau) = \lim_{T \to \infty} \frac{1}{2T} \int_{-T}^{T} s(x)s(x+\tau)dx \qquad (5.2\text{-}35)$$

For signals $s(t)$ with finite energy, this last function will be zero in the limit due to the multiplier $1/2T$. It is convenient, therefore, to define a finite time autocorrelation function $P_F(\tau)$, when it exists, by

$$P_F(\tau) = \int_{-\infty}^{t_1-\tau} s(x)s(x+\tau)dx, \quad \tau \geq 0 \qquad (5.2\text{-}36)$$

A comparison of this definition with Eq. (5.2-34) shows that the signal component of the output of the matched filter is

$$s_o(t) = P_F(t_1-t) \qquad (5.2\text{-}37)$$

This function $P_F(\tau)$ has a maximum at the origin (just as does the ordinary correlation function) and this maximum is

$$s_o(t_1) = P_F(0) = \int_{-\infty}^{t_1} s^2(x)dx = E \qquad (5.2\text{-}38)$$

the total input signal energy in the interval $(-\infty,t_1)$. For a normalized input noise spectral density of $N_o/2 = 1$, this last expression is just R_o, the output signal-to-noise power ratio.

Let $f(t)$ be all signals with energy E in the interval $(-\infty,t_1)$; that is, all signals for which

$$\int_{-\infty}^{t_1} f^2(t)dt = E \qquad (5.2\text{-}39)$$

Then it is easily shown that the matched filter output at time t_1 due to the signal alone is greatest for $f(t)=s(t)$, where $s(t)$ is the signal to which the filter is matched. From Eq. (5.2-2), the signal output at t_1 due to the signal $f(t)$ is

$$s_o(t_1) = \int_{-\infty}^{t_1} f(\tau)h^\dagger(t_1-\tau)d\tau \qquad (5.2\text{-}40)$$

If the filter is matched to $s(t)$, then it follows form Eq. (5.2-31) that

$$s_o(t_1) = \int_{-\infty}^{t_1} f(\tau)s(\tau)d\tau \qquad (5.2\text{-}41)$$

From the Schwarz inequality of Appendix E, we have

$$\left| \int_{-\infty}^{t_1} f(\tau)s(\tau)d\tau \right|^2 \leq \int_{-\infty}^{t_1} f^2(\tau)d\tau \int_{-\infty}^{t_1} s^2(\tau)d\tau = E^2 \qquad (5.2\text{-}42)$$

or

$$s_o(t_1) \leq E \qquad (5.2\text{-}43)$$

The equality is satisfied when $f(t) = s(t)$ and has already been given by Eq. (5.2-38).

Example 5.1

We consider the following signal corrupted by additive white noise:

$$s(t) = \begin{cases} Be^{bt} & , t < 0, \quad B, b > 0 \\ \\ 0 & , t \geq 0 \end{cases}$$

The transform $S(\omega)$ of this signal is

$$S(\omega) = \int_{-\infty}^{0} Be^{bt} e^{-j\omega t} \, dt = \frac{B}{b-j\omega}$$

The matched filter transfer function $H^{\dagger}(\omega)$ is given by Eq. (5.2-29) as

$$H^{\dagger}(\omega) = S(-\omega)e^{-j\omega t_1} = \frac{B}{b+j\omega} e^{-j\omega t_1}$$

and the impulse response by Eq. (5.3-21):

$$h^{\dagger}(t) = s(t_1 - t) = \begin{cases} Be^{b(t_1-t)} & , \quad t \geq t_1 \\ \\ 0 & , \quad t < t_1 \end{cases}$$

The physical realizability requirement that $h^{\dagger}(t)$ vanish for negative time is satisfied by taking $t_1 \geq 0$. The simplest choice is $t_1 = 0$ so that

$$H^{\dagger}(\omega) = \frac{B}{b+j\omega}$$

and

$$h^{\dagger}(t) = \begin{cases} Be^{-bt} & t \geq 0 \\ \\ 0 & t < 0 \end{cases}$$

This is just an *RC* low-pass filter preceded by an ideal amplifier. The transform $S_o(\omega)$ of the signal output is

$$S_o(\omega) = S(\omega)H^{\dagger}(\omega) = \frac{B^2}{b^2+\omega^2}$$

and

$$s_o(t) = \frac{1}{2\pi} \int\limits_{-\infty}^{\infty} \frac{B^2}{b^2 + \omega^2} e^{j\omega t} \; d\omega$$

or

$$s_o(t) = \begin{cases} \dfrac{B^2}{2b} e^{bt} & , \; t < 0 \\[3em] \dfrac{B^2}{2b} e^{-bt} & , \; t > 0 \end{cases}$$

by contour integration or by Pair No. 5 of Table 3.1. The same results could be obtained from Eq. (5.2-34), of course. The various quantities relating to the problem are illustrated in Fig. 5.2.

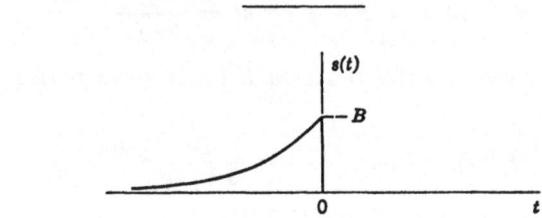

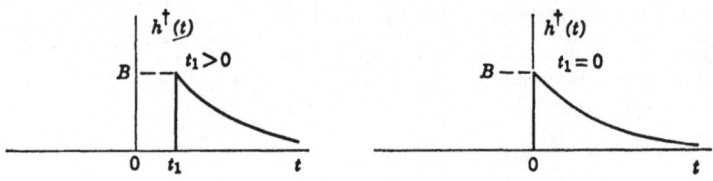

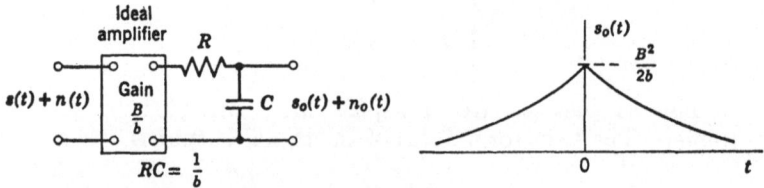

Fig. 5.2 - A matched filter example in continuous time

The choice of signal has made the foregoing a rather trivial example. For arbitrary signals, however, the construction of the matched filter will not be simple and may involve complicated approximation techniques. Also it may not be possible to choose the time t_1 large enough so that all of the signal will have passed through the filter by t_1. In such cases, the output signal $s_o(t_1)$ will be less than the total signal energy E.

Example 5.2

Let the signal $s(t)$ be given by

$$s(t) = \begin{cases} 0 & , t < 0 \\ Be^{-bt} & , t \geq 0, \quad B, b > 0 \end{cases}$$

and let the noise be white and additive. The impulse response of the matched filter is

$$h^{\dagger}(t) = \begin{cases} s(t_1 - t), & t \geq 0 \\ \\ 0 & , t < 0 \end{cases} = \begin{cases} 0 & , t > t_1 \\ Be^{-b(t_1 - t)}, & 0 \leq t \leq t_1 \\ 0 & , t < 0 \end{cases}$$

Both $s(t)$ and $h^{\dagger}(t)$ are plotted in Fig. 5.3. For any finite t_1, all of the signal will not have passed through the filter; therefore the output signal is a function of t_1 and is given by

$$s_o(t_1) = \int_{-\infty}^{t_1} s(\tau) h^{\dagger}(t_1 - \tau) d\tau = \int_0^{t_1} s^2(\tau) d\tau$$

or

$$s_o(t_1) = B^2 \left[\frac{1 - e^{-2bt_1}}{2b} \right]$$

This function is zero when $t_1 = 0$ and a maximum when $t_1 \to \infty$ as shown in Fig. 5.3.

It should be emphasized that there is a noise component $N_o(t)$ present in the output of the filter as given by Eq. (5.2-3). This component tends to obscure the signal output, and, if the noise power N_o is large enough, the signal may be undetectable in the output waveform. The actual output signal-to-noise power for the realizable matched filter is given by Eq. (5.2-21) and by Eq. (5.2-23) as

$$\frac{1}{2\pi} \frac{2}{N_o} \int_{-\infty}^{\infty} |H^{\dagger}(\omega)|^2 d\omega = \frac{2}{N_o} \int_{-\infty}^{t_1} s^2(t) dt \qquad (5.2\text{-}44)$$

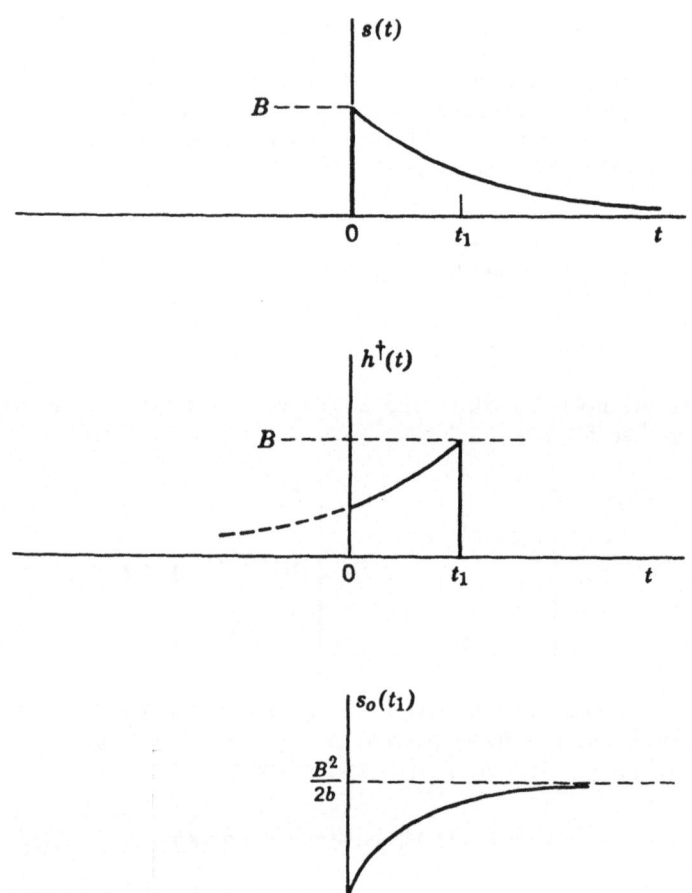

Fig. 5.3 - Another matched filter example

5.3 *The Unrealizable Matched Filter in Continuous Time* - In this section we consider the solution of Eq. (5.2-14) without regard to the realizability of the resulting filter $h^\dagger(t)$ and without restricting the input noise to be white. This assumption is equivalent to rewriting Eq. (5.2-14) as

$$\int_{-\infty}^{\infty} R_n(\tau-\sigma) \, h^\dagger(t_1-\sigma)d\sigma = s(\tau) \quad , \quad -\infty < \tau < \infty \quad (5.3\text{-}1)$$

Here we have replaced the upper limit t_1 on σ and on τ by ∞. In terms of the original convolutions of Eqs. (5.2-3) and (5.2-4), we are allowing contributions from the future; that is, we do not require that

$$h(t) = 0 \quad , \quad t < 0$$

Now the Fourier transform can be taken of both sides of Eq. (5.3-1) to yield

$$\int\limits_{-\infty}^{\infty} \int\limits_{-\infty}^{\infty} R_n(\tau-\sigma) \, h^\dagger(t_1-\sigma) \, d\sigma \, e^{-j\omega\tau} d\tau = \int\limits_{-\infty}^{\infty} s(\tau)e^{-j\omega\tau} d\tau \quad (5.3\text{-}2)$$

or, on rearranging,

$$\int\limits_{-\infty}^{\infty} h^\dagger(t_1-\sigma)e^{-j\omega\sigma} \, \phi_n(\omega)d\sigma = S(\omega) \quad (5.3\text{-}3)$$

After a change of variable $x = t_1 - \sigma$, this becomes

$$e^{-j\omega t_1}\phi_n(\omega) \int\limits_{-\infty}^{\infty} h^\dagger(x)e^{j\omega x} \, dx = S(\omega) \quad (5.3\text{-}4)$$

Thus

$$H^\dagger(-\omega) = \frac{S(\omega)}{\phi_n(\omega)} e^{j\omega t_1} \quad (5.3\text{-}5)$$

or, finally,

$$H^\dagger(\omega) = \frac{S(-\omega)}{\phi_n(\omega)} e^{-j\omega t_1} \quad (5.3\text{-}6)$$

since $\phi_n(\omega) = \phi_n(-\omega)$. Equation (5.3-6) specifies the unrealizable matched filter for arbitrary (but time-invariant) noise spectrum $\phi_n(\omega)$. For the white noise case where $\phi_n(\omega)$ is a constant, this equation reduces to

$$H^\dagger(\omega) = S(-\omega)e^{-j\omega t_1} \quad (5.3\text{-}7)$$

as before.

It should be emphasized again that the filter given by Eq. (5.3-6) is not necessarily realizable as the following two examples show.

Example 5.3

Let the signal $s(t)$ be given by

$$s(t) = u(-t)[e^{t/2} - e^{3t/2}]$$

where $u(t)$ is the unit-step function and let the noise be wide-sense stationary with power spectral density

$$\phi_n(\omega) = \frac{4}{1+4\omega^2}$$

and autocorrelation function

$$R_n(\tau) = e^{-|\tau|/2}$$

The signal $s(t)$ and the noise autocorrelation function $R_n(\tau)$ are shown in Fig. 5.4(a). The transform $S(\omega)$ of the signal is

$$S(\omega) = \frac{2}{1-j\,2\omega} - \frac{2}{3-j\,2\omega} = \frac{4}{(1-j\,2\omega)(3-j\,2\omega)}$$

It follows from Eq. (5.3-6) that the matched filter has a transfer function given by

$$H^\dagger(\omega) = \frac{1-j\,2\omega}{3+j\,2\omega}\, e^{-j\omega t_1} = \left[-1 + \frac{2}{(3/2)+j\,\omega}\right] e^{-j\omega t_1}$$

and an impulse response

$$h^\dagger(t) = -\delta(t-t_1) + 2e^{-(3/2)(t-t_1)}\, u(t-t_1)$$

This filter is realizable for all $t_1 \geq 0$ as shown in Fig. 5.4(b) and (c). The transform $S_0(\omega)$ of the output signal $s_o(t)$ for $t_1 = 0$ is given by

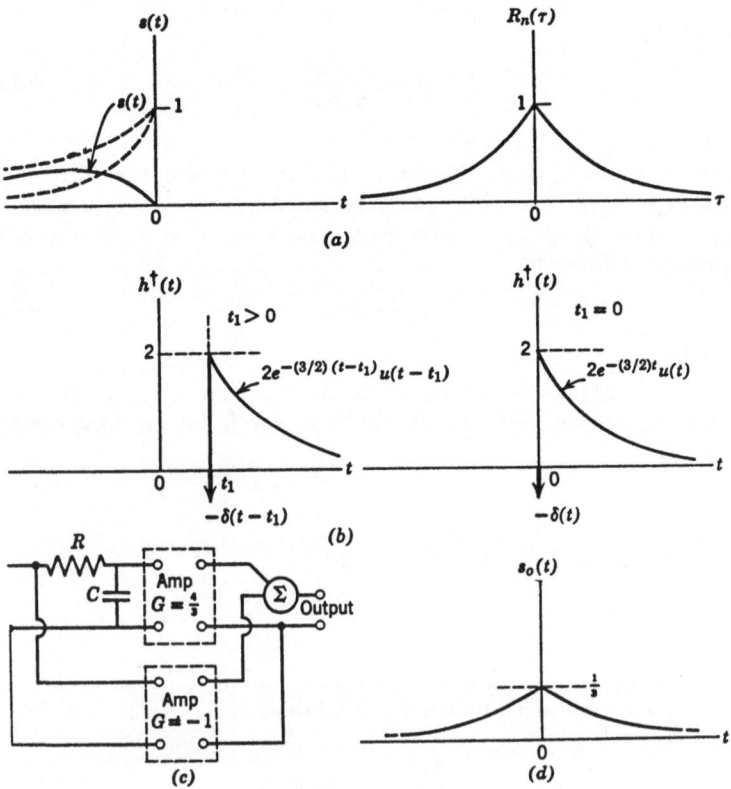

(a)

(b)

(c)

(d)

Fig. 5.4 - A matched filter example for non-white noise

$$S_0(\omega) = H^\dagger(\omega) \, S(\omega)$$

or

$$S_0(\omega) = \frac{1-j\,2\omega}{3+j\,2\omega} \, \frac{4}{(1-j\,2\omega)(3-j\,2\omega)} = \frac{1}{(9/4)+\omega^2}$$

The output signal is then

$$s_0(t) = (1/3)e^{-(3/2)\,|\,t\,|}$$

as shown in Fig. 5.4(d). For this example the generalization to non-white noise has produced little additional complication in filter realization.

————

Example 5.4

We consider the input signal $s(t)$ given by

$$s(t) = u(t)\left[e^{-t/2} - e^{-3t/2}\right]$$

The input noise is the same as in the previous example; that is,

$$\phi_n(\omega) = \frac{4}{1+4\omega^2}$$

and

$$R_n(\tau) = e^{-|\tau|/2}$$

As before, $s(t)$ and $R_n(\tau)$ are shown in Fig. 5.5(a). The transform $S(\omega)$ of the signal is now

$$S(\omega) = \frac{2}{1+j\,2\omega} - \frac{2}{3+j\,2\omega} = \frac{4}{(1+j\,2\omega)(3+j\,2\omega)}$$

Hence the matched filter transfer function is

$$H^\dagger(\omega) = \frac{1+j\,2\omega}{3-j\,2\omega} \, e^{-j\,\omega t_1} = \left[-1 + \frac{2}{(3/2)-j\,\omega}\right] e^{-j\,\omega t_1}$$

and the matched filter impulse response is

$$h^\dagger(t) = -\delta(t-t_1) + 2e^{(3/2)(t-t_1)} u(t_1-t)$$

It is apparent from Fig. 5.5(b) that this latter expression is not physically realizable; furthermore the best choice of a physically realizable filter is not obvious. This problem will be solved later. Meanwhile let us take as a realizable approximation the filter given by the impulse response

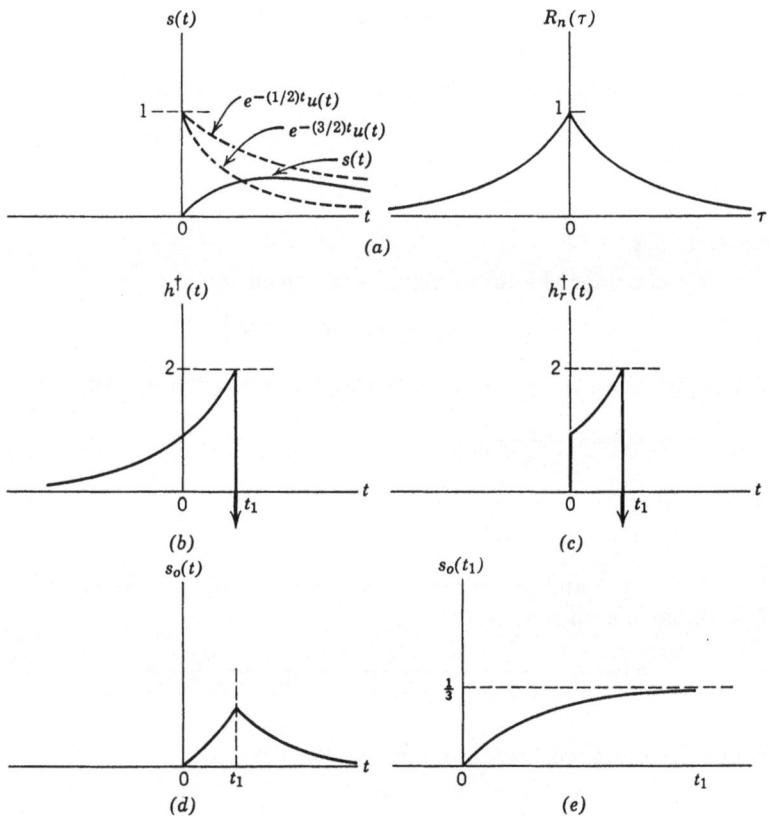

Fig. 5.5 - Another matched filter example for non-white noise

$$h_r{}^\dagger(t) = h^\dagger(t)\, u\,(t)$$

as shown in Fig. 5.5(c). The transform $S_0(\omega)$ of the output signal $s_o(t)$ due to this filter is

$$S_0(\omega) = S\,(\omega)\, H_r{}^\dagger(\omega)$$

where $H_r{}^\dagger(\omega)$ is the Fourier transform of $h_r{}^\dagger(t)$. Note that $h_r{}^\dagger(t)$ can be rewritten as

$$h_r{}^\dagger(t) = -\delta(t-t_1) + 2e^{(3/2)(t-t_1)}[u(t-t_1) - u(-t)]$$

The transform of this expression is

$$H_r{}^\dagger(\omega) = \int_0^{t_1+0} h_r{}^\dagger(t)\, e^{-j\omega t}\, dt$$

or

$$H_r{}^\dagger(\omega) = -e^{-j\omega t_1} + \frac{2}{(3/2)-j\omega}[e^{-j\omega t_1} - e^{-(3/2)t_1}]$$

Since $S(\omega)$ is already given, we have, after some rearranging

$$S_0(\omega) = e^{-(3/2)t_1}\left[\frac{-1}{(1/2)+j\omega} + \frac{2/3}{(3/2)+j\omega} + \frac{-(1/3)}{(3/2)-j\omega}\right]$$

$$+ e^{-j\omega t_1}\left[\frac{1/3}{(3/2)+j\omega} + \frac{1/3}{(3/2)-j\omega}\right]$$

This expression may be transformed term-by-term to yield the output signal

$$s_0(t) = -e^{-(3/2)t_1}\, e^{-(1/2)t}\, u(t) + (2/3)e^{-(3/2)(t_1+t)}u(t)$$
$$- (1/3)e^{-(3/2)(t_1-t)}u(-t) + (1/3)e^{-(3/2)|t_1-t|}$$

This signal is plotted in Fig. 5.5(d) and does have a maximum value at $t = t_1$; nevertheless there is no guarantee that the procedure which has been used to find a realizable filter is optimum. Suppose we take $t = t_1 \geq 0$; then we have

$$s_0(t_1) = -e^{-2t_1} + (2/3)\, e^{-3t_1} + 1/3$$

This function is plotted in Fig. 5.5(e) vs. the delay t_1. As the delay approaches infinity, we would expect the filter to become optimum since, in this case, the impulse response $h^\dagger(t)$ of Fig. 5.5(b) can be realized exactly. For this case the output signal $s_0(t_1)$ is given by Eq. (5.2-38) as

$$\lim_{t_1 \to \infty} s_0(t_1) = \int_{-\infty}^{\infty} s^2(x)\, dx = \int_0^{\infty} [e^{-(1/2)x} - e^{-(3/2)x}]^2 dx$$
$$= [-e^{-x} - (1/3)e^{-3x} + e^{-2x}]_0^{\infty} = 1/3$$

as expected from Fig. 5.5(e).

5.4 *Spectral Factorization for Continuous-Parameter Random Processes* - We are interested ultimately in the solution of Eq. (5.2-14) for the realizable matched filter. In this section we consider a topic which is a necessary preliminary to such a solution, namely the factorization of spectral densities.

The power spectral density $\phi(\omega)$ of a random process is given by

$$\phi(\omega) = \int_{-\infty}^{\infty} R(\tau) e^{-j\omega\tau} d\tau \qquad (5.4\text{-}1)$$

where $R(\tau)$ is the autocorrelation function of the random process (which is assumed to be at least wide sense stationary). It was pointed out in Section 3.8 that $\phi(\omega)$ is real, even, and non-negative for real values of ω. In most problems that we will encounter, this power spectral density $\phi(\omega)$ can be assumed to be a rational function of ω; that is, the ratio of two polynomials in ω. This will always be true if $\phi(\omega)$ arises from the filtering of a white noise by a lumped-parameter linear time-invariant filter.

Example 5.5

Suppose that $\phi(\omega)$ is the result of passing white noise of spectral density unity through an RC low pass filter with transfer function

$$H(\omega) = \frac{1}{1+j\omega RC}$$

The power spectral density $\phi(\omega)$ is given by

$$\phi(\omega) = |H(\omega)|^2 = H(\omega)H(-\omega)) = \frac{1}{1+\omega^2 R^2 C^2}$$

which is a rational function of ω.

Example 5.6

Suppose that $\phi(\omega)$ is the result of passing white noise of spectral density unity through the lumped-parameter linear time-invariant filter with transfer function

$$H(\omega) = K\frac{\omega^n + a_{n-1}\omega^{n-1} + \cdots + a_1\omega + a_0}{\omega^d + b_{d-1}\omega^{d-1} + \cdots + b_1\omega + b_0}$$

Since the variable ω occurs only as power of $j\omega$ in the determination of this transfer function, we can write

$$H(\omega) = A(\omega^2) + j\omega B(\omega^2)$$

where $A(\omega^2)$ and $B(\omega^2)$ are real polynomials in ω^2. Thus

$$\phi(\omega) = |H(\omega)|^2 = H(\omega)H(-\omega) = A^2(\omega^2) + \omega^2 B^2(\omega^2)$$

The output spectral density $\phi(\omega)$ can now be written as

$$\phi(\omega) = K^2 \frac{N_n(\omega^2)}{D_d(\omega^2)}$$

where $N_n(\omega^2)$ and $D_d(\omega^2)$ are polynomials of degree n and d, respectively, in ω^2.

In many cases not covered by the previous examples, the correlation function $R(\tau)$ can be closely approximated by a sum of exponentials or damped sinusoids

$$R(\tau) \approx \sum_n A_n\, e^{-a_n |\tau|} \cos b_n\, \tau \tag{5.4-2}$$

where some or all of the b_n may be zero. The corresponding power spectral density is given approximately by the transform of the right side of Eq. (5.4-2):

$$\phi(\omega) \approx \sum_n a_n A_n\, \frac{(a_n^2 + b_n^2) + \omega^2}{\omega^4 + 2(a_n^2 - b_n^2)\,\omega^2 + (a_n^2 + b_n^2)^2} \tag{5.4-3}$$

which is of the form

$$\phi(\omega) = K^2 \frac{N(\omega^2)}{D(\omega^2)} \tag{5.4-4}$$

and is a rational function of ω. For the remainder of this discussion it will be assumed that power spectral densities can be written in the form of Eq. (5.4-4).

Since $\phi(\omega)$ is real, then the polynomials $N(\omega^2)$ and $D(\omega^2)$ must have real coefficients (It has been assumed that common factors have been deleted). The roots of polynomials with real coefficients are either real or occur in complex conjugate pairs. At this point it is convenient to introduce the complex variable λ where

$$\lambda = \omega + j\,\sigma \tag{5.4-5}$$

so that we can consider the complex λ-plane as shown in Fig. 5.6. In this plane roots of $N(\lambda^2)$ and $D(\lambda^2)$ must occur in pairs (λ_k, λ_k^*) where

$$\lambda_k = \omega_k + j\,\sigma_k \tag{5.4-6}$$

$$\lambda_k^* = \omega_k - j\,\sigma_k \tag{5.4-7}$$

as illustrated in Fig. 5.6. Furthermore, since the polynomials $N(\lambda^2)$ and $D(\lambda^2)$ are even, for each root

$$\lambda_k = \omega_k + j\,\sigma_k \tag{5.4-8}$$

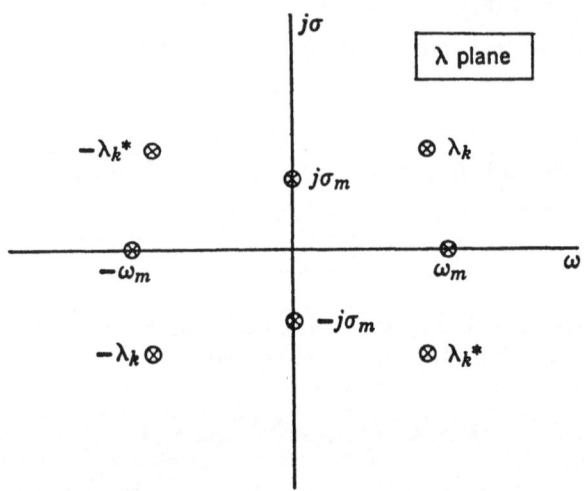

Fig. 5.6 - The complex λ - plane and possible root locations

there is a root

$$-\lambda_k = -\omega_k - j \, \sigma_k \qquad (5.4\text{-}9)$$

as shown in Fig. 5.6. Thus purely real and purely imaginary roots occur in pairs $(\omega_k, -\omega_k)$ and $(j\sigma_k, -j\sigma_k)$ while complex roots occur in sets of four $(\lambda_k, \lambda_k^*, -\lambda_k, -\lambda_k^*)$ placed symmetrically with respect to both real and imaginary axes.

If we restrict our attention to random processes $X(t)$ with finite energy, then

$$E\{X^2(t)\} = R(0) = \frac{1}{2\pi} \int\limits_{-\infty}^{\infty} \phi(\omega) \, d\omega < \infty \qquad (5.4\text{-}10)$$

In this case $D(\lambda^2)$ can have no real roots since then $\phi(\omega)$ would possess a pole on the real axis in violation of Eq. (5.4-10). Also, since $\phi(\omega)$ is non-negative for all real ω, the polynomial $N(\omega^2)$ cannot change sign and any real root of $N(\lambda^2)$ must be a multiple root of even order arising from a set of factors of the form

$$(\lambda - \omega_k)^{2r} (\lambda + \omega_k)^{2r} \qquad (5.4\text{-}11)$$

From the previous discussion it is apparent that the denominator polynomial $D(\lambda^2)$ can be factored into the form

$$D(\lambda^2) = \pi_k (\lambda - \lambda_k)(\lambda - \lambda_k^*)(\lambda + \lambda_k)(\lambda + \lambda_k^*)$$

$$\times \pi_m (\lambda - j\,\sigma_m)(\lambda + j\,\sigma_m) \qquad (5.4\text{-}12)$$

while the numerator polynomial can be written as

$$N(\lambda^2) = \pi_p(\lambda-\lambda_p)(\lambda-\lambda_p^*)(\lambda+\lambda_p)(\lambda+\lambda_p^*)$$

$$\times \; \pi_q(\lambda-j\,\sigma_q\,)(\lambda+j\,\sigma_q\,) \qquad (5.4\text{-}13)$$

$$\times \; \pi_r(\lambda-\omega_r\,)^{n_r}(\lambda-\omega_r\,)^{n_r}(\lambda+\omega_r\,)^{n_r}(\lambda+\omega_r\,)^{n_r}$$

where n_r is an integer whose value depends on the index r. For purposes of notation, let us define $\lambda_k, \lambda_p, j\sigma_q, j\sigma_m$ to be in the upper half λ-plane (UHP). Note that $-\lambda_k^*$ and $-\lambda_p^*$ will also be in the UHP. Now we factor $D(\lambda^2)$ into two terms $D^+(\lambda)$ and $D^-(\lambda)$ such that

$$D(\lambda^2) = D^+(\lambda)D^-(\lambda) \qquad (5.4\text{-}14)$$

where $D^+(\lambda)$ has only UHP roots and $D^-(\lambda)$ has only lower half λ-plane (LHP) roots. In the same way

$$N(\lambda^2) = N^+(\lambda)\,N^-(\lambda) \qquad (5.4\text{-}15)$$

If $N(\lambda^2)$ has real roots, they occur in even multiples and half are assigned to each half-plane. It is now apparent that

$$D^+(\lambda) = \pi_k(\lambda-\lambda_k)(\lambda+\lambda_k^*)\pi_m(\lambda-j\,\sigma_m) \qquad (5.4\text{-}16)$$

$$D^-(\lambda) = \pi_k(\lambda-\lambda_k^*)(\lambda+\lambda_k)\pi_m(\lambda+j\,\sigma_m) \qquad (5.4\text{-}17)$$

$$N^+(\lambda) = \pi_p(\lambda-\lambda_p)(\lambda+\lambda_p^*)\pi_q(\lambda-j\,\sigma_q)\;\pi_r(\lambda-\omega_r\,)^{n_r}(\lambda+\omega_r\,)^{n_r} \qquad (5.4\text{-}18)$$

and

$$N^-(\lambda) = \pi_p(\lambda-\lambda_p^*)(\lambda+\lambda_p)\pi_q(\lambda+j\,\sigma_q)\;\pi_r(\lambda-\omega_r\,)^{n_r}(\lambda+\omega_r\,)^{n_r} \qquad (5.4\text{-}19)$$

We have now factored the power spectral density $\phi(\lambda)$ into the form

$$\phi(\lambda) = \phi^+(\lambda)\phi^-(\lambda) \qquad (5.4\text{-}20)$$

where

$$\phi^+(\lambda) = \frac{N^+(\lambda)}{D^+(\lambda)} \qquad (5.4\text{-}21)$$

has all the zeroes and poles of $\phi(\lambda)$ that are in the UHP and where

$$\phi^-(\lambda) = \frac{N^-(\lambda)}{D^-(\lambda)} \qquad (5.4\text{-}22)$$

has all the zeroes and poles of $\phi(\lambda)$ that are in the LHP. Furthermore, $\phi^+(\lambda)$ and $\phi^-(\lambda)$ are conjugates so that

$$\phi^+(\lambda) = [\phi^-(\lambda)]^* \qquad (5.4\text{-}23)$$

$$\phi^-(\lambda) = [\phi^+(\lambda)]^* \qquad (5.4\text{-}24)$$

and

$$\phi(\lambda) = \phi^+(\lambda)\phi^-(\lambda) = |\phi^\pm(\lambda)|^2$$

Example 5.7

Let the random process $X(t)$ be given as the sum of the two independent random processes $X_1(t)$ and $X_2(t)$, each with zero mean and spectral densities respectively of

$$\phi_1(\omega) = 2$$

and

$$\phi_2(\omega) = \frac{1}{1+\omega^2}$$

In this case the power spectral density of the process $X(t)$ is

$$\phi(\omega) = \phi_1(\omega) + \phi_2(\omega) = \frac{2\omega^2 + 3}{\omega^2 + 1}$$

which may be written as

$$\phi(\lambda) = \phi^+(\lambda)\phi^-(\lambda)$$

where

$$\phi^+(\lambda) = \sqrt{2}\,\frac{\lambda - j\sqrt{3/2}}{\lambda - j}$$

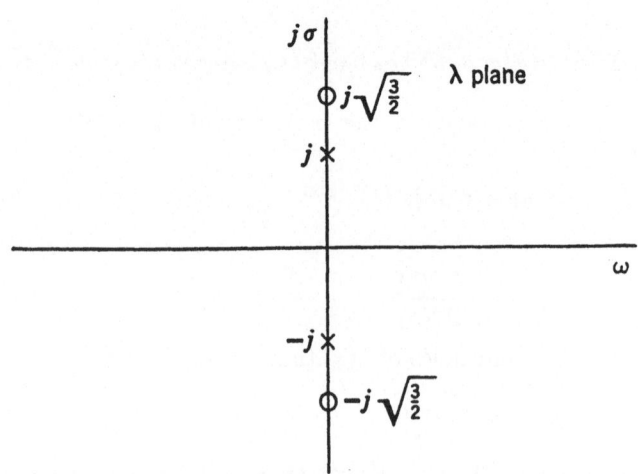

Fig. 5.7 - Pole and zero locations of a power spectral density of a continuous-parameter random process

and

$$\phi^-(\lambda) = \sqrt{2}\ \frac{\lambda + j\sqrt{3/2}}{\lambda + j}$$

Thus ϕ^+ has a zero at $\lambda = j\sqrt{3/2}$ and a pole at $\lambda = j$ while ϕ^- has a zero at $\lambda = -j\sqrt{3/2}$ and a pole at $\lambda = -j$ as shown in Fig. 5.7.

We have succeeded in factoring the power spectral density in the form

$$\phi(\omega) = \phi^+(\omega)\phi^-(\omega)$$

We now wish to relate the location for the poles of the functions $\phi^+(\omega)$ and $\phi^-(\omega)$ to the time-domain behavior of their inverse Fourier transforms $R^-(t)$ and $R^+(t)$ where

$$R^+(t) = \frac{1}{2\pi} \int\limits_{-\infty}^{\infty} \phi^+(\omega)e^{j\omega t}\ d\omega \qquad (5.4\text{-}26)$$

and

$$R^-(t) = \frac{1}{2\pi} \int\limits_{-\infty}^{\infty} \phi^-(\omega)e^{j\omega t}\ d\omega \qquad (5.4\text{-}27)$$

Suppose $f(t)$ is a function which vanishes on the negative half-line, that is,

$$f(t) = 0, \quad t < 0 \qquad (5.4\text{-}28)$$

If $f(t)$ is absolutely integrable; that is, if

$$\int\limits_{-\infty}^{\infty} |f(t)|\ dt = \int\limits_{0}^{\infty} |f(t)|\ dt < \infty \qquad (5.4\text{-}29)$$

then the Fourier transform

$$F(\omega) = \int\limits_{-\infty}^{\infty} f(t)e^{-j\omega t}\ dt = \int\limits_{0}^{\infty} f(t)e^{-j\omega t}\ dt \qquad (5.4\text{-}30)$$

exists for all ω in the range $(-\infty, \infty)$. Now $F(\omega)$ may be written in complex form as

$$F(\lambda) = \int\limits_{0}^{\infty} f(t)e^{-j\lambda t}\ dt = \int\limits_{0}^{\infty} [f(t)e^{\sigma t}]e^{-j\omega t}\ dt \qquad (5.4\text{-}31)$$

where λ is the complex variable

$$\lambda = \omega + j\ \sigma \qquad (5.4\text{-}32)$$

Clearly, if Eq. (5.4-29) holds, then

$$\int\limits_{0}^{\infty} |f(t)e^{\sigma t}| \, dt < \infty \qquad (5.4\text{-}33)$$

for all $\sigma \leq 0$, that is, for all values of λ in the LHP. Thus $F(\lambda)$ exists for all λ in the LHP and, hence, has no singularities in the LHP including the ω-axis. Since any function $F(\lambda)$ of a complex variable must have at least one singularity, unless the function is a constant or zero, the singularities of $F(\lambda)$ are in the UHP.

A function $f(t)$, which vanishes on the negative half line, may not be absolutely integrable but it may happen that

$$\int\limits_{0}^{\infty} |f(t)\,e^{\sigma t}| \, dt < \infty \quad , \quad \sigma < 0 \qquad (5.4\text{-}34)$$

In this case $F(\lambda)$ may have some of its singularities on the ω-axis, although it will have none in the LHP. An example is the time function

$$f(t) = t^n \, u(t) \quad , \quad n = 1,2,... \qquad (5.4\text{-}35)$$

where $u(t)$ is the unit-step function. The transform $F(\lambda)$ of this function is readily calculated to be

$$F(\lambda) = \int\limits_{0}^{\infty} t^n \, e^{\sigma t} \, e^{-j\omega t} \, dt = \frac{n!}{\lambda^{n+1}} \qquad (5.4\text{-}36)$$

for $\sigma < 0$. Thus $F(\lambda)$ has a pole of order $n+1$ at the origin.

Suppose $g(t)$ is a function which vanishes on the positive half-line; that is,

$$g(t) = 0 \quad , \quad t > 0 \qquad (5.4\text{-}37)$$

If $g(t)$ is absolutely integrable so that

$$\int\limits_{-\infty}^{0} |g(t)| \, dt < \infty \qquad (5.4\text{-}38)$$

then the Fourier transform

$$G(\omega) = \int\limits_{-\infty}^{0} g(t)e^{-j\omega t} \, dt \qquad (5.4\text{-}39)$$

exists for all ω in the range $(-\infty,\infty)$. This transform may be written in complex form as

$$G(\lambda) = \int\limits_{-\infty}^{0} [g(t)e^{\sigma t}] \, e^{-j\omega t} \, dt \qquad (5.4\text{-}40)$$

Again it is clear from Eq. (5.4-38) that

$$\int\limits_{-\infty}^{0} |g(t)e^{\sigma t}| \, dt < \infty \qquad (5.4\text{-}41)$$

for all $\sigma \geq 0$; that is for all values of λ in the UHP including the ω-axis. The function $G(\lambda)$ has no singularities in the UHP including the ω-axis, and its singularities must be in the LHP. As before, if $g(t)$ is not absolutely integrable but

$$\int_{-\infty}^{0} |g(t)e^{\sigma t}| \, dt < \infty \qquad (5.4\text{-}42)$$

for $\sigma > 0$, then $G(\lambda)$ may have singularities on the ω-axis as well as in the LHP.

Thus we have shown that time domain behavior is related to the location of frequency domain singularities in the following two ways:

1) A function $\phi^+(\lambda)$ of the complex variable $\lambda = \omega + j\sigma$ having all of its singularities in the upper half λ-plane (UHP) has an inverse Fourier transform $R^+(t)$ which vanishes for negative time.

2) A function $\phi^-(\lambda)$ of the complex variable $\lambda = \omega + j\sigma$ having all of its singularities in the lower half λ-plane (LHP) has an inverse Fourier transform $R^-(t)$ which vanishes for positive time.

We note in passing that, since $\phi^+(\lambda)$ and $\phi^-(\lambda)$ were defined previously to have their zeroes as well as their poles in the UHP and LHP respectively, the reciprocal functions $1/\phi^+(\lambda)$ and $1/\phi^-(\lambda)$ also have transforms which vanish for negative time and positive time respectively. Thus both $\phi^+(\omega)$ and $1/\phi^+(\omega)$ can be considered as the complex transfer functions of physically realizable filters since the corresponding impulse responses vanish for negative time.

The process of *pre-whitening* or *spectral shaping* consists of filtering a random process so that the filter output is a normalized white noise; that is, has a constant power spectral density of unity. If the input process $X(t)$ is at least wide sense stationary with power spectral density $\phi(\omega)$, then the pre-whitening filter $H(\omega)$ is given by

$$1 = \phi(\omega) |H(\omega)|^2 \qquad (5.4\text{-}43)$$

The situation is illustrated in Fig. 5.8. Equation (5.4-43) may be rewritten as

$$1 = [\phi^+(\omega)H(\omega)][\phi^-(\omega)H(-\omega)] \qquad (5.4\text{-}44)$$

This equation is satisfied if

$$1 = \phi^+(\omega)H(\omega) \qquad (5.4\text{-}45)$$

since

$$\phi^+(\omega) \, H(\omega) = [\phi^-(\omega)H(-\omega)]^*$$

where the asterisk indicates the complex conjugate. The

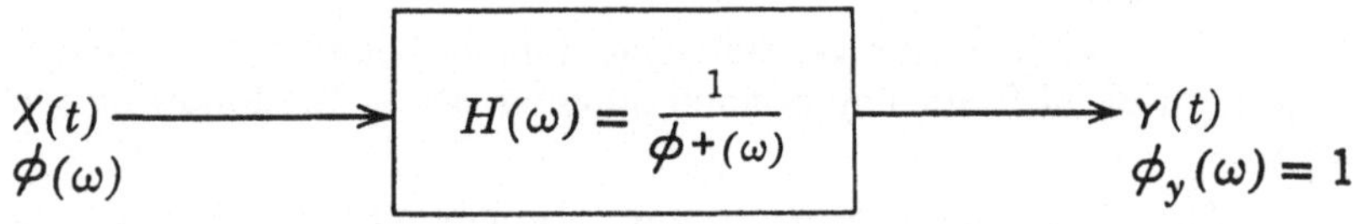

$$X(t) \longrightarrow \boxed{H(\omega) = \frac{1}{\phi^+(\omega)}} \longrightarrow \begin{array}{l} Y(t) \\ \phi_y(\omega) = 1 \end{array}$$

$$\begin{array}{l} X(t) \\ \phi(\omega) \end{array}$$

Fig. 5.8 - The pre-whitening filter

pre-whitening filter $H(\omega)$ is given by Eq. (5.4-45) as

$$H(\omega) = 1/\phi^+(\omega) \tag{5.4-46}$$

and is physically realizable.

5.5 *Solution of the Integral Equation for the Continuous-Time Matched Filter* - It was pointed out earlier that the difficulty in the solution of the integral equation (5.2-19) for the matched filter arises because the upper limits on σ and τ in this equation are both t_1 instead of ∞. The equation is not valid for $\sigma, \tau > t_1$. We begin the solution of Eq. (5.2-14) by requiring that $h^\dagger(t)$ be physically realizable so that

$$h^\dagger(t) = 0 \quad , \quad t < 0 \tag{5.5-1}$$

Then $h(t_1 - \sigma)$ is zero for $\sigma > t_1$ and Eq. (5.2-14) becomes

$$\int_{-\infty}^{\infty} R_n(\tau-\sigma) h^\dagger(t_1-\sigma) d\sigma = s(\tau) \quad , \quad -\infty < \tau \le t_1 \tag{5.5-2}$$

After a linear change of variable $t_1 - \sigma = u$ and $t_1 - \tau = t$, this equation becomes

$$\int_{-\infty}^{\infty} R_n(u - t) h^\dagger(u) du = s(t_1-t) \quad , \quad 0 \le t < \infty \tag{5.5-3}$$

which is a slightly more convenient form for our purposes.

Let us now define the function $g(t)$ from Eq. (5.5-3) as

$$g(t) = s(t_1-t) - \int_{-\infty}^{\infty} R_n(u-t) \, h^\dagger(u) \, du \quad , \quad -\infty < t < \infty \qquad (5.5\text{-}4)$$

It is apparent that $g(t)$ vanishes for positive t; that is,

$$g(t) = 0 \qquad t \geq 0 \qquad (5.5\text{-}5)$$

Thus the complex Fourier transform $G(\lambda)$ of $g(t)$ has no singularities in the upper half λ-plane (UHP). We take the complex Fourier transform of Eq. (5.5-4) and obtain

$$G(\lambda) = S(-\lambda) \, e^{-j\lambda t_1} - \phi_n(\lambda) \, H^\dagger(\lambda) \qquad (5.5\text{-}6)$$

where $\lambda = \omega + j\sigma$. It follows from Eq. (5.4-20) that the noise power spectral density $\phi_n(\lambda)$ may be factored and the equation rearranged to yield

$$\frac{G(\lambda)}{\phi_n^-(\lambda)} = \frac{S(-\lambda)}{\phi_n^-(\lambda)} e^{-j\lambda t_1} - \phi_n^+(\lambda) \, H^\dagger(\lambda) \qquad (5.5\text{-}7)$$

Let us consider the inverse transform of this equation term by term and write

$$a(t) = b(t) - c(t) \qquad (5.5\text{-}8)$$

where

$$a(t) = \frac{1}{2\pi} \int_{-\infty}^{\infty} \frac{G(\omega)}{\phi_n^-(\omega)} \, e^{j\omega t} \, d\omega \qquad (5.5\text{-}9)$$

and

$$b(t) = \frac{1}{2\pi} \int_{-\infty}^{\infty} \frac{S(-\omega)}{\phi_n^-(\omega)} \, e^{-j\omega t_1} e^{j\omega t} \, d\omega \qquad (5.5\text{-}10)$$

and

$$c(t) = \frac{1}{2\pi} \int_{-\infty}^{\infty} \phi_n^+(\omega) \, H^\dagger(\omega) \, e^{j\omega t} \, d\omega \qquad (5.5\text{-}11)$$

It is clear that the term $G(\lambda)/\phi_n^-(\lambda)$ has no UHP singularities and, hence, that its inverse Fourier transform $a(t)$ must vanish for $t \geq 0$. Therefore, for $t \geq 0$, it follows from Eq. (5.5-8) that

$$c(t) = b(t) \quad , \quad t \geq 0 \qquad (5.5\text{-}12)$$

Furthermore the function $\phi_n^+(\lambda) \, H^\dagger(\lambda)$ is free of singularities in the lower half λ-plane (LHP) so that

$$c(t) = 0 \quad , \quad t \leq 0 \qquad (5.5\text{-}13)$$

or

$$\phi_n^+(\omega) \, H^\dagger(\omega) = \int_0^{\infty} c(t) \, e^{-j\omega t} \, dt \qquad (5.5\text{-}14)$$

On replacing $c(t)$ by $b(t)$ as given by Eq. (5.5-10), we obtain

$$H^\dagger(\omega) = \frac{1}{2\pi\phi_n^+(\omega)} \int_0^\infty e^{-j\omega t} \int_{-\infty}^\infty \frac{S(-\nu)}{\phi_n^-(\nu)} e^{j\nu(t-t_1)} d\nu \, dt \quad (5.5\text{-}15)$$

which is an explicit solution for the realizable matched filter.

Notice that this last equation can be rearranged as

$$H^\dagger(\omega) = H_1(\omega)H_2(\omega) \qquad (5.5\text{-}16)$$

where

$$H_1(\omega) = \frac{1}{\phi_n^+(\omega)} \qquad (5.5\text{-}17)$$

is a prewhitening filter for the input noise process $N(t)$. The output of this filter consists of two components. One is the signal $s_1(t)$ with transform $S_1(\omega)$ given by

$$S_1(\omega) = H_1(\omega)S(\omega) = \frac{S(\omega)}{\phi_n^+(\omega)} \qquad (5.5\text{-}18)$$

where $S(\omega)$ is the transform of the input signal s(t). The other component is the white noise $N_1(t)$ with a power spectral density of unity. The filter $H_2(\omega)$ is given by

$$H_2(\omega) = \frac{1}{2\pi} \int_0^\infty e^{-j\omega t} \int_{-\infty}^\infty \frac{S(-\nu)}{\phi_n^-(\nu)} e^{j\nu(t-t_1)} d\nu \, dt \qquad (5.5\text{-}19)$$

and is physically realizable. That such is the case may be seen by noting that

$$h_2(t) = h_3(t)u(t) \qquad (5.5\text{-}20)$$

where

$$h_3(t) = \frac{1}{2\pi} \int_{-\infty}^\infty \frac{S(-\nu)}{\phi_n^-(\nu)} e^{j\nu(t-t_1)} d\nu \qquad (5.5\text{-}21)$$

In terms of the original unrealizable filter of Eq. (5.3-6), we may write Eq. (5.5-15) as

$$H^\dagger(\omega) = \frac{1}{\phi_n^+(\omega)} \left[\frac{S(-\omega)e^{-j\omega t_1}}{\phi_n^-(\omega)} \right]_+ \qquad (5.5\text{-}22)$$

where the plus sign on the term in brackets is used to indicate the operation of Eq. (5.5-20). The situation is illustrated in Fig. 5.9. The first filter $H_1(\omega)$ is a prewhitener for the input noise. The second filter is the remainder of the unrealizable filter of Eq. (5.3-6) made realizable by truncating its impulse response $h_3(t)$ for negative time. Note, however, that Eq. (5.5-22) is not equivalent to truncating the impulse response $h_s(t)$ of the unrealizable matched filter; prewhitening must be accomplished first.

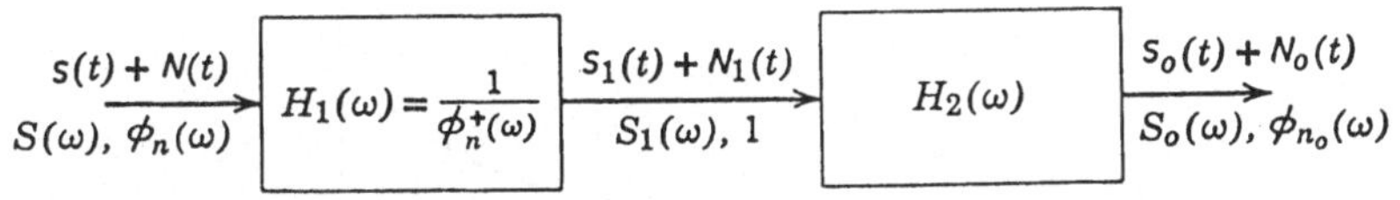

$$\begin{array}{|c|c|c|} \hline \text{s(t) + N(t)} & H_1(\omega) = \frac{1}{\phi_n^+(\omega)} & \text{s}_1(t) + N_1(t) \\ S(\omega), \, \phi_n(\omega) & & S_1(\omega), \, 1 \\ \hline \end{array} \quad \boxed{H_2(\omega)} \quad \begin{array}{c} S_o(t) + N_o(t) \\ S_o(\omega), \, \phi_{n_o}(\omega) \end{array}$$

Fig. 5.9 - Solution for the matched filter by prewhitening techniques

Example 5.8

Let us refer again to Example 5.4 and consider the same input signal

$$s(t) = u(t)(e^{-t/2} - e^{-3t/2})$$

and the same input noise $N(t)$ with spectral density

$$\phi_n(\omega) = \frac{1}{\frac{1}{4} + \omega^2}$$

and autocorrelation function $R_n(\tau) = e^{-|\tau|/2}$. Both $s(t)$ and $R_n(\tau)$ are shown in Figure 5.10(a). The noise spectral density may be factored as

$$\phi_n(\omega) = \phi_n^+(\omega)\phi_n^-(\omega) = \frac{1}{\frac{1}{2} + j\omega} \quad \frac{1}{\frac{1}{2} - j\omega}$$

Now the prewhitening filter has a complex transfer function

$$H_1(\omega) = \frac{1}{\phi_n^+(\omega)} = \frac{1}{2} + j\omega$$

The signal $s_1(t)$ after this filter has a transform

$$S_1(\omega) = S(\omega)H_1(\omega) = \left[\frac{2}{1 + j\,2\omega} - \frac{2}{3 + j\,2\omega} \right] \frac{1 + j\,2\omega}{2}$$

or

$$S_1(\omega) = 1 - \frac{1 + j\,2\omega}{3 + j\,2\omega} = \frac{2}{3 + j\,2\omega}$$

Thus, at the output of the prewhitening filter, the signal $s_1(t)$ is given by

$$s_1(t) = u(t)\, e^{-3t/2}$$

and exists in white noise of unity spectral density as shown in Figure 5.10(b). The unrealizable matched filter for this input has a transfer function

$$H_3(\omega) = S_1(-\omega)\, e^{-j\omega t_1} = \frac{2}{3 - j\,2\omega}\, e^{-j\omega t_1}$$

and an impulse response

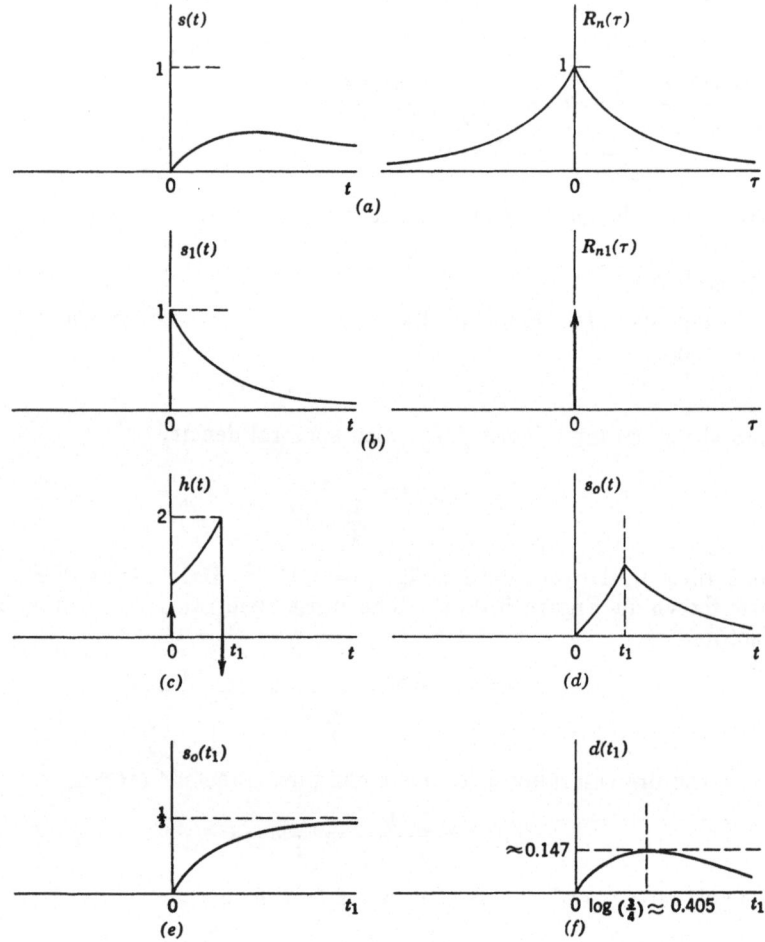

Fig. 5.10 - A matched filter example solved by prewhitening techniques

$$h_3(t) = u(t_1 - t)\, e^{(3/2)(t - t_1)}$$

We take as the realizable filter

$$h_2(t) = u(t)h_3(t)$$

with corresponding transfer function

$$H_2(\omega) = \int_0^{t_1} e^{(3/2)(t - t_1)}\, e^{-j\omega t}\, dt = \frac{2}{3 - j\,2\omega}\left(e^{-j\omega t_1} - e^{-3t_1/2}\right)$$

Now we can combine $H_1(\omega)$ and $H_2(\omega)$ to form the overall matched filter

$$H(\omega) = H_1(\omega)H_2(\omega) = \frac{1 + j\,2\omega}{3 - j\,2\omega}\left(e^{-j\omega t_1} - e^{-3t_1/2}\right)$$

$$= \left(-1 + \frac{4}{3 - j\,2\omega} \right) (e^{-j\omega t_1} - e^{-3t_1/2}).$$

We may take the inverse transform of $H(\omega)$ term by term to obtain the impulse response

$$h(t) = -\delta(t - t_1) + e^{-3t_1/2}\,\delta(t) + 2e^{(3/2)(t - t_1)}[u(t_1 - t) - u(-t)]$$

as shown in Figure 5.10(c).

The output signal $s_0(t)$ has a Fourier transform given by

$$S_0(\omega) = H(\omega)S(\omega) = \frac{4}{(3 - j\,2\omega)(3 + j\,2\omega)}\,[e^{-j\omega t_1} - e^{-3t_1/2}]$$

with inverse transform

$$s_0(t) = \frac{1}{3}[e^{-(3/2)|t - t_1|} - e^{-3t_1/2}\,e^{-(3/2)|t|}]$$

as shown in Figure 5.10(d). For $t = t_1 \geq 0$ this becomes

$$s_0(t_1) = \frac{1}{3}(1 - e^{-3t_1})$$

and is plotted in Figure 5.10(e) versus t_1. It is interesting to compare the quantity $s_0(t_1)$ with the same quantity obtained in Example 5.4. Call them $s_{01}(t_1)$ and $s_{02}(t_1)$ respectively. Recall, from Eq.(5.2-23), that each of these is the actual signal-to-noise output power ratio at time t_1 as given by Eq. (5.2-1) or (5.2-4). This quantity is a maximum for the matched filter. On forming the difference

$$d(t_1) = s_{01}(t_1) - s_{02}(t_1) = e^{-2t_1} - e^{-3t_1},$$

we see that this difference is nonnegative, or

$$d(t_1) \geq 0, \quad \text{all} \quad t_1 \geq 0,$$

as shown in Figure 5.10(f). Note that this difference has a maximum of approximately 0.147 at $t_1 = log(\frac{3}{2}) = 0.40547... $. It is apparent that the prewhitening realization procedure used in this example yields a larger output signal-to-noise power ratio than the procedure used in Example 5.4 where the unrealizable impulse response corresponding to Eq.(5.3-6) was simply truncated to yield a realizable filter.

The solution of this example has corresponded exactly to a step-by-step solution of Eq.(5.5-15).

Part II - The Matched Filter For Discrete-Time Inputs

5.6 *Derivation* - We consider here the same problem treated in Section 5.2 except that we take the input signal s_i and the input noise process N_i to be defined in discrete time as shown in Fig. 5.11.

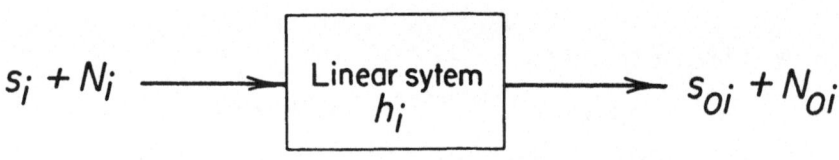

$$s_i + N_i \longrightarrow \boxed{\begin{array}{c} \text{Linear sytem} \\ h_i \end{array}} \longrightarrow s_{oi} + N_{oi}$$

Fig. 5.11 - The matched filter in discrete time

The corresponding output signal s_{oi} and noise N_{oi} at time i are given by

$$s_{oi} = \sum_{k=-\infty}^{i} h_{i-k}\, s_k \qquad (5.6\text{-}1)$$

and

$$N_{oi} = \sum_{k=-\infty}^{i} h_{i-k}\, N_k \qquad (5.6\text{-}2)$$

where h_k is the impulse response of the linear, time-invariant matched filter. As before, the criterion of optimization is the maximization of the output signal-to-noise power ratio R_o at some time q where

$$R_o = \frac{s_{oq}^2}{E\{N_{oq}^2\}} \qquad (5.6\text{-}3)$$

For the optimum filter, we take $R_{o\ max} = 1/\alpha$. For an arbitrary filter, Eq. (5.6-3) becomes

$$E\{N_{oq}^2\} - \alpha \, s_{oq}^2 = K \geq 0 \qquad (5.6\text{-}4)$$

where the equality holds only for the optimum filter $h_k{}^\dagger$. After substituting Eqs. (5.6-1) and (5.6-2) into the last expression, we have

$$\sum_{k=-\infty}^{q} \sum_{j=-\infty}^{q} R_n(k-j)h_{q-k} h_{q-j} - \alpha \mid \sum_{k=-\infty}^{q} s_k \, h_{q-k} \mid^2 = K \geq 0 \quad (5.6\text{-}5)$$

where $R_n(k-j) = R(j-k) = E\{N_k N_j\}$; that is, we have assumed that the input noise process N_k is at least wide-sense stationary. We now wish to find the matched filter $h_k{}^\dagger$ such that the left side of Eqs. (5.6-5) is zero.

We proceed with the same variational technique described in Section 5.2; that is, we replace $h^\dagger$ by $h^\dagger + \gamma \delta h$ where γ is a real variable and δh is an *arbitrary* variation in $h^\dagger$. Following the steps outlined in Eqs. (5.2-8) thru (5.2-18), we finally obtain

$$\sum_{j=-\infty}^{q} R_n(k-j)h_{q-j}^\dagger = s_k \quad , \quad k = \ldots -1,0,1,\ldots,q \qquad (5.6\text{-}6)$$

This is the equation which must be solved to find the matched filter and is the discrete-time equivalent to Eq. (5.2-19).

Consider the case where the input noise process N_k is white with autocorrelation function

$$R_n(k) = \begin{cases} N_o/2 \ , & k = 0 \\ \\ 0 & , \ k \neq 0 \end{cases} \qquad (5.6\text{-}7)$$

and corresponding power spectral density

$$\phi_n(z) = \sum_{k=-\infty}^{\infty} R_n(k) \, z^{-k} = N_o/2 \qquad (5.6\text{-}8)$$

It is clear that the solution to Eq. (5.6-6) is just

$$h_{q-k}^\dagger = s_k \qquad (5.6\text{-}9)$$

or

$$h_l{}^\dagger = s_{q-l} \qquad (5.6\text{-}10)$$

We may transform Eq. (5.6-10) to obtain

$$H^\dagger(z) = \sum_{l=-\infty}^{\infty} h_l{}^\dagger z^{-l} = \sum_{l=-\infty}^{\infty} s_{q-l} \, z^{-l} \qquad (5.6\text{-}11)$$

or, after changing the index of summation,

$$H^\dagger(z) = \sum_{k=-\infty}^{\infty} s_k \, z^k \, z^{-q} = S(1/z)z^{-q} \qquad (5.6\text{-}12)$$

where $S(z)$ is the transform

$$S(z) = \sum_{k=-\infty}^{\infty} s_k z^{-k} \qquad (5.6\text{-}13)$$

Recall that Eq. (5.6-12) gives the matched filter for input signal s_k and white input noise and without consideration of realizability.

The output signal-to-noise power ratio $R_{0\,\text{max}}$ can be found for the general case by using Eq. (5.6-6) to rewrite Eq. (5.6-3) as

$$R_{0\,\text{max}} = \left| \sum_{k=-\infty}^{q} s_k\, h_{q-k} \right| = |s_{0q}| = \frac{1}{\alpha} \qquad (5.6\text{-}14)$$

Exactly as before, Eq. (5.6-6) is a sufficient, as well as necessary, condition for R_0 to be a maximum.

Example 5.9

Let a signal s_k be given by

$$s_k = \begin{cases} Be^{bk} & , \quad k \leq 0 , \quad B, b > 0 \\ 0 & , \quad k > 0 \end{cases}$$

and let the noise be white and additive with power spectral density $N_0/2$. The transform $G_s(z)$ of this signal is

$$G_s(z) = B \sum_{k=-\infty}^{0} e^{bk} z^{-k} = B \sum_{k=0}^{\infty} (e^b z^{-1})^{-k} = \frac{Be^b}{e^b - z}$$

The impulse response of the matched filter is obtained from Eq. (5.6-10) as

$$h_k^\dagger = s_{q-k} = \begin{cases} Be^{b(q-k)} & , \quad k \geq q \\ 0 & , \quad k < q \end{cases}$$

The physical realizability requirement that $h_k^\dagger$ vanish for negative k is satisfied by taking $q \geq 0$. The simplest choice is $q = 0$ so that

$$h_k^\dagger = \begin{cases} Be^{-bk} & , \quad k \geq 0 \\ 0 & , \quad k < 0 \end{cases}$$

and

$$H(z) = \sum_{k=0}^{\infty} h_k^\dagger z^{-k} = B \sum_{k=0}^{\infty} (ze^b)^{-k} = \frac{Be^b}{e^b - z^{-1}}$$

The output signal at time l is given by

$$s_{0l} = \sum_{k=-\infty}^{l} s_k \, h_{l-k}$$

and we must distinguish two cases. When $l \geq 0$, we have

$$s_{0l} = \sum_{k=-\infty}^{0} B e^{bk} \, B e^{b(k-l)} = B^2 e^{-bl} \sum_{k=0}^{\infty} e^{-2bk} = \frac{B^2}{1-e^{2b}} \, e^{-bl}$$

When $l < 0$, we have

$$s_{0l} = \sum_{k=-\infty}^{l} B e^{bk} \, B e^{b(k-l)} = B^2 e^{-bl} \sum_{k=-l}^{\infty} e^{-2bk}$$

$$= B^2 e^{-bl} \, \frac{e^{2bl}}{1-e^{2b}} = \frac{B^2}{1-e^{2b}} \, e^{bl}$$

or, for all l,

$$s_{0l} = \frac{B^2}{1-e^{2b}} \, e^{-b|l|}$$

This last expression [see Eq.(5.6-14)] is also the output S/N as a function of l. It is a maximum at $q = l = 0$, as expected. The appropriate quantities are shown in Fig. 5.12.

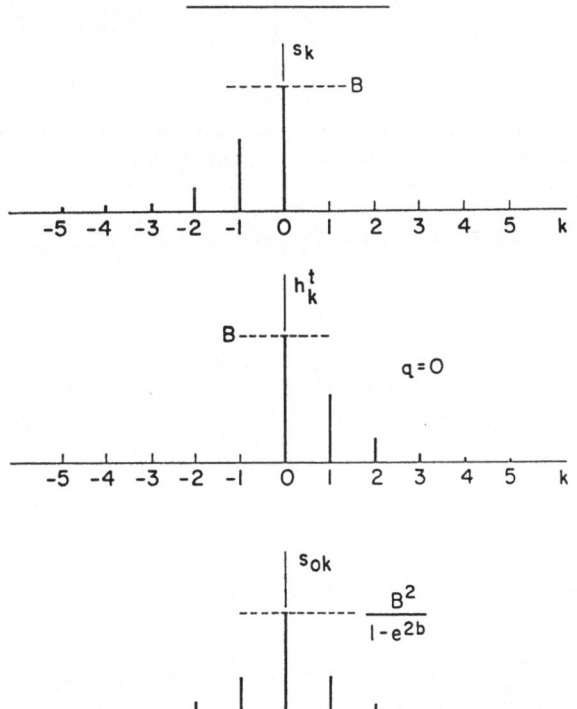

Fig. 5.12 - A matched filter example in discrete time

As in the continuous-time case [compare Examples 5.1 and 5.2], this last example was somewhat trivial, at least in so far as concerned realizability. For many signals, the choice of q to obtain realizability may not be so simple.

Example 5.10

Consider the signal

$$s_k = \begin{cases} 0 & , \quad k < 0 \\ Be^{-bk} & , \quad k \geq 0 \end{cases} , \quad B, b > 0$$

corrupted by additive white noise with power spectral density $N_0/2$. The transform of the signal is

$$G_s(z) = B \sum_{k=0}^{\infty} (e^b z)^{-k} = \frac{Be^b}{e^b - z^{-1}}$$

as worked out in detail in Example 4.2. The impulse response of the realizable matched filter is given by

$$h_k^\dagger = \begin{cases} s_{q-k} & , \quad k \geq 0 \\ 0 & , \quad k < 0 \end{cases} = \begin{cases} 0 & , \quad k > q \\ Be^{-b(q-k)} & , \quad 0 \leq k \leq q \\ 0 & , \quad k < 0 \end{cases}$$

For any finite value of q, all of the signal will not have passed through the filter; therefore the output signal is a function of q and is given by

$$s_{0q} = \sum_{k=0}^{q} s_k^2 = B^2 \sum_{k=0}^{q} e^{-2bk} = B^2 \frac{1 - e^{-2b(q+1)}}{1 - e^{-2b}}$$

In the limit as $q \rightarrow \infty$, we have

$$\lim_{q \rightarrow \infty} s_{0q} = \frac{B^2}{1 - e^{2b}}$$

as shown in Fig. 5.13.

5.7 *The Unrealizable Matched Filter in Discrete Time* - It is easy to solve Eq. (5.6-6) if it is rewritten as

$$\sum_{j=-\infty}^{\infty} R_n(k-j)h_{q-j}^\dagger = s_k \quad , \quad k = 0, \pm 1, \pm 2, \dots \quad (5.7\text{-}1)$$

We have replaced the limit q by ∞ and, as a consequence, have ignored realizability. On transforming, we have

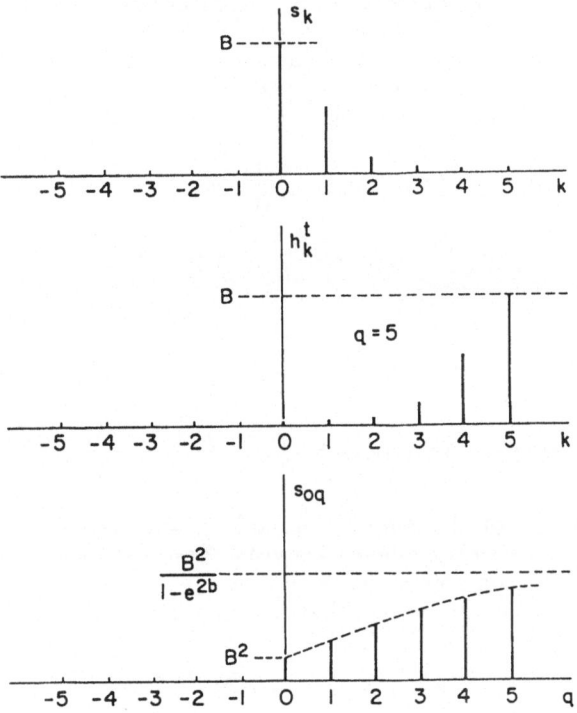

Fig. 5.13 - Another matched filter example in discrete time

$$\sum_{k=-\infty}^{\infty} \sum_{j=-\infty}^{\infty} R_n(k-j)h_{q-j}^{\dagger} z^{-k} = \sum_{k=-\infty}^{\infty} s_k z^{-k} \qquad (5.7\text{-}2)$$

or

$$\sum_{j=-\infty}^{\infty} h_{q-j}^{\dagger} z^{-j} \phi_n(z) = z^{-q} \phi_n(z) \sum_{l=-\infty}^{\infty} h_l z^l = S(z) \qquad (5.7\text{-}3)$$

so that

$$H^{\dagger}(1/z) = \frac{S(z)}{\phi_n(z)} z^q \qquad (5.7\text{-}4)$$

or

$$H^{\dagger}(z) = \frac{S(1/z)}{\phi_n(z)} z^{-q} \qquad (5.7\text{-}5)$$

since $\phi_n(z) = \phi_n(1/z)$.

As in the continuous-time case, the matched filter specified by Eq. (5.7-5) will not be realizable for arbitrary signal and noise since $h_k^{\dagger}$ will not vanish for negative values of the index k.

5.8 *Spectral Factorization for Discrete-Parameter Random Processes* - In Sections 3.8 and 4.2, we developed two alternative transform pairs to describe the autocorrelation function and power spectral density of a wide-sense stationary process with discrete parameter. These pairs were

$$R(n) = E\{X_m X_{m+n}\} = \frac{1}{2\pi} \int_{-\pi}^{\pi} \phi(\omega) e^{jn\omega} d\omega \qquad (5.8\text{-}1a)$$

$$\phi(\omega) = \sum_{n=-\infty}^{\infty} R(n) e^{-jn\omega} \quad , \quad -\pi \le \omega \le \pi \qquad (5.8\text{-}1b)$$

and

$$R(n) = E\{X_m X_{m+n}\} = \frac{1}{2\pi j} \int_{c} \phi(z) z^{n-1} dz \qquad (5.8\text{-}2a)$$

$$\phi(z) = \sum_{n=-\infty}^{\infty} R(n) z^{-n} \qquad (5.8\text{-}2b)$$

In terms of the complex variable $\lambda = \omega + j\sigma$ discussed in Section 5.4, the correspondence between these two pairs involves $e^{j\lambda}$ and another complex variable $z = x + jy$. A comparison of Eqs. (5.8-1b) and (5.8-2b) shows that (at least formally),

$$z = x + jy = e^{j\lambda} = e^{j(\omega + j\sigma)} = e^{-\sigma} e^{j\omega} \qquad (5.8\text{-}3)$$

The situation is illustrated in Fig. 5.14 which shows the complex

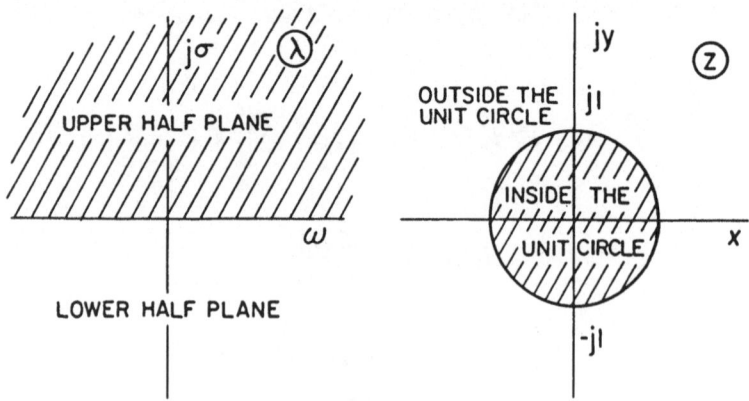

Fig. 5.14 - The λ-plane and the z-plane

λ-plane and the equivalent complex z-plane. The upper half λ-plane (UHP), defined by $\sigma > 0$, corresponds in the z-plane to the region inside the unit circle (IUC), defined by $|z| < 1$.

We now wish to factor the power spectral density $\phi(z)$ into two factors

$$\phi(z) = \phi^+(z)\phi^-(z) \qquad (5.8\text{-}4)$$

where $\phi^+(z)$ has all of the poles and zeroes of $\phi(z)$ that are inside the unit circle (IUC) and $\phi^-(z)$ has all the poles and zeroes of $\phi(z)$ that are outside the unit circle (OUC). It should be clear that

$$\phi^+(z) = \phi^-(1/z) \qquad (5.8\text{-}5)$$

and

$$\phi^-(z) = \phi^+(1/z) \qquad (5.8\text{-}6)$$

Suppose f_k is a discrete-time function which vanishes on the negative half-line; that is,

$$f_k = 0 \quad , \quad k < 0 \qquad (5.8\text{-}7)$$

If f_k is absolutely summable; that is, if

$$\sum_{k=-\infty}^{\infty} |f_k| = \sum_{k=0}^{\infty} |f_k| < \infty \qquad (5.8\text{-}8)$$

then the generating function

$$F(z) = \sum_{k=-\infty}^{\infty} f_k z^{-k} = \sum_{k=0}^{\infty} f_k z^{-k} \qquad (5.8\text{-}9)$$

exists everywhere on the unit circle $|z| = 1$ and everywhere *outside* the unit circle; that is, where $|z| > 1$. Hence the poles of $F(z)$ will all be *inside the unit circle* (IUC).

Suppose g_k is a function which vanishes on the positive half-line so that

$$g_k = 0 \quad , \quad k \geq 0 \qquad (5.8\text{-}10)$$

If g_k is absolutely summable; that is, if

$$\sum_{k=-\infty}^{\infty} |g_k| = \sum_{k=-\infty}^{-1} |g_k| < \infty \qquad (5.8\text{-}11)$$

then the generating function

$$G(z) = \sum_{k=-\infty}^{-1} g_k z^{-k} \qquad (5.8\text{-}12)$$

exists everywhere *on* the unit circle $|z| = 1$ and everywhere *inside* the unit circle; that is, where $|z| < 1$. Hence the poles of $G(z)$ will all be *outside the unit circle* (OUC).

In the discrete-time case, a prewhitening filter $H(z)$ for the process N_k with power spectral density $\phi_n(z)$ must satisfy

$$1 = \phi_n(z)H(z)H(1/z) \qquad (5.8\text{-}13)$$

or

$$1 = [\phi_n^+(z)H(z)][\phi_n^-(z)H(1/z)] \qquad (5.8\text{-}14)$$

From this last expression, we conclude that the prewhitening filter is

$$H(z) = \frac{1}{\phi_n^+(z)} \qquad (5.8\text{-}15)$$

The impulse response h_k of this filter will vanish for $k < 0$; hence $H(z)$ is physically realizable.

Example 5.11

In Example 4.2, we encountered the power spectral density

$$\phi_{zz}(z) = \sqrt{\frac{N_0}{2}\frac{e^\alpha}{e^\alpha - z^{-1}}}\sqrt{\frac{N_0}{2}\frac{e^\alpha}{e^\alpha - z}} = \phi_{zz}^+(z)\phi_{zz}^-(z)$$

where

$$\phi_{zz}^+(z) = \sqrt{\frac{N_0}{2}\frac{e^\alpha}{e^\alpha - z^{-1}}}$$

This factor has a pole at $z = e^{-\alpha}$, a real number which is less than unity since we assumed $\alpha > 0$. In the same way, we have

$$\phi_{zz}^-(z) = \sqrt{\frac{N_0}{2}\frac{e^\alpha}{e^\alpha - z}}$$

with a pole at $z = e^\alpha$ which is a real number greater than unity. These pole locations are shown in Fig. 5.15.

5.9 *Solution of the Integral Equation for the Discrete-Time Matched Filter* - We now proceed to solve Eq. (5.6-6) under a realizability constraint. We begin by changing the indices of summation to obtain a more convenient form. Let $q - j = m$ and $q - k = n$ so that Eq. (5.6-6) becomes

$$\sum_{m=0}^{\infty} R_n(n-m)h_m^\dagger = s_{q-n} \quad , \quad n = 0,1,\dots \qquad (5.9\text{-}1)$$

If $h_m^\dagger$ were physically-realizable, this expression would be

$$\sum_{m=-\infty}^{\infty} R_n(n-m)h_m^\dagger = s_{q-n} \quad , \quad n = 0,1,\dots \qquad (5.9\text{-}2)$$

Let us now define a discrete-time function g_n by

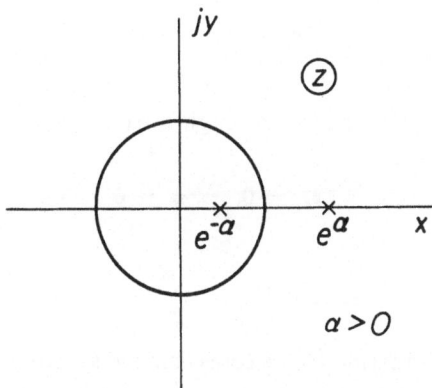

Fig. 5.15 - Pole locations of a power spectral density
for a discrete-parameter random process

$$g_n = s_{q-n} - \sum_{m=-\infty}^{\infty} R_n(n-m)h_m^\dagger \qquad (5.9\text{-}3)$$

It is clear that

$$g_n = 0 \ , \quad n \geq 0 \qquad (5.9\text{-}4)$$

and that the generating function

$$G(z) = \sum_{n=-\infty}^{-1} g_n z^{-n} \qquad (5.9\text{-}5)$$

has no singularities *inside* the unit circle. We take the generating
function of Eq. (5.9-3) term-by-term to obtain

$$G(z) = S(1/z)z^{-q} - \phi_n(z)H^\dagger(z) \qquad (5.9\text{-}6)$$

After factoring $\phi_n(z) = \phi_n^+(z)\phi_n^-(z)$ and rearranging we get

$$\frac{G(z)}{\phi_n^-(z)} = \frac{S(1/z)}{\phi_n^-(z)} z^{-q} - \phi_n^+(z)H^\dagger(z) \qquad (5.9\text{-}7)$$

We now consider the inverse transform of this equation term-by-
term to get

$$a_n = b_n - c_n \qquad (5.9\text{-}8)$$

where

$$a_n = \frac{1}{2\pi j} \int_c \frac{G(z)}{\phi_n^-(z)} z^{n-1} dz \qquad (5.9\text{-}9)$$

$$b_n = \frac{1}{2\pi j} \int_c \frac{S(1/z)}{\phi_n^-(z)} z^{-q} z^{n-1} dz \qquad (5.9\text{-}10)$$

and

$$c_n = \frac{1}{2\pi j} \int_c \phi_n^+(z) H^\dagger(z) z^{n-1} dz \qquad (5.9\text{-}11)$$

Since $a_n = 0$ for $n \geq 0$, we have

$$b_n = c_n \quad , \quad n \geq 0 \qquad (5.9\text{-}12)$$

Also, since $\phi_n^+(z) H^\dagger(z)$ has no singularities *outside* the unit circle, it follows that

$$c_n = 0 \quad , \quad n < 0 \qquad (5.9\text{-}13)$$

or

$$\phi_n^+(z) H^\dagger(z) = \sum_{n=0}^{\infty} c_n z^{-n} \qquad (5.9\text{-}14)$$

On replacing c_n in this last expression by b_n from Eq. (5.9-10), we have

$$H^\dagger(z) = \frac{1}{\phi_n^+(z)} \sum_{n=0}^{\infty} z^{-n} \frac{1}{2\pi j} \int_c \frac{S(1/z)}{\phi_n^-(z)} z^{-q} z^{n-1} dz \qquad (5.9\text{-}15)$$

as the explicit solution to the matched filter. Notice that this last expression could be written as

$$H^\dagger(z) = \frac{1}{\phi_n^+(z)} \left[\frac{S(1/z)}{\phi_n^-(z)} z^{-q} \right]_+ \qquad (5.9\text{-}16)$$

Here we have separated the sequence b_k into two parts so that

$$\sum_{k=-\infty}^{\infty} b_k z^{-k} = \sum_{k=-\infty}^{-1} b_k z^{-k} + \sum_{k=0}^{\infty} b_k z^{-k} \qquad (5.9\text{-}17)$$

with corresponding transform

$$\frac{S(1/z)}{\phi_n^-(z)} z^{-q} = \left[\frac{S(1/z)}{\phi_n^-(z)} z^{-q} \right]_- + \left[\frac{S(1/z)}{\phi_n^-(z)} z^{-q} \right]_+ \qquad (5.9\text{-}18)$$

where the first term on the right side contains all singularities outside the unit circle and the second term contains all singularities inside the unit circle.

As in the continuous-time case, the solution of Eq. (5.9-16) can be regarded as two linear systems in tandem. The first is a prewhitener for the input noise. The second is the remainder of the unrealizable matched filter transformed back into the time domain and made realizable by throwing away that part which does not vanish for negative time.

Part III - The Linear Least-Mean-Squared-Error Filter For Continuous-Time Inputs

5.10 *Formulation of the Linear Least-Mean-Squared-Error (LLMSE) Filtering and Prediction Problem in Continuous Time* - Many of the fundamental ideas of statistical optimization theory must be credited to Norbert Wiener and to his book *Extrapolation, Interpolation, and Smoothing of Stationary Time Series* [9]. The material of this book was first published in 1942 as a classified defense report and had a far-reaching impact on communications and control. Specifically, he considered least mean-squared-error linear filtering and prediction, although the influence of his work has extended over a much wider area.

We formulate the least mean-squared-error linear filtering and prediction problem in terms of the system of Fig. 5.16(a).

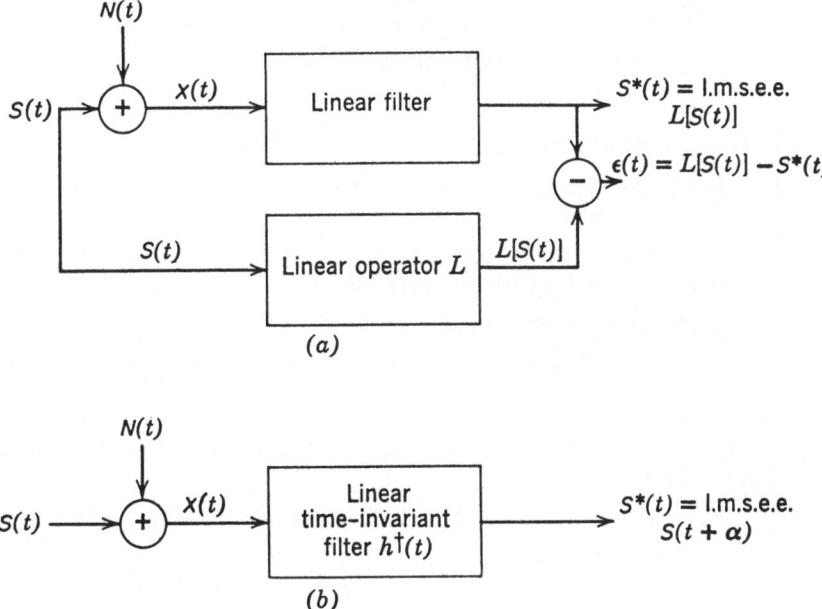

(a)

(b)

Fig. 5.16 - Linear least-mean-squared-error filtering and prediction in continuous time

Here a random signal $S(t)$ in an additive random noise $N(t)$ is to be sent through a linear filter designed such that its output $S^*(t)$ is an approximation to some desired linear operation $L[S(t)]$ on the input signal $S(t)$. The filter is to be designed so that the least mean-squared-error

$$E\{\epsilon^2(t)\} = E\{\,|\,L\,[S(t)] - S^*(t)\,|^{\,2}\} \qquad (5.10\text{-}1)$$

is as small as possible. We say that $S^*(t)$ is a linear least mean-squared- error estimate (LLMSEE) of $L\,[S(t)]$. In this treatment, we will restrict the problem somewhat further as shown in Fig. 5.16(b) although these restrictions are not necessary [10]. We shall require that the desired linear operator $L\,[S(t)]$ have a complex transfer function

$$L(\omega) = e^{\,j\omega\alpha} \qquad (5.10\text{-}2)$$

so that the desired filter output is $S(t + \alpha)$. It is convenient to distinguish three cases:

1. $\alpha > 0$ and $N(t) = 0$; the problem is one of the *pure prediction* of $S(t)$.

2. $\alpha > 0$ and $N(t) \neq 0$; the problem is one of the *prediction and estimation* of $S(t)$.

3. $\alpha \leq 0$ and $N(t) \neq 0$; the problem is one of the *estimation* of $S(t)$ with or without a time delay α in the estimate.

The specific restrictions on the input

$$X(t) = S(t) + N(t) \qquad (5.10\text{-}3)$$

and on the filter $h^\dagger(t)$ are:

1. Both the signal $S(t)$ and the noise $N(t)$ are zero-mean random processes which are at least wide-sense stationary.

2. The filter, with impulse response $h^\dagger(t)$, is to be time-invariant and physically realizable.

3. The mean-squared error

$$E\{\epsilon^2\} = E\{\,|\,S(t + \alpha) - S^*(t)\,|^{\,2}\} \qquad (5.10\text{-}4)$$

is to be a minimum.

We proceed now to find the filter $h^\dagger(t)$ which satisfies the condition of Eq. (5.10-4).

For any linear filter $h(t)$, Eq. (5.10-4) can be written as

$$E\{\epsilon^2\} = E\{\,|\,S(t + \alpha) - \int_0^\infty h(\tau)X(t-\tau)d\tau\,|^{\,2}\} \qquad (5.10\text{-}5)$$

This expression may be expanded and the expectation operator taken inside the integral to yield

$$E\{\epsilon^2\} = R_{ss}(0) - 2\int_0^\infty R_{xs}(\tau + \alpha)h(\tau)d\tau$$

$$\qquad (5.10\text{-}6)$$

$$+ \int_0^\infty h(\sigma)\int_0^\infty R_{xx}(\sigma - \tau)h(\tau)\,d\tau\,d\sigma$$

where $R_{ss}(\tau)$ is the autocorrelation function of the signal $s(t)$ and is given by

$$R_{ss}(\tau) = E\{S(t)\,S(t+\tau)\} \tag{5.10-7}$$

$R_{xs}(\tau)$ is the cross correlation function of the total input $X(t)$ and the input signal $S(t)$ and is given by

$$R_{xs}(\tau) = E\{X(t)S(t+\tau)\} = E\{[S(t)+N(t)]\,S(t+\tau)\} \tag{5.10-8}$$

$$= R_{ss}(\tau) + R_{ns}(\tau)$$

and $R_{xx}(\tau)$ is the autocorrelation function of the total input $X(t)$:

$$R_{xx}(\tau) = E\{X(t)X(t+\tau)\}$$

$$= E\{[S(t)+N(t)][S(t+\tau)+N(t+\tau)]\}$$

$$= R_{ss}(\tau) + R_{ns}(\tau) + R_{sn}(\tau) + R_{nn}(\tau) \tag{5.10-9}$$

The problem is to choose $h^{\dagger}(t) = h(t)$ such that Eq. (5.10-6) is a minimum.

We follow the procedure previously outlined in Section 5.2 and discussed in Appendix F. We assume that the filter $h^{\dagger}(t)$ to make Eq. (5.10-6) a minimum is known. We replace $h^{\dagger}(\bullet)$ by $h^{\dagger}(\bullet) + \gamma\delta h(\bullet)$ where γ is a real variable and $\delta h(\bullet)$ is an arbitrary variation in $h^{\dagger}(\bullet)$. With this substitution the mean-squared-error given by Eq. (5.10-6) becomes a function of γ and can be written as

$$E\{\epsilon^2(\gamma)\} = J(\gamma) \tag{5.10-10}$$

where

$$J(0) = E\{\epsilon^2(0)\} = E\{\epsilon^2\} \tag{5.10-11}$$

If the function $J(\gamma)$ has a minimum, it will occur when $h(\bullet) = h^{\dagger}(\bullet)$. A necessary condition for this minimum to occur is that

$$\frac{\partial}{\partial\gamma}\,J(\gamma)\,|_{\gamma=0} = 0 \tag{5.10-12}$$

Let us use Eq. (5.10-6) to write

$$J(\gamma) = R_{ss}(0) - 2\int_0^\infty R_{xs}(\tau+\alpha)[h(\tau)+\gamma\delta h(\tau)]\,d\tau \tag{5.10-13}$$

$$+ \int_0^\infty [h(\sigma)+\gamma\delta h(\sigma)]\int_0^\infty R_{xx}(\sigma-\tau)[h(\tau)+\gamma\delta h(\tau)]\,d\tau\,d\sigma$$

On noting that $R(\sigma-\tau) = R(\tau-\sigma)$, we can rearrange this expression to give

$$J(\gamma) = J(0) + 2\gamma C + \gamma^2 D \qquad (5.10\text{-}14)$$

where $J(0)$ has been given by Eq. (5.10-6) and where

$$C = \int\limits_0^\infty \delta h\,(\tau)[-R_{zs}\,(\tau + \alpha) + \int\limits_0^\infty R_{zz}\,(\tau - \sigma)h\,(\sigma)d\,\sigma]\ d\,\tau \quad (5.10\text{-}15)$$

and

$$D = \int\limits_0^\infty \delta h\,(\tau) \int\limits_0^\infty R_{zz}\,(\sigma - \tau)\delta h\,(\sigma)\ d\,\sigma\,d\,\tau \qquad (5.10\text{-}16)$$

On applying Eq. (5.10-12) to Eq. (5.10-14), we have that

$$C = 0 = \int\limits_0^\infty \delta h\,(\tau)[-R_{zs}\,(\tau + \alpha) + \int\limits_0^\infty R_{zz}\,(\tau - \sigma)h\,(\sigma)d\,\sigma]d\,\tau \qquad (5.10\text{-}17)$$

As before we argue that the term in square brackets must be zero for this relation to be satisfied for arbitrary variation $\delta h\,(\tau)$; that is,

$$\int\limits_0^\infty R_{zz}\,(\tau - \sigma)h^\dagger(\sigma)d\,\sigma = R_{zs}\,(\tau + \alpha) \quad , \quad 0 \le \tau < \infty \quad (5.10\text{-}18)$$

is the integral equation that must be solved to obtain the least mean-squared-error filter $h^\dagger(t)$.

We have shown that Eq. (5.10-18) is a necessary condition for the mean-squared error $J(\gamma)$ to be a minimum. It is easily shown that it is also sufficient. Note that the quantity D is non-negative since Eq. (5.10-16) can be written as

$$D = E\{|\int\limits_0^\infty X(\tau)\,\delta h\,(\tau)d\,\tau|^2\} \ge 0 \qquad (5.10\text{-}19)$$

Since $C = 0$ if Eq. (5.10-18) is satisfied, it follows that

$$J(\gamma) \ge J(0) \qquad (5.10\text{-}20)$$

In other words the mean-squared-error cannot be decreased when $h^\dagger(t)$ is allowed to vary from its value determined from Eq. (5.10-18).

It is clear from Eq. (5.10-18) that the optimum filter will depend only on the correlation functions or power spectral densities of the input signal and noise. Two sets of inputs with the same power spectral densities will require the same optimum filters. This fact is a result of the mean-squared-error criterion and the restriction to a linear filter.

5.11 *The Unrealizable LLMSE Filter in Continuous Time* -
Let us temporarily ignore the physical-realizability requirement
and replace the bottom limit of Eq. (5.10-18) by $-\infty$ without however insuring that $h^\dagger(t) = 0$, $t < 0$. This is the same procedure
as that followed in Section 5.3 for the matched filter. We may
now transform both sides of Eq. (5.10-8) and solve for $H^\dagger(\omega)$ to
obtain

$$H_u^\dagger(\omega) = \frac{\phi_{zs}(\omega)}{\phi_{zz}(\omega)} e^{j\omega\alpha} \qquad (5.11\text{-}1)$$

where the subscript u is used to denote the fact that $H_u^\dagger(\omega)$ is
optimum but unrealizable; that is, it possesses, in general, singularities in the lower half λ-plane and, hence, it inverse Fourier
transform does not vanish for negative time. This equation is the
equivalent of Eq. (5.3-6) for the matched filter.

A special case which occurs frequently in that where the
input signal $S(t)$ and the input noise $N(t)$ are uncorrelated so
that

$$R_{zz}(\tau) = R_{ss}(\tau) + R_{nn}(\tau) \qquad (5.11\text{-}2)$$

$$\phi_{zz}(\omega) = \phi_{ss}(\omega) + \phi_{nn}(\omega) \qquad (5.11\text{-}3)$$

$$R_{zs}(\tau) = R_{ss}(\tau) \qquad (5.11\text{-}4)$$

$$\phi_{zs}(\omega) = \phi_{ss}(\omega) \qquad (5.11\text{-}5)$$

In this case, Eq. (5.11-1) reduces to

$$H_u^\dagger(\omega) = \frac{\phi_{ss}(\omega)}{\phi_{ss}(\omega) + \phi_{nn}(\omega)} e^{j\omega\alpha} \qquad (5.11\text{-}6)$$

For the noiseless case $[\phi_{nn}(\omega) = 0]$, both Eqs. (5.11-1) and (5.11-6)
become

$$H_u^\dagger(\omega) = e^{j\omega\alpha} \qquad (5.11\text{-}7)$$

a pure advance or delay, depending on whether α is positive or
negative. This result is certainly to be expected.

The expression for the mean-squared error for the optimum
unrealizable filter is given by Eq. (5.10-6) with $h(\bullet) = h_u^\dagger(\bullet)$. This
equation may be simplified by the use of Eq. (5.10-18) to yield

$$\min E\}\epsilon^2\}_u = R_{ss}(0) - \int_{-\infty}^{\infty} R_{zs}(\tau + \alpha) h_u^\dagger(\tau) d\tau \qquad (5.11\text{-}8)$$

or, on transforming,

$$\min E\{\epsilon^2\}_u = \frac{1}{2\pi} \int_{-\infty}^{\infty} [\phi_{ss}(\omega) - \phi_{zs}(\omega)H_u^\dagger(-\omega) e^{j\omega\alpha}] d\omega \qquad (5.11\text{-}9)$$

If the expression for $H_u^\dagger(\omega)$, as given by Eq. (5.11-1), is substituted
into this last expression, the result is

$$\min E\{\epsilon^2\}_* = \frac{1}{2\pi} \int\limits_{-\infty}^{\infty} \frac{\phi_{ss}(\omega)\phi_{zz}(\omega) - |\phi_{zs}(\omega)|^2}{\phi_{zz}(\omega)} \, d\omega \qquad (5.11\text{-}10)$$

For the special case where $S(t)$ and $N(t)$ are uncorrelated, we have

$$\min E\{\epsilon^2\}_* = \frac{1}{2\pi} \int\limits_{-\infty}^{\infty} \frac{\phi_{ss}(\omega)\,\phi_{nn}(\omega)}{\phi_{ss}(\omega) + \phi_{nn}(\omega)} \, d\omega \qquad (5.11\text{-}11)$$

Example 5.12

Let the input signal $S(t)$ be a wide-sense stationary process with autocorrelation function

$$R_{ss}(\tau) = \frac{a^2}{2b} \, e^{-|\tau|/b}$$

and power spectral density

$$\phi_{ss}(\omega) = \frac{a^2}{b^2\omega^2 + 1}$$

where a and b are positive numbers. Let the input noise $N(t)$ be independent and white with autocorrelation function

$$R_{nn}(\tau) = c^2\delta(\tau)$$

and power spectral density

$$\phi_{nn}(\omega) = c^2$$

Then

$$\phi_{zz}(\omega) = \phi_{ss}(\omega) + \phi_{nn}(\omega) = \frac{(a^2 + c^2) + b^2c^2\omega^2}{b^2\omega^2 + 1}$$

The optimum unrealizable filter has a complex transfer function given by Eq. (5.11-6) or

$$H_*^\dagger(\omega) = \frac{a^2}{b^2c^2} \frac{e^{j\omega\alpha}}{\dfrac{a^2 + c^2}{b^2c^2} + \omega^2}$$

which is easily inverted to yield

$$h_*^\dagger(t) = \frac{a^2}{2bc\sqrt{a^2+c^2}} \, e^{-\frac{\sqrt{a^2+c^2}}{bc}|t + \alpha|}$$

the impulse response for the least-mean-squared-error unrealizable filter. This function is plotted in Fig. 5.17 for various values of the prediction time α. It is apparent that the filter is unrealizable unless $\alpha \to -\infty$ (the infinite lag filter).

The mean-squared error can be found from Eq. (5.11-11) to be

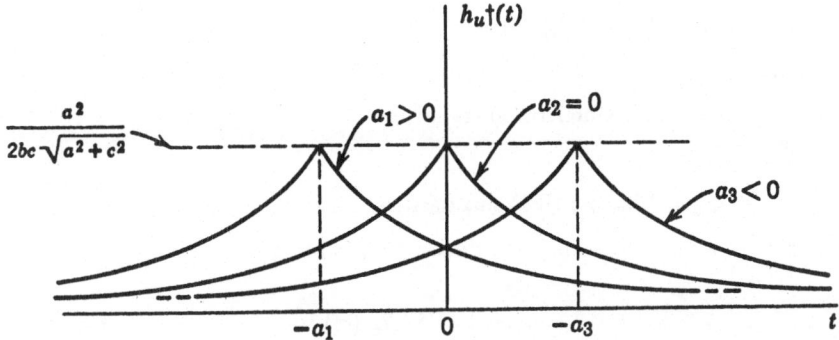

Fig. 5.17 - An unrealizable impulse response

$$\min E\{\epsilon^2\}_* = \frac{1}{2\pi} \int_{-\infty}^{\infty} \frac{a^2 c^2}{(a^2+c^2)+b^2 c^2 \omega^2} \, d\omega = \frac{a^2 c}{2b \sqrt{a^2+c^2}}$$

since

$$\int_{-\infty}^{\infty} \frac{a}{a^2+x^2} \, dx = \begin{cases} \pi & , \ a > 0 \\ 0 & , \ a = 0 \\ -\pi & , \ a < 0 \end{cases}$$

5.12 *Solution of the Integral Equation for the Continuous-Time LLMSE Filter* - The solution of Eq. (5.10-8) is readily apparent from the work already done with the matched filter. Notice that Eq. (5.10-18) is of the same form as Eq. (5.2-19) which specified the matched filter. This similarity may be more obvious if we examine Eq. (5.5-3) which results after a linear change of variables in Eq. (5.2-19). The form of this latter expression is identical to that of Eq. (5.10-18) where $R_{xx}(\tau - \sigma)$ corresponds to $R_n(t - u)$, and $R_{ss}(\tau + \alpha)$ corresponds to $s(t_1 - t)$. In Eq. (5.5-3) it has already been required that $h^\dagger(\bullet)$ be physically realizable so that the lower limit on the integral can be made $-\infty$ instead of zero, but this assumption must also be made for Eq. (5.10-18). It follows, then, that the solution for the least-mean-squared-error filter is given by Eq. (5.9-15) or, after changing notation, by

$$H^\dagger(\omega) = \frac{1}{2\pi\,\phi_{zz}^+(\omega)} \int_0^\infty e^{-j\omega t} \int_{-\infty}^\infty \frac{\phi_{zs}(\nu)}{\phi_{zz}^-(\nu)}\, e^{j\nu(\alpha+t)} d\nu\, dt \quad (5.12\text{-}1)$$

Here the input power spectral density $\phi_{zz}(\omega)$ is given by

$$\phi_{zz}(\omega) = \mathbf{F}\{R_{zz}(\tau)\} = \phi_{zz}^+(\omega)\,\phi_{zz}^-(\omega) \quad (5.12\text{-}2)$$

where

$\phi_{zz}^+(\omega)$ has no LHP singularities

and

$\phi_{zz}^-(\omega)$ has no UHP singularities

and where the cross spectral density $\phi_{zs}(\omega)$ is given by

$$\phi_{zs}(\omega) = \mathbf{F}\{R_{zs}(\tau)\} \quad (5.12\text{-}3)$$

Example 5.13

We consider the same inputs as for Example 5.12 except that we now find the least-mean-squared-error *realizable* filter. Recall that the input spectral density was

$$\phi_{zz}(\omega) = \frac{b^2 c^2 \omega^2 + (a^2+c^2)}{b^2 \omega^2 + 1}$$

This spectral density may be factored as

$$\phi_{zz}(\omega) = \phi_{zz}^+(\omega)\phi_{zz}^-(\omega)$$

where

$$\phi_{zz}^+(\lambda) = \frac{bc\,\lambda - j\sqrt{a^2+c^2}}{b\,\lambda - j}$$

is analytic, bounded, and free of poles and zeroes in the lower half λ-plane (LHP) and where

$$\phi_{zz}^-(\lambda) = \frac{bc\,\lambda + j\sqrt{a^2+c^2}}{b\,\lambda + j}$$

is analytic, bounded, and free of poles and zeroes in the upper half λ-plane (UHP). We form the function

$$\frac{\phi_{zs}(\nu)}{\phi_{zz}^-(\nu)} = \frac{\phi_{ss}(\nu)}{\phi_{zz}^-(\nu)} = \frac{K}{k_1+j\,\nu} + \frac{K}{k_2-j\,\nu}$$

where

$$K = \frac{a^2}{b\,(c+\sqrt{a^2+c^2})}$$

$$k_1 = 1/b$$

$$k_2 = \sqrt{a^2 + c^2}/bc$$

We are now in a position to evaluate the inner integral of Eq. (5.12-1). We have

$$\frac{1}{2\pi} \int_{-\infty}^{\infty} \frac{1}{k_1 + j\nu} \, e^{j\nu(t+\alpha)} d\nu = u(t+\alpha)e^{-k_1(t+\alpha)}$$

and

$$\frac{1}{2\pi} \int_{-\infty}^{\infty} \frac{1}{k_2 - j\nu} \, e^{j\nu(t+\alpha)} d\nu = u(-t-\alpha)e^{k_2(t+\alpha)}$$

from Pair No. 3 of Table 3.1 or from contour integration techniques. Hence

$$\frac{1}{2\pi} \int_{-\infty}^{\infty} \frac{\phi_{zs}(\nu)}{\phi_{zz}^-(\nu)} \, e^{j\nu(t+\alpha)} d\nu = u(-t-\alpha)e^{k_2(t+\alpha)}$$

$$\text{(A)}$$

$$+ \, u(t+\alpha)e^{-k_1(t+\alpha)}$$

This expression is similar to Fig. 5.17 except that the two sides of the exponential have different time constants $1/k_1$ and $1/k_2$. We must now perform the last integration in Eq. (5.12-1). It is clear from Fig. 5.17 that we may distinguish three cases, namely those where $\alpha = 0$, $\alpha < 0$, and $\alpha > 0$.

Case I - Zero-Lag Filter ($\alpha = 0$)

We need only consider the term

$$u(t)e^{-k_1 t}$$

on the right side of Eq. (A) since the other term

$$u(-t)e^{k_2 t}$$

vanishes for positive time. The integration is easily carried out to give

$$\int_0^{\infty} e^{-j\omega t} e^{-k_1 t} \, dt = \frac{1}{k_1 + j\omega}$$

or finally

$$H^\dagger(\omega) = \frac{1}{\phi_{rr}^+(\omega)} \frac{K}{k_1 + j\omega} = \frac{k_1 + j\omega}{c(k_2 + j\omega)} \frac{K}{k_1 + j\omega} = \frac{K}{(k_2 + j\omega)c}$$

$$H^\dagger(\omega) = \frac{a^2}{c + \sqrt{a^2 + c^2}} \frac{1}{\sqrt{a^2 + c^2} + j\omega bc}$$

This last expression is the complex transfer function of an $R-C$ low pass filter preceded by an ideal attenuator as shown in Fig. 5.18.

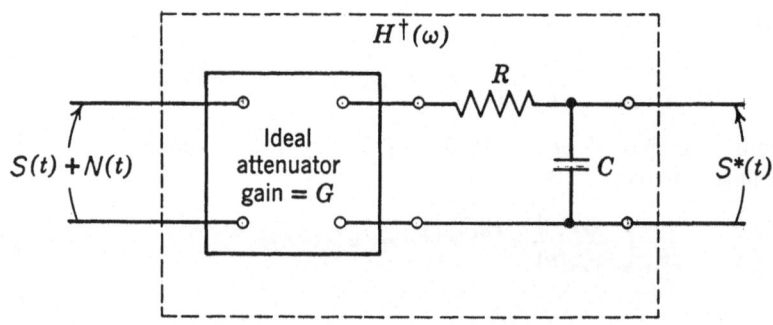

$$H^\dagger(\omega) = \frac{G}{1+j\,\omega RC}$$

$$G = \frac{a^2}{a^2+c^2+c\,\sqrt{a^2+c^2}} \leq 1$$

$$RC = \frac{bc}{\sqrt{a^2+c^2}}$$

Fig. 5.18 - A least mean-squared-error zero-lag filter

Case II - Lag Filter ($\alpha < 0$)

On examining Eq. (A), we conclude that the t-integral in Eq. (5.12-1) can be written as

$$\int_0^{-\alpha} e^{k_2(t+\alpha)}e^{-j\omega t}\,dt + \int_{-\alpha}^{\infty} e^{-k_1(t+\alpha)}e^{-j\omega t}\,dt \tag{B}$$

It should be kept in mind that $\alpha < 0$. Perhaps it would be clearer to define a positive number β by

$$\beta = -\alpha > 0$$

Now expression (B) becomes

$$\int_0^{\beta} e^{(k_2-j\omega)t}\,e^{-k_2\beta}\,dt + \int_{\beta}^{\infty} e^{-(k_1+j\omega)t}\,e^{k_1\beta}\,dt$$

$$= \frac{(k_1+k_2)e^{-j\omega\beta}-(k_1+j\omega)e^{-k_2\beta}}{(k_2-j\omega)(k_1+j\omega)}$$

Finally, the optimum filter is obtained by multiplying this last expression by $K/\phi_{zz}^+(\omega)$. After simplification, we obtain

$$H^\dagger(\omega) = \frac{a^2}{c^2}\frac{k_1^2}{k_1+k_2}\frac{(k_1+k_2)e^{-j\omega\beta}-(k_1+j\omega)e^{-k_2\beta}}{k_2^2+\omega^2}$$

where β is a positive number and is the desired delay. This filter is not a rational function of ω and cannot be synthesized exactly with a finite number of lumped, linear elements, although it may be approximated arbitrarily closely. For $\beta=\alpha=0$, the expression reduces to that obtained in Case I.

Case III - Prediction Filter ($\alpha > 0$)

It is clear from Eq. (A) that only the term

$$e^{-k_1(t+\alpha)}$$

exists for positive time. Consequently we have,

$$\int_0^\infty e^{-k_1(t+\alpha)}e^{-j\omega t}\,dt = \frac{e^{-k_1\alpha}}{k_1+j\omega}$$

The transfer function of the least-mean-squared-error prediction filter is

$$H^\dagger(\omega) = \frac{K}{\phi_{zz}^+(\omega)}\frac{e^{-k_1\alpha}}{k_1+j\omega}$$

or

$$H^\dagger(\omega) = \frac{K}{c}\frac{k_1+j\omega}{k_2+j\omega}\frac{e^{-k_1\alpha}}{k_1+j\omega}$$

$$= e^{-k_1\alpha}\frac{a^2}{a^2+c^2+c\sqrt{a^2+c^2}}\frac{1}{1+j\omega\dfrac{bc}{\sqrt{a^2+c^2}}}$$

It is convenient to rewrite this last expression as

$$H^\dagger(\omega) = G_1(\alpha)\frac{G}{1+j\omega RC}$$

where $G_1(\alpha)$ is an attenuation factor given by

$$G_1(\alpha) = e^{-k_1\alpha}$$

and where G and RC have been defined in Fig. 5.18. Note that $G_1(0)$ is unity and that the optimum prediction filter is just the zero-lag filter of Case I and Fig. 5.18 with another series attenuator $G_1(\alpha)$. As $\alpha \to \infty$, the transfer function approaches zero.

5.13 *The Mean-Squared Error for the Continuous-Time Case* - A general expression for the mean-squared error for the least-mean-squared-error filter is given by Eq. (5.10-6) when $h(\bullet)$ is replaced by $h^\dagger(\bullet)$ as given by Eq. (5.10-18). The result is

$$\min E\{\epsilon^2\} = R_{ss}(0) - \int_0^\infty R_{zs}(\tau + \alpha)h^\dagger(\tau)d\tau \qquad (5.13\text{-}1)$$

corresponding to Eq. (5.11-8) for the unrealizable filter. After transforming, Eq. (5.13-1) becomes

$$\min E\{\epsilon^2\} = \frac{1}{2\pi}\int_{-\infty}^\infty [\phi_{ss}(\omega)-\phi_{zs}(\omega)H^\dagger(-\omega)e^{j\omega\alpha}]d\omega \qquad (5.13\text{-}2)$$

corresponding to Eq. (5.11-9). For the case where the signal $S(t)$ and the noise $N(t)$ are uncorrelated so that Eq. (5.10-25) and (5.10-27) apply, the last expression may be written as

$$\min E\{\epsilon^2\} = \frac{1}{2\pi}\int_{-\infty}^\infty \phi_{ss}(\omega)[1-H^\dagger(-\omega)e^{j\omega\alpha}]d\omega \qquad (5.13\text{-}3)$$

Actually, in some ways, it is more instructive to transform Eq. (5.10-6) term-by-term without using Eq. (5.10-18). The result, in the general case, is

$$\min E\{\epsilon^2\} = \qquad (5.13\text{-}4)$$

$$\frac{1}{2\pi}\int_{-\infty}^\infty [\phi_{ss}(\omega)-2\phi_{zs}(\omega)H^\dagger(-\omega)e^{j\omega\alpha} +\phi_{zz}(\omega)\,|\,H^\dagger(\omega)\,|^2]d\omega$$

and, for uncorrelated signal and noise,

$$\min E\{\epsilon^2\} = \qquad (5.13\text{-}5)$$

$$\frac{1}{2\pi}\int_{-\infty}^\infty [\phi_{ss}(\omega)\,|\,1-H^\dagger(-\omega)e^{j\omega\alpha}\,|^2 + \phi_{nn}(\omega)\,|\,H^\dagger(\omega)\,|^2]d\omega$$

These last two equations are just slightly disguised forms of Eqs. (5.13-2) and (5.13-3), respectively. However, in Eq. (5.13-5), two distinct error terms are readily identified. The first part of the integral, involving the integrand

$$\phi_{ss}(\omega)\,|\,1-H^\dagger(-\omega)e^{j\omega\alpha}\,|^2$$

gives the error that results from the filter's distortion of the signal. The second part, involving the integrand

$$\phi_{nn}(\omega)\,|\,H^\dagger(\omega)\,|^2$$

gives the error that results from the input noise.

Let us return to Eq. (5.13-1) for the minimum mean-squared error. We will call the last term $G(\alpha)$ and show that it is non-negative; that is,

$$G(\alpha) = \int\limits_0^\infty R_{ze}(\tau+\alpha)h^\dagger(\tau)d\tau \geq 0 \qquad (5.13\text{-}6)$$

for all α. This is easily seen by comparing Eqs. (5.13-1) and (5.10-6). From these equations, it is clear that

$$G(\alpha) = \int\limits_0^\infty R_{ze}(\tau+\alpha)h^\dagger(\tau)d\tau = \int\limits_0^\infty h^\dagger(\sigma)\int\limits_0^\infty R_{zz}(\sigma-\tau)h^\dagger(\tau)d\tau\,d\sigma$$

$$= E\left\{ \left[\int\limits_0^\infty X(\tau)h^\dagger(\tau)d\tau\right]^2 \right\} \geq 0 \qquad (5.13\text{-}7)$$

which proves Eq. (5.13-6). Suppose we now rewrite Eq. (5.13-1) with a linear change of variable $x = \tau + \alpha$ and note that $\min E\{\epsilon^2\}$ must be non-negative by definition:

$$\min E\{\epsilon^2\} = R_{ss}(0) - \int\limits_\alpha^\infty R_{ze}(x)h(x-\alpha)dx \geq 0 \qquad (5.13\text{-}8)$$

Now this expression is non-negative and is the difference of two non-negative terms, $R_{ss}(0)$ and $G(\alpha)$; therefore it will be a minimum when the term $G(\alpha)$ is a maximum and a maximum when $G(\alpha)$ is a minimum. It will be an absolute maximum when $G(\alpha)$ is zero:

$$abs \ \max\left[\min\ E\{\epsilon^2\}\right] = R_{ss}(0) \qquad (5.13\text{-}9)$$

Note that $R_{ss}(0)$ is just the normalized signal power. It is clear from Eq. (5.13-6) that this absolute maximum can always be obtained by making $h(t)$ identically zero for all $t \in (0,\infty)$ or by allowing $\alpha \to \infty$.

Although it is not obvious, the function $G(\alpha)$ is monotone non-increasing in α, that is,

$$G(\alpha_2) \leq G(\alpha_1) \ , \quad \alpha_2 > \alpha_1 \qquad (5.13\text{-}10)$$

It follows then from Eq. (5.13-1) that the minimum mean-squared error $\min E\{\epsilon^2\}$ is monotone non-decreasing in α since the term $R_{ss}(0)$ is a constant. In other words, increasing α (the prediction time) increases the error, and the minimum value of the minimum mean-squared error is obtained when $\alpha \to -\infty$ (the infinite lag filter):

$$\lim_{\alpha \to -\infty} \min E\{\epsilon^2\} = \min\left[\min\ E\{\epsilon^2\}\right] \qquad (5.13\text{-}11)$$

$$\lim_{\alpha \to \infty} \min E\{\epsilon^2\} = \max\left[\min\ E\{\epsilon^2\}\right] = R_{ss}(0) \qquad (5.13\text{-}12)$$

The relationship of Eq. (5.13-10) may be proved as follows. We write $G(\alpha)$ from Eq. (5.13-6) in the form

$$G(\alpha) = \int_0^\infty R_{ze}(\tau+\alpha)h^\dagger(\tau)d\tau = \int_{-\infty}^\infty R_{ze}(\tau+\alpha)h^\dagger(\tau)d\tau$$

$$= \int_0^\infty R_{ze}(\tau+\alpha)\frac{1}{2\pi}\int_{-\infty}^\infty H^\dagger(\omega)e^{j\omega\tau}d\omega d\tau \qquad (5.13\text{-}13)$$

Now we substitute Eq. (5.12-1) for $H^\dagger(\omega)$:

$$G(\alpha) = \int_{-\infty}^\infty R_{ze}(\tau+\alpha)\frac{1}{2\pi}\int_{-\infty}^\infty e^{j\omega\tau}\frac{1}{\phi_{zz}^+(\omega)}\int_0^\infty e^{-j\omega t}$$

$$\times \frac{1}{2\pi}\int_{-\infty}^\infty \frac{\phi_{ze}(\nu)}{\phi_{zz}^-(\nu)}e^{j\nu(t+\alpha)}\,d\nu dt d\omega d\tau \qquad (5.13\text{-}14)$$

Let us now define the function $K(\nu)$ by

$$K(\nu) = \frac{\phi_{ze}(\nu)}{\phi_{zz}^-(\nu)} \qquad (5.13\text{-}15)$$

with inverse Fourier transform

$$k(t) = \frac{1}{2\pi}\int_{-\infty}^\infty K(\nu)e^{j\nu t}\,d\nu \qquad (5.13\text{-}16)$$

We see now that the integral with respect to ν becomes $k(t+\alpha)$. We may interchange the order of integration to obtain

$$G(\alpha) = \int_0^\infty k(t+\alpha)\frac{1}{2\pi}\int_{-\infty}^\infty \frac{e^{-j\omega t}}{\phi_{zz}^+(\omega)}\int_{-\infty}^\infty R_{ze}(\tau+\alpha)e^{j\omega\tau}\,d\tau d\omega dt \qquad (5.13\text{-}17)$$

or, since the inner integral is $\phi_{ze}(-\omega)e^{-j\omega\alpha}$,

$$G(\alpha) = \int_0^\infty k(t+\alpha)\frac{1}{2\pi}\int_{-\infty}^\infty K(-\omega)e^{-j\omega(t+\alpha)}d\omega\,dt \qquad (5.13\text{-}18)$$

Finally, we have

$$G(\alpha) = \int_0^\infty [k(t+\alpha)]^2 dt \qquad (5.13\text{-}19)$$

or

$$G(\alpha) = \int_\alpha^\infty [k(t)]^2 dt \qquad (5.13\text{-}20)$$

Since the integrand in Eq. (5.13-20) is non-negative for all t, it follows that $G(\alpha)$ is monotone non-increasing in α as previously stated.

Let us now consider more closely the minimum value of the mean-squared error. This minimum is obtained when $\alpha \to -\infty$, where the function $G(\alpha)$ becomes

$$G(-\infty) = \int_{-\infty}^{\infty} [k(t)]^2 dt = \frac{1}{2\pi} \int_{-\infty}^{\infty} |K(\omega)|^2 d\omega \quad (5.13\text{-}21)$$

The last relationship follows from Parseval's Theorem or from interchanging the order of integration in Eq. (5.13-18). Since $K(\omega)$ is defined by Eq. (5.13-15), we have

$$\min[\min E\{\epsilon^2\}] = R_{ss}(0) - G(-\infty)$$

$$= \frac{1}{2\pi} \int_{-\infty}^{\infty} \left[\phi_{ss}(\omega) - \frac{|\phi_{zs}(\omega)|^2}{\phi_{zz}(\omega)} \right] d\omega$$

$$= \frac{1}{2\pi} \int_{-\infty}^{\infty} \frac{\phi_{ss}(\omega)\phi_{zz}(\omega) - |\phi_{zs}(\omega)|^2}{\phi_{zz}(\omega)} d\omega \quad (5.13\text{-}22)$$

This last expression is identical to Eq. (5.11-10) for the minimum mean-squared error in the unrealizable case. The result is to be expected since ignoring realizability and allowing the lag to become infinite are equivalent.

Example 5.14

We return to the filter problem of Examples 5.12 and 5.13, where the function $K(\omega)$ is

$$K(\omega) = \frac{K}{k_1 + j\omega} + \frac{K}{k_2 - j\omega}$$

and where K, k_1, and k_2 have been defined in Example 5.13. The inverse transform of $K(\omega)$ is

$$k(t) = Ke^{k_2 t} u(-t) + Ke^{-k_1 t} u(t)$$

and its square is

$$[k(t)]^2 = K^2 e^{2k_2 t} u(-t) + K^2 e^{-2k_1 t} u(t)$$

as shown in Fig. 5.19(a). Thus the function $G(\alpha)$ is given by

$$G(\alpha) = \int_{\alpha}^{\infty} [k(t)]^2 dt$$

or

$$G(\alpha) = \begin{cases} (K^2/2k_2) \left[1 - e^{2k_2 \alpha}\right] + K^2/2k_1 & , \quad \alpha < 0 \\ K^2/2k_1 & , \quad \alpha = 0 \\ (K^2/2k_1) e^{-2k_1 \alpha} & , \quad \alpha > 0 \end{cases}$$

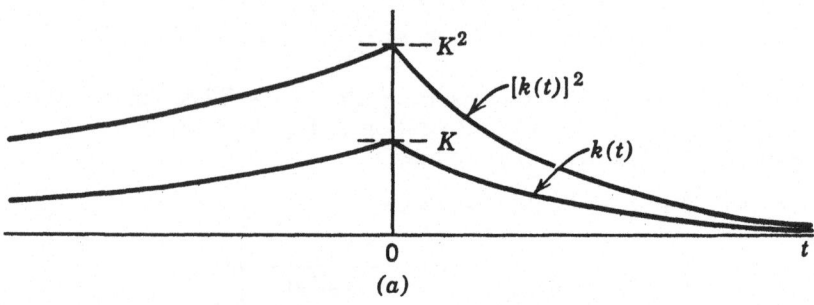

(a)

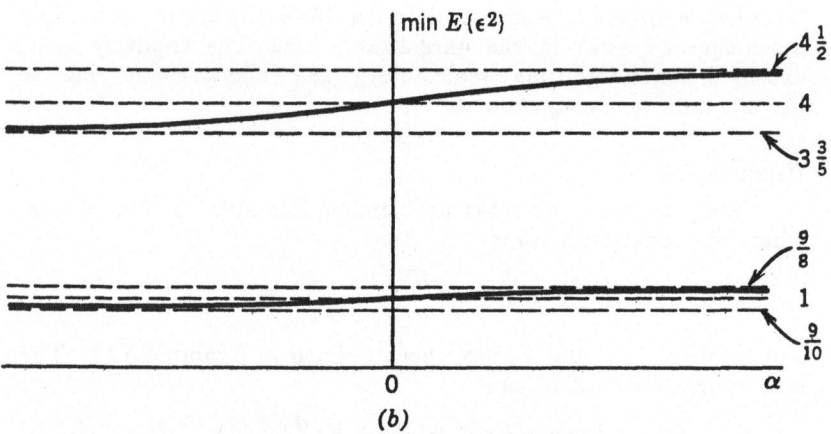

(b)

Fig. 5.19 - A least-mean-squared error calculation: (a) plot of $k(t)$
and $k^2(t)$; (b) minimum mean-squared error as a function of
prediction time α: $a=3$, $b=1$, $c=4$, and $a=3$, $b=4$, $c=4$

The minimum mean-squared error is

$$\min E\{\epsilon^2\} = R_{ss}(0) - G(\alpha) = \frac{a^2}{2b} - G(\alpha)$$

This expression is rather complicated. Note that, for $\alpha \to -\infty$, it
becomes

$$\min E\{\epsilon^2\}_{\alpha\to-\infty} = \frac{a^2}{2b} - \frac{K^2(k_1+k_2)}{2k_1k_2} = \frac{a^2c}{2b\sqrt{a^2+c^2}}$$

in agreement with the results obtained in Example 5.12. Also, for
$\alpha = 0$, we have

$$\min E\{\epsilon^2\}_{\alpha=0} = \frac{a^2}{2b} - \frac{K^2}{2k_1} = \frac{a^2c}{bc + b\sqrt{a^2+c^2}}$$

For convenience in plotting let us assume that
$$a = 3 \quad ; \quad c = 4$$
so that
$$K = k_1 = 1/b \quad ; \quad k_2 = 5/4b$$
Then the error is

$$\min E\{\epsilon^2\} = \begin{cases} (8/2b) - (2/5b)[1 - e^{\frac{5a}{2b}}] & , \quad \alpha < 0 \\[2em] 8/2b & , \quad \alpha = 0 \\[2em] (1/2b)[9 - e^{\frac{-2\alpha}{b}}] & , \quad \alpha > 0 \end{cases}$$

and is plotted in Fig. 5.19(b) for $b = 1$ and $b = 4$.

5.14 *The Pure Prediction Problem for the Continuous-Time Case* - In pure prediction, the input noise $N(t)$ is identically zero and

$$\phi_{zs}(\omega) = \phi_{ss}(\omega)$$
$$\phi_{zz}(\omega) = \phi_{ss}(\omega)$$

Hence Eq. (5.12-1) for the least-mean-squared-error filter becomes

$$H^\dagger(\omega) = \frac{1}{2\pi\phi_{ss}^+(\omega)} \int_0^\infty e^{-j\omega t} \int_{-\infty}^\infty \phi_{ss}^+(\nu)e^{j\nu(t+\alpha)}d\nu dt \quad (5.14\text{-}1)$$

The mean-squared error is given by Eqs. (5.13-1) and (5.13-20) as

$$\min E\{\epsilon^2\} = R_{ss}(0) - \int_\alpha^\infty [k(t)]^2 \, dt \quad (5.14\text{-}2)$$

where $k(t)$ is given by Eq. (5.13-16), and, for the noiseless case, is just

$$k(t) = \frac{1}{2\pi} \int_{-\infty}^\infty \phi_{ss}^+(\omega)e^{j\omega t} \, d\omega \quad (5.14\text{-}3)$$

Notice that $k(t)=0$ for $t < 0$ since $\phi_{ss}^+(\omega)$ has no LHP singularities. Also $R_{ss}(0)$ can be written as

$$R_{ss}(0) = \frac{1}{2\pi} \int_{-\infty}^\infty \phi_{ss}^+(\omega)\phi_{ss}^-(\omega)d\omega = \int_0^\infty [k(t)]^2 dt \quad (5.14\text{-}4)$$

Now Eq. (5.14-2) becomes

$$\min E\{\epsilon^2\} = \begin{cases} \int\limits_0^\alpha [k(t)]^2 dt & , \alpha > 0 \\ 0 & , \alpha \le 0 \end{cases} \tag{5.14-5}$$

Example 5.15

We consider Example 5.14 for the noiseless case. The signal power spectral density $\phi_{ss}(\omega)$ may be factored as

$$\phi_{ss}(\omega) = \phi_{ss}^+(\omega)\phi_{ss}^-(\omega) = \frac{a/b}{k_1+j\omega}\frac{a/b}{k_1-j\omega}$$

and

$$\frac{1}{2\pi}\int\limits_{-\infty}^\infty \frac{a/b}{k_1+j\nu} e^{j\nu(t+\alpha)}d\nu = \frac{a}{b}e^{-k_1(t+\alpha)}u(t+\alpha)$$

where $k_1 = 1/b$ has been defined previously. Now the optimum filter is

$$H^\dagger(\omega) = (k_1+j\omega)\int\limits_0^\infty e^{-j\omega t}\, e^{-k_1(t+\alpha)}\, u(t+\alpha)dt$$

For the case where $\alpha = -\beta \le 0$, this becomes

$$H^\dagger(\omega) = (k_1+j\omega)e^{k_1\beta}\int\limits_\beta^\infty e^{-(k_1+j\omega)t}\, dt = e^{-j\omega\beta}$$

a pure delay, as expected. For the prediction case where $\alpha > 0$, we have

$$H^\dagger(\omega) = (k_1+j\omega)e^{-k_1\alpha}\int\limits_0^\infty e^{-(k_1+j\omega)t}\, dt = e^{-k_1\alpha}$$

a constant attenuator. As $\alpha \to \infty$, the filter becomes an open-circuit. Of course the same results could be obtained by allowing $c \to 0$ in Example 5.13.

The mean-squared error is zero for $\alpha \le 0$ and, for $\alpha > 0$, it is given by

$$\min E\{\epsilon^2\} = \int\limits_0^\alpha [k(t)]^2 dt = \frac{a^2}{b^2}\int\limits_0^\alpha e^{-2k_1 t}\, dt$$

$$= \frac{a^2}{2b}\left[1 - e^{-2k_1\alpha}\right]$$

since

$$k(t) = \frac{a}{b}e^{-k_1 t}u(t)$$

At $\alpha = 0$, the error is zero; as $\alpha \rightarrow \infty$, the error becomes

$$\frac{a^2}{2b} = R_{ss}(0)$$

the total signal energy.

———————

Part IV - The Linear Least-Mean-Squared-Error Filter For Discrete-Time Inputs

5.15 *Formulation of the LLMSE Filtering and Prediction Problem in Discrete Time* - We formulate the problem to be considered here as shown in Fig. 5.20. We could, of course treat the more general case shown in Fig. 5.16(a), for continuous time, but we shall restrict ourselves to the discrete-time equivalent of Fig. 5.16(b).

$$S_k + N_k \longrightarrow \boxed{h_k} \longrightarrow S_k^* = S_{ok} + N_{ok}$$

$$E\{[S_{k+\alpha} - S_k^*]^2\} = a \text{ min}$$

Fig. 5.20 - Linear least-mean-squared-error filtering and prediction in discrete time

Here the random signal S_k in an additive noise N_k is sent through the linear time-invariant filter h_k designed so that the least-mean-squared error

$$E\{\epsilon^2\} = E\{[S_{k+\alpha} - S_k^*]^2\} \qquad (5.15\text{-}1)$$

is a minimum. As in the continuous-time case of Section 5.10, it will be assumed that the input

$$X_k = S_k + N_k \qquad (5.15\text{-}2)$$

is composed of two random processes $\{S_k; k = 0,\pm 1,\pm 2...\}$ and $\{N_k; k = 0,\pm 1,\pm 2...\}$ which are independent of each other and at least wide-sense stationary with zero means and autocorrelation functions

$$R_{ss}(m) = E\{S_k S_{k+m}\} \qquad (5.15\text{-}3)$$

and

$$R_{nn}(m) = E\{N_k N_{k+m}\} \qquad (5.15\text{-}4)$$

This assumption of a w.s.s input is a sufficient condition for the optimal filter $h_k{}^\dagger$ to be time-invariant.

Since the filter h_k is linear, Eq. (5.15-1) may be rewritten as

$$E\{\epsilon^2\} = E\{[S_{k+\alpha} - \sum_{n=0}^{\infty} h_n R_{k-n}]^2\} \qquad (5.15\text{-}5)$$

or, after squaring,

$$E\{\epsilon^2\} = \qquad (5.15\text{-}6)$$

$$R_{ss}(0) - 2 \sum_{n=0}^{\infty} R_{ss}(n+\alpha)h_n + \sum_{m=0}^{\infty} h_m \sum_{n=0}^{\infty} R_{xx}(m-n)h_n$$

where we have written

$$R_{ss}(0) = E\{S_{k+\alpha}^2\} = E\{S_k^2\} \qquad (5.15\text{-}7)$$

$$R_{xs}(n+\alpha) = E\{X_{k-n} S_{k+\alpha}\} = R_{ss}(n+\alpha) \qquad (5.15\text{-}8)$$

and

$$R_{xx}(m-n) = E\{X_n X_m\} = R_{ss}(m-n) + R_{nn}(m-n) \qquad (5.15\text{-}9)$$

We now apply exactly the same procedure to Eq. (5.15-6) as was applied to Eq. (5.10-6); that is, we replace h by $h + \gamma \delta h$ to form

$$E\{\epsilon^2(\gamma)\} = J(\gamma) \qquad (5.15\text{-}10)$$

as in Eq. (5.10-13). The result, after differentiating $J(\gamma)$ with respect to γ and setting the result equal to zero in the limit as $\gamma \to 0$, is the equation specifying the optimal filter. In analogy with Eq. (5.10-18), we obtain

$$\sum_{n=0}^{\infty} R_{xx}(m-n)h_n{}^\dagger = R_{ss}(m+\alpha) \quad , \quad m = 0,1,... \quad (5.15\text{-}11)$$

This last equation must be solved to find $h_k{}^\dagger$ the LLMSE filter in discrete time. As before, the equation is both a necessary and sufficient condition for $h_k{}^\dagger$ to be a least-mean-squared-error filter.

5.16 *The Unrealizable LLMSE Filter in Discrete Time* - As in Section 5.7, we may rewrite Eq. (5.15-11) as

$$\sum_{n=-\infty}^{\infty} R_{zz}(m-n)h_n{}^{\dagger} = R_{ss}(m+\alpha), \quad m = 0,1,2,\dots \quad (5.16\text{-}1)$$

where we have changed the bottom limit on the sum from zero to $-\infty$, and, of course, are ignoring realizability. Now, on transforming, we have

$$\sum_{m=-\infty}^{\infty} \sum_{n=-\infty}^{\infty} R_{zz}(m-n)h_n{}^{\dagger}z^{-m} = \sum_{m=-\infty}^{\infty} R_{ss}(m+\alpha)z^{-m} \quad (5.16\text{-}2)$$

or

$$\sum_{n=-\infty}^{\infty} h_n{}^{\dagger}z^{-n}\phi_{zz}(z) = H^{\dagger}(z)\phi_{zz}(z) = \phi_{ss}(z)z^{\alpha} \quad (5.16\text{-}3)$$

Now, finally, the unrealizable LLMSE filter is

$$H^{\dagger}(z) = \frac{\phi_{ss}(z)}{\phi_{zz}(z)}z^{\alpha} = \frac{\phi_{ss}(z)}{\phi_{ss}(z)+\phi_{nn}(z)}z^{\alpha} \quad (5.16\text{-}4)$$

In general, if we transform $H^{\dagger}(z)$, as determined by this last equation, the resulting impulse response $h_k{}^{\dagger}$ will not vanish for negative values of the index k.

Example 5.16

Let the input signal S_k be a wide-sense stationary random process with autocorrelation function

$$R_{ss}(n) = ae^{-b|n|}, \quad a,b > 0$$

and power spectral density

$$\phi_{ss}(z) = \frac{a(e^{2b}-1)}{(e^{b}-z)(e^{b}-z^{-1})}$$

Let the additive input noise N_k be white so that

$$R_{nn}(n) = \begin{cases} c^2 & , \quad n = 0 \\ 0 & , \quad n \neq 0 \end{cases}$$

and

$$\phi_{nn}(n) = c^2$$

We could now substitute these expressions to find the (unrealizable) optimal complex transfer function $H^{\dagger}(z)$. We could use the results of Example 4.2 to find the corresponding impulse response $h_k{}^{\dagger}$ and could plot $h_k{}^{\dagger}$ for $\alpha < 0$, $\alpha = 0$, and $\alpha > 0$ as was done in Example 5.12 and Fig. 5.17 for the continuous-time case. We suggest that this example be completed as Problem 5.21.

5.17 *Solution of the Integral Equation for the Discrete-Time LLMSE Filter* - We now solve Eq. (5.15-11) under a realizability constraint. If $h_n{}^\dagger$ were physically realizable, this equation would become

$$\sum_{n=-\infty}^{\infty} R_{zz}(m-n)h_n{}^\dagger = R_{ss}(m+\alpha) \quad , \quad m = 0,1,2,\dots \quad (5.17\text{-}1)$$

Let us define a discrete-time function g_m by

$$g_m = R_{ss}(m+\alpha) - \sum_{n=-\infty}^{\infty} R_{zz}(m-n)h_n{}^\dagger \qquad (5.17\text{-}2)$$

It is clear that

$$g_m = 0 \quad , \quad m \geq 0 \qquad (5.17\text{-}3)$$

and that the generating function

$$G(z) = \sum_{m=-\infty}^{-1} g_m z^{-n} \qquad (5.17\text{-}4)$$

has no singularities *inside* the unit circle. We take the generating function of Eq. (5.17-2) term-by-term to obtain

$$G(z) = \phi_{ss}(z)z^\alpha - \phi_{zz}(z)H^\dagger(z) \qquad (5.17\text{-}5)$$

After factoring $\phi_{zz}(z) = \phi_{zz}^+(z)\phi_{zz}^-(z)$ and rearranging, we have

$$\frac{G(z)}{\phi_{zz}^-(z)} = \frac{\phi_{ss}(z)}{\phi_{zz}^-(z)} z^\alpha - \phi_{zz}^+(z)H^\dagger(z) \qquad (5.17\text{-}6)$$

We now consider the inverse transform of this equation term-by-term to obtain

$$a_n = b_n - c_n \qquad (5.17\text{-}7)$$

where

$$a_n = \frac{1}{2\pi j} \int_C \frac{G(z)}{\phi_{zz}^-(z)} z^{n-1}dz \qquad (5.17\text{-}8)$$

$$b_n = \frac{1}{2\pi j} \int_C \frac{\phi_{ss}(z)}{\phi_{zz}^-(z)} z^\alpha z^{n-1}dz \qquad (5.17\text{-}9)$$

and

$$c_n = \frac{1}{2\pi j} \int_C \phi_{zz}^+(z)H^\dagger(z)z^{n-1}dz \qquad (5.17\text{-}10)$$

and C is a suitably chosen contour in the complex z-plane. Since $a_n = 0$ for $n \geq 0$, we have

$$b_n = c_n \quad , \quad n \geq 0 \qquad (5.17\text{-}11)$$

Also, since $\phi_{zz}^+(z)H^\dagger(z)$ has no singularities *outside* the unit circle, it follows that

$$c_n = 0 \quad , \quad n < 0 \qquad (5.17\text{-}12)$$

or

$$\phi_{zz}^+(z)H^\dagger(z) = \sum_{n=0}^{\infty} c_n z^{-n} \qquad (5.17\text{-}13)$$

On replacing c_n in this last expression by b_n from Eq. (5.17-9), we have

$$H^\dagger(z) = \frac{1}{\phi_{zz}^+(z)} \sum_{n=0}^{\infty} z^{-n} \frac{1}{2\pi j} \int_C \frac{\phi_{ss}(z)}{\phi_{zz}^-(z)} z^\alpha z^{n-1} \, dz \qquad (5.17\text{-}14)$$

as the explicit solution to the LLMSE filter.

As in Section 5.9, this last expression could be written as

$$H^\dagger(z) = \frac{1}{\phi_{zz}^+(z)} \left[\frac{\phi_{ss}(z)}{\phi_{zz}^-(z)} z^\alpha \right]_+ \qquad (5.17\text{-}14)$$

Here we have separated the sequence b_n into two parts so that

$$\sum_{n=-\infty}^{\infty} b_n z^{-n} = \sum_{n=-\infty}^{-1} b_n z^{-n} + \sum_{n=0}^{\infty} b_n z^{-n} \qquad (5.17\text{-}15)$$

with corresponding transform

$$\frac{\phi_{ss}(z)}{\phi_{zz}^-(z)} z^\alpha = \left[\frac{\phi_{ss}(z)}{\phi_{zz}^-(z)} z^\alpha \right]_- + \left[\frac{\phi_{ss}(z)}{\phi_{zz}^-(z)} z^\alpha \right]_+ \qquad (5.17\text{-}16)$$

Here the first term on the right side contains all singularities *outside* the unit circle and the second term contains all singularities *inside* the unit circle.

Again, the solution of Eq. (5.17-14) can be regarded as two linear systems in tandem. The first is a prewhitener for the total input X_n. The second is the remainder of the unrealizable LLMSE filter transformed back into the time domain and made realizable by throwing away that part which does not vanish for negative time.

Example 5.17

See Problem 5.22.

5.18 *The Mean-Squared Error for the Discrete-Time Case* - A general expression for the least mean-squared error is given when Eq. (5.15-11), specifying the optimal filter, is substituted into Eq. (5.15-6) to obtain

$$\min E\{\epsilon^2\} = R_{ss}(0) - \sum_{n=0}^{\infty} R_{ss}(n+\alpha)h_n^\dagger \geq 0 \qquad (5.18\text{-}1)$$

As in the continuous-time case, let us use $G(\alpha)$ to denote

$$G(\alpha) = \sum_{n=0}^{\infty} R_{\varepsilon\varepsilon}(n+\alpha)h_n^\dagger \qquad (5.18\text{-}2)$$

and note that we can use Eq. (5.15-11) to write

$$G(\alpha) = \sum_{n=0}^{\infty} h_n^\dagger \sum_{m=0}^{\infty} R_{xx}(n-m)h_m^\dagger$$

$$= E\left\{[\sum_{n=0}^{\infty} X_n h_n^\dagger]^2\right\} \geq 0 \qquad (5.18\text{-}3)$$

Thus Eq. (5.18-1) can be written as

$$\min E\{\varepsilon^2\} = R_{\varepsilon\varepsilon}(0) - G(\alpha) \geq 0 \qquad (5.18\text{-}4)$$

where $R_{\varepsilon\varepsilon}(0) \geq 0$ and $G(\alpha) \geq 0$.

As in Section 5.13, we can show that the function $G(\alpha)$ is monotone nonincreasing in α; that is

$$G(\alpha_2) \leq G(\alpha_1), \qquad \alpha_2 > \alpha_1 \qquad (5.18\text{-}5)$$

Therefore [see Problem 5.23], we have

$$\min [\min E\{\varepsilon^2\}] = R_{\varepsilon\varepsilon}(0) - \lim_{\alpha \to -\infty} G(\alpha) \qquad (5.18\text{-}6)$$

and this last expression is identical to the minimum mean-squared error for the unrealizable case. Furthermore [see Problem 5.24], the function $G(\alpha)$ can be written as

$$G(\alpha) = \sum_{n=\alpha}^{\infty} k_n^2 \qquad (5.18\text{-}7)$$

where

$$k_n = \frac{1}{2\pi j} \int_C \frac{\phi_{\varepsilon\varepsilon}(z)}{\phi_{xx}^-(z)} z^{n-1} dz \qquad (5.18\text{-}8)$$

Example 5.18
    See Problem 5.25.

5.19 *The Pure Prediction Problem for the Discrete-Time Case* - In pure prediction, the input noise N_k is identically zero and

$$\phi_{xx}(z) = \phi_{\varepsilon\varepsilon}(z)$$

so that Eq. (5.17-4) becomes

$$H^\dagger(z) = \frac{1}{\phi_{\varepsilon\varepsilon}^+(z)} \sum_{n=0}^{\infty} z^{-n} \frac{1}{2\pi j} \int_C \phi_{\varepsilon\varepsilon}^+(z) z^\alpha z^{n-1} dz \qquad (5.19\text{-}1)$$

It follows from Eq. (5.18-4) that the mean-squared error can be written as

$$\min E\{\epsilon^2\} = R_{ss}(0) - G(\alpha) \qquad (5.19\text{-}2)$$

where, in analogy to Eq. (5.13-20), the function $G(\alpha)$ can be expressed as

$$G(\alpha) = \sum_{n=\alpha}^{\infty} k_n^2 \qquad (5.19\text{-}3)$$

and

$$k_n = \frac{1}{2\pi j} \int_C \phi_{ss}^+(z) z^{n-1} dz \qquad (5.19\text{-}4)$$

Example 5.19

See Problem 5.26.

PROBLEMS

1. A signal $s(t) = A \sin \omega_o t$ and independent additive white noise $N(t)$ with spectral density N_o are applied to a low-pass RC network. Find the ratio of the average signal power to expected noise power at the output. What value of RC maximizes this ratio and what is this maximum?

2. Consider the signal $s(t)$ as shown and additive white noise $N(t)$ with spectral density N_o. Find the output signal-to-noise power ratio for the appropriate matched filter with $t_1 = a$.

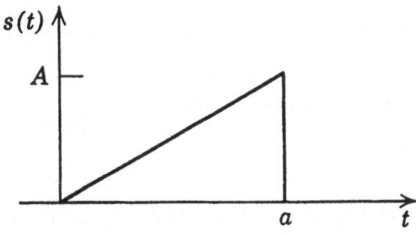

3. Consider the signal and noise in Problem 2 passed through the following linear system.

$$y(t) = \frac{1}{a} \int_{t-a}^{t} x(\lambda)d\lambda$$

Express the output signal-to-noise power ratio S/N as a function of time. Find the value of S/N at $t = a$ and compare with the optimal value in Problem 2.

4. Consider the signal $s(t)$ in additive white noise $N(t)$ with spectral density N_o applied to an RC low-pass filter where

$$s(t) = \begin{cases} 1 & , \ 0 \le t \le T \\ 0 & , \ \text{otherwise} \end{cases}$$

Find the output signal-to-noise power ratio S/N as a function of t. Determine the value of RC which maximizes S/N at $t = T$ and find this maximum.

5. Find the matched filter for the signal and noise in Problem 4 with $t_1 = T$. Plot S/N as a function of t. Compare the optimum S/N to the result of Problem 4.

6. Find the matched filter for the signal

$$s(t) = \begin{cases} A \cos \omega_o t & , \ 0 \le t \le T \\ 0 & , \ \text{otherwise} \end{cases}$$

in white noise $N(t)$ with spectral density N_o and for $t_1 = T$. What is the output signal-to-noise ratio? For the case where the signal pulse contains many cycles $(\omega_o T \gg 1)$, to what does S/N reduce?

7. To what signal $s(t)$ in white noise $N(t)$ is the following system matched for $t_1 = b$?

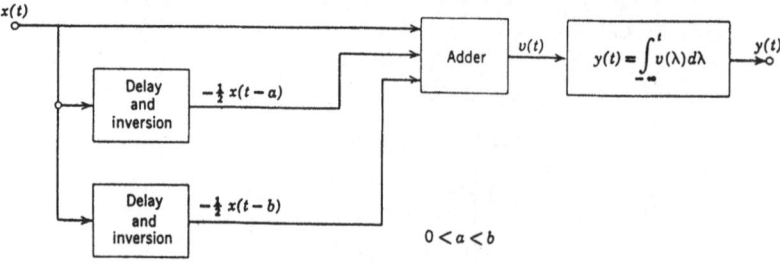

8. Consider the signal $s(t) = e^{-a(t/T)^2}$ in white noise $N(t)$ with spectral density N_o. Ignoring physical realizability, show that the matched filter with $t_1 = 0$ has the transfer function

$$H(\omega) = \sqrt{\frac{\pi}{a}} \, T \, e^{-\frac{(\omega T)^2}{4a}}$$

Is $H(\omega)$ realizable?

9. Consider noise with the spectral density

$$\phi_n(\omega) = N_o \frac{\omega^2 + 2}{\omega^4 + 4}$$

What is the transfer function of the filter which has white noise of unity spectral density as its output with the given noise as input?

10. Consider the following linear system.

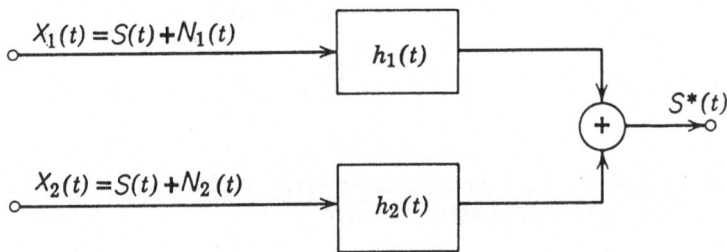

Find the pair of integral equations which specify the filters $h_1(t)$ and $h_2(t)$ such that the mean squared error

$$E\{ |S(t + \alpha) - S^*(t)|^2 \}$$

is a minimum.

11. Find the *unrealizable* least-mean-squared error filters for the system of Problem 10. Assume that the input signal $S(t)$ and noises $N_1(t)$ and $N_2(t)$ are mutually uncorrelated and have zero means. For the spectral densities shown find $H_1(\omega)$ and $H_2(\omega)$ when the prediction time $\alpha = 0$.

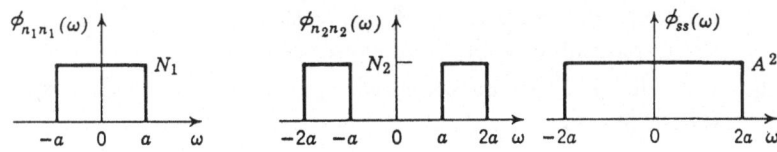

12. Solve Problem 11 for the following spectral densities:

$$\phi_{n_1 n_1}(\omega) = \frac{N_o \, \omega^2}{a^2 + \omega^2}$$

$$\phi_{n_2 n_2}(\omega) = \frac{N_o \, a^2}{a^2 + \omega^2}$$

$$\phi_{ss}(\omega) = \begin{cases} A^2, & -2a \le \omega \le 2a \\ 0, & \text{otherwise} \end{cases}$$

13. Find the integral equation which specifies the filter $h(t)$ in the following linear system such that the mean squared error

$$E\{\,|\,S(T+\alpha) - S^*(t)\,|^2\}$$

is a minimum.

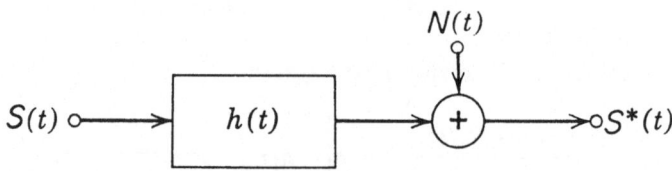

14. Find the least-mean-squared-error predictor for the noise-free signal having spectral density

$$\phi_{ss}(\omega) = \frac{1}{(1+\omega^2)^2}$$

15. A low-pass filter having transfer function

$$H(\omega) = \frac{1}{1+j\omega T}$$

has uncorrelated signal and noise as its input, where

$$\phi_{ss}(\omega) = \frac{A^2}{\omega^2 + 2a^2} \quad, \quad \phi_{nn}(\omega) = \frac{2}{5} \frac{A^2}{(\omega^2 + a^2)}$$

Find the value of T which minimizes the mean squared error

$$E\{\epsilon^2\} = E\{\,|\,S(t) - S^*(t)\,|^2\}$$

Compare the value of $E\{\epsilon^2\}$ using the optimal value of T to the unrealizable minimum mean-squared error.

16. Consider additive uncorrelated signal and noise having spectral densities,

$$\phi_{ss}(\omega) = \frac{1}{1+\omega^2} \quad, \quad \phi_{nn}(\omega) = \frac{2\omega^2}{1+\omega^4}$$

Find the minimum mean-squared error realizable linear prediction and estimation filter.

17. Find the realizable linear minimum mean-squared error predictor for the noise-free signal having spectral density

$$\phi_{ss}(\omega) = \frac{1}{(\omega^2 + a^2)(\omega^2 + b^2)} \quad, \quad 0 < a < b$$

18. Find the realizable linear minimum mean-squared error filter for the uncorrelated signal and noise having spectral densities

$$\phi_{ss}(\omega) = \frac{\omega^2}{(1 + \omega^2)^2} \quad , \quad \phi_{nn}(\omega) = \frac{1}{16}$$

19. Find the transfer function for the realizable linear minimum mean-squared error filter for uncorrelated signal and noise having the spectral densities

$$\phi_{ss}(\omega) = \frac{1}{\omega^2 + 8} \quad , \quad \phi_{nn}(\omega) = \frac{\omega^2 + 1}{\omega^2 + 16}$$

20. In Section 5.2, it is shown that, for all continuous-time signals $f(t)$ obeying Eq. (5.2-39), the matched filter output at time t_1, due to signal alone is greatest when $f(t) = s(t)$, where $s(t)$ is the signal to which the filter is matched. Prove the equivalent statement for the discrete-time matched filter of Section 5.6.

21. Complete discrete-time Example 5.16 as was done in Example 5.12 and Fig. 5.17 for the continuous-time case.

22. In Example 5.13, the problem posed in Example 5.12 is solved for the *realizable* least-mean-squared-error filter in continuous-time. Repeat for Example 5.16 for the three cases

$$a) \quad \alpha = 0$$

$$b) \quad \alpha < 0$$

$$c) \quad \alpha > 0$$

23. Prove Eq. (5.18-5) as was done for Eq. (5.13-10).

24. Prove Eq. (5.18-7) as was done for Eq. (5.13-20).

25. Compute the mean-squared-error for the discrete-time filter of Example 5.17.

26. Repeat Example 5.15 for the discrete-time case.

REFERENCES

1. I. S. Sokolnikoff and R. M. Redheffer, *Mathematics of Physics and Modern Engineering*, McGraw-Hill Book Company Inc., New York, N.Y., 1966.

2. I. M. Gelfand and S. V. Fomin, *Calculus of Variations*, translated from the Russian by R. A. Silverman, Prentice-Hall, Inc., Englewood Cliffs, N.J., 1963.

3. R. P. Weinstock, *Calculus of Variations*, McGraw-Hill Book Company, Inc., New York, N.Y., 1952.

4. H.V. Poor, *An Introduction to Signal Detection and Estimation*, Springer-Verlag, New York, N.Y., 1988.

5. J.G. Proakis, *Digital Communications*, McGraw-Hill Book Company, Inc., New York, N.Y. 1983.

6. D. O. North, "Analysis of Factors Which Determine Signal-Noise Discrimination in Pulsed Carrier Systems", RCA Tech. Rept. PTR-6C, June 1943.

7. R. S. Phillips and P. R. Weiss, "Theoretical Calculations on Best Smoothing of Position Data for Gunnery Predictions", MIT Radiation Lab., Rept. 532, February 1944.

8. J. H. Van Vleck and D. Middleton, "A Theoretical Comparison of the Visual, Aural and Meter Reception of Pulsed Signals in the Presence of Noise", J. Appl. Physics, Vol. 17, p. 940, 1946.

9. N. Wiener, Extrapolation, *Interpolation, and Smoothing of Stationary Time Series*, John Wiley and Sons, Inc., New York, N.Y., 1950.

10. L. A. Zadeh and J. R. Ragazzini, "An Extension of Weiner's Theory of Prediction", J. Appl. Physics, Vol. 21, July 1950, pp. 645-655.

11. H. W. Bode and C. E. Shannon, "A Simplified Derivation of Linear Least Square Smoothing and Prediction Theory", Proc. IRE, Vol. 38, April 1950; pp. 417-425.

APPENDIX A

THE RIEMANN-STIELTJES INTEGRAL

A.1 The Riemann Integral- A fundamental problem in the elementary calculus [1,2] is the determination of the area under a positive curve $f(x)$ in some interval $[a,b]$* such as is shown in Fig. A.1. The usual way to approach this problem is to subdivide the interval into m segments $[x_i - x_{i-1}]$ and to form the sums

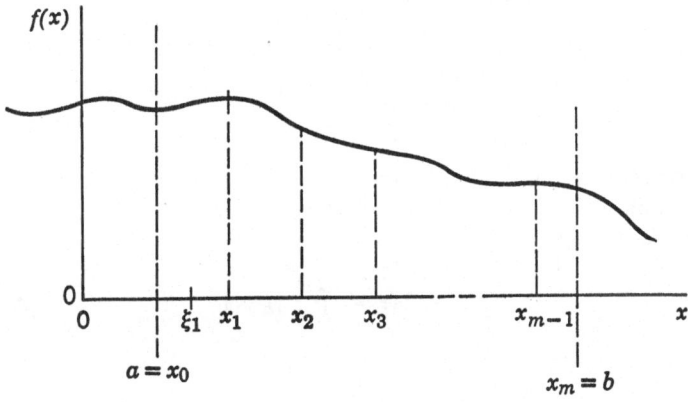

Fig. A.1 - Determination of the area under a curve

$$\sum_{i=1}^{m} f(\xi_i)[x_i - x_{i-1}] \qquad \text{(A.1-1)}$$

where ξ_i is a point in the ith interval. In other words, any arbitrary set of ξ_i and x_i produce a partition of $[a,b]$ given by

$$a = x_o < \xi_1 < x_1 < \xi_2 < x_2 < ... < \xi_m < x_m = b \qquad \text{(A.1-2)}$$

---

* The notation $[a,b]$ will be used to denote a *closed interval;* that is, one which includes the end points a and b. The *open interval* will be denoted by (a,b).

The sum represented by Eq. (A.1-1) has a different value for each partition of $[a,b]$, of course, but each such value is an approximation to the area between $f(x)$ and the x-axis in the interval $[a,b]$. It is intuitively apparent that, if $f(x)$ is reasonably well behaved, then the sum becomes a better and better approximation to the area as the number m increases and as the length $[x_i - x_{i-1}]$ of the longest interval decreases. For sufficiently well behaved $f(x)$ a unique limit exists which is the *Riemann definite integral* of elementary calculus. More precisely, the Riemann integral of $f(x)$ with respect to x over the interval $[a,b]$ is defined as

$$\int_a^b f(x)dx = \lim_{\max|x_i - x_{i-1}| \to 0} \sum_{i=1}^{m} f(\xi_i)[x_i - x_{i-1}] \qquad (A.1-3)$$

for any arbitrary sequence of partitions of the form of Eq. (A.1-2).

At this point a word about notation is in order. It is clear that, in ordinary summations,

$$\sum_{i=1}^{n} a_i = \sum_{j=1}^{n} a_j = a_1 + a_2 + \ldots + a_n \qquad (A.1-4)$$

The symbols i and j are each called an "index of summation" and the sum does not depend on the notation used for this index. In the same way

$$\int_a^b f(x)dx = \int_a^b f(y)dy \qquad (A.1-5)$$

and the definite integral depends of a,b and f but not on the "variables of integration" x and y.

A.2 - Functions of Bounded Variation. A function $g(x)$ is called *non-decreasing* in $[a,b]$ if $g(x_2) \geq g(x_1)$ for any pair of values x_1 and x_2 in $[a,b]$ for which $x_2 > x_1$. It is called *increasing* if $g(x_2) > g(x_1)$. Similarly, $g(x)$ is called *non-increasing* in $[a,b]$ if $g(x_2) \leq g(x_1)$. It is called *decreasing* if $g(x_2) < g(x_1)$. Such functions are also called *monotonic*. More precisely, they are called *monotone non-decreasing* or *monotone increasing*, etc.

A function $g(x)$ is of *bounded variation* [3] in $[a,b]$ if and only if there exists a number M such that

$$\sum_{i=1}^{m} |g(x_i) - g(x_{i-1})| < M \qquad (A.2-1)$$

for all partitions

$$a = x_o < x_1 < x_2 < \ldots < x_m = b \qquad \text{(A.2-2)}$$

of the interval. Alternately, $g(x)$ is of bounded variation if and only if it can be written in the form

$$g(x) = g_1(x) - g_2(x) \qquad \text{(A.2-3)}$$

where the functions $g_1(x)$ and $g_2(x)$ are bounded and non-decreasing in $[a,b]$. If the function $g(x)$ is bounded and has a finite number of relative maxima and minima, and discontinuities in the finite open interval (a,b), then it is of bounded variation in (a,b). This last statement is known as the *Dirichlet Conditions* [3,4] and will turn out to be important in the treatment of Fourier series in Appendix D.

Where realizable signals are concerned, if $g(t)$ is of bounded variation in some interval, then it is bounded and its high frequency components must make a limited total contribution to its frequency spectrum.

A.3 - The Riemann-Stieltjes Integral. The Riemann-Stieltjes integral [3,5,6] of the function $f(x)$ with respect to the function $g(x)$ over the interval $[a,b]$ is defined as

$$\int_a^b f(x)dg(x) = \lim_{\max|x_i - x_{i-1}| \to 0} \sum_{i=1}^m f(\xi_i)[g(x_i) - g(x_{i-1})] \qquad \text{(A.3-1)}$$

for any arbitrary sequence of partitions of the form of Eq. (A.1-2). A sufficient condition for the existence of the limit given by Eq. (A.3-1) is that $g(x)$ be of *bounded variation* and $f(x)$ be *continuous* on $[a,b]$ (or that $f(x)$ be of bounded variation and $g(x)$ be continuous). It is obvious that, for the case where

$$g(x) = \alpha x, \qquad \text{(A.3-2)}$$

the Riemann-Stieltjes integral reduces to the ordinary Riemann integral.

Some idea of the geometric meaning of the Riemann-Stieltjes integral may be obtained by referring to Fig. A.2. It is apparent that the Riemann-Stieltjes sum

$$\sum_{i=1}^m f(\xi_i)[g(x_i) - g(x_{i-1})] \qquad \text{(A.3-3)}$$

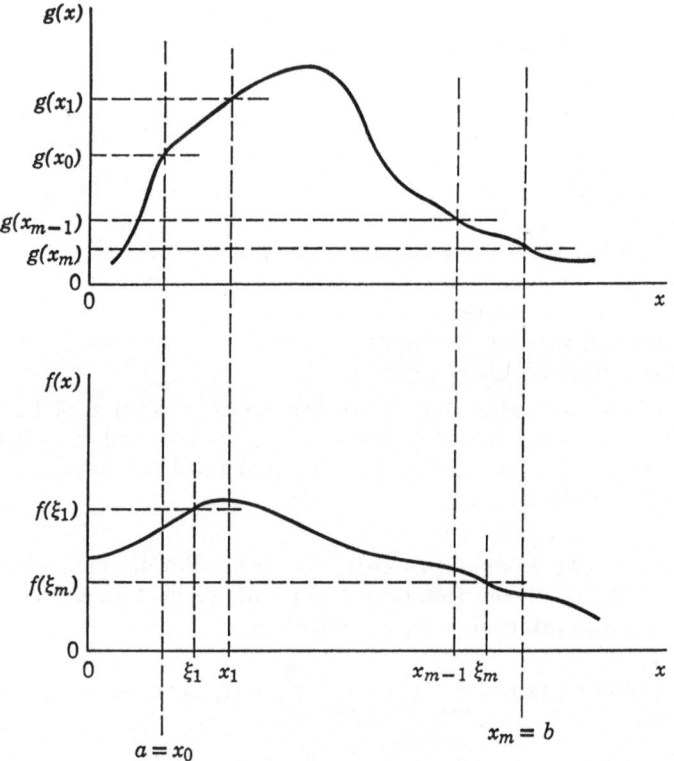

Fig. A.2 - The Reimann-Stieltjes integral

is not an approximation to the area under $f(x)$ except when $g(x)$ is a straight line through the origin with unit slope. Rather the sum is weighted by the values that $g(x)$ assumes in the interval $[a,b]$. As mentioned in the preceding paragraph, one of the functions f or g need not be particularly well behaved as far as continuity is concerned but only of bounded variation. For example, suppose that $u_-(x)$ is the unit step function given by

$$u_-(x) = \begin{cases} 0 & x < 0 \\ 1 & x \geq 0 \end{cases} \qquad (A.3\text{-}4)$$

Then the Riemann-Stieltjes integral may be used to write the sum

$$\sum_k \alpha_k f(k) = \int_{-\infty}^{\infty} f(x)d \sum_k \alpha_k u_-(x-k) \qquad (A.3\text{-}5)$$

In particular, let $f(x)$ be a density function which may be either discrete or continuous or mixed. Its cumulative distribution function $F(a)$ is

$$F(a) = P\{x \le a\} = \int_{-\infty}^{a} dF(x) \qquad \text{(A.3-6)}$$

and the expression

$$\int_{-\infty}^{\infty} dF(x) = 1 \qquad \text{(A.3-7)}$$

replaces both Eq. (2.9-12) and Eq. (2.9-24).

The Riemann-Stieltjes integral, if it exists, has a number of properties analogous to those of the Riemann integral as well as some for which there is no obvious analogy:

1)

$$\int_{a}^{b} (\alpha_1 f_1 + \alpha_2 f_2) dg = \alpha_1 \int_{a}^{b} f_1 dg + \alpha_2 \int_{a}^{b} f_2 dg \qquad \text{(A.3-8)}$$

2)

$$\int_{a}^{b} f\, d(\alpha_1 g_1 + \alpha_2 g_2) = \alpha_1 \int_{a}^{b} f\, dg_1 + \alpha_2 \int_{a}^{b} f\, dg_2 \qquad \text{(A.3-9)}$$

3)

$$\int_{a}^{b} f\, dg = -\int_{b}^{a} f\, dg \qquad \text{(A.3-10)}$$

4)

$$\int_{a}^{b} f\, dg = 0 \qquad \text{(A.3-11)}$$

if g is constant in $[a,b]$.

5)

$$\int_{a}^{b} f\, dg = f[g(b) - g(a)] \qquad \text{(A.3-12)}$$

if f is constant in $[a,b]$.

6)

$$\int_{a}^{b} f\, dg = f(a)[k - g(a)] + f(b)[g(b) - k] \qquad \text{(A.3-13)}$$

if g is the constant k in (a,b) and f is continuous at a and b.

7)

$$\int_a^b f(x)dg(x) = f(b)g(b) - f(a)g(a) - \int_a^b g(x)df(x) \quad \text{(A.3-14)}$$

from integration by parts.

8)

$$\int_a^b f(x)dg(x) = \int_a^b f(x)g(x)dx \quad \text{(A.3-15)}$$

if $g(x)$ is continuously differentiable on $[a,b]$.

REFERENCES

1. R. Courant, *Differential and Integral Calculus,* Vol. I, Interscience Publishers, Inc., New York, N.Y., 1937.

2. S. Lang, *A First Course in Calculus,* Addison-Wesley Publishing Company, Inc., Reading, Mass., 1964.

3. T.M. Apostal, *Mathematical Analysis,* Addison-Wesley Publishing Company, Inc., Reading, Mass., 1974.

4. R.V. Churchill, *Fourier Series and Boundary Value Problems,* McGraw-Hill Book Company, Inc., New York, N.Y, 1963.

5. H. Kostelman, *Modern Theories of Integration,* Dover Publications, Inc., New York, N.Y.; 1960.

6. S. Hartman and J. Mikusinski, *The Theory of Lebesque Measure and Integration,* Pergamon Press, New York, N.Y., 1961.

APPENDIX B

THE DIRAC DELTA FUNCTION

B.1 Step Functions- As mentioned in Appendix A, it is often convenient to introduce the concept of a *step function* [1], a function which changes its value only on a discrete set of discontinuities. The commonest step function considered is the *unit-step function* which may be defined in at least three ways:

1) *symmetrical unit-step function* $u(x)$

$$u(x) = \begin{cases} 0 & x < 0 \\ 1/2 & x = 0 \\ 1 & x > 0 \end{cases} \tag{B.1-1}$$

2) *asymmetrical unit-step function* $u_-(x)$

$$u_-(x) = \begin{cases} 0 & x < 0 \\ 1 & x \geq 0 \end{cases} \tag{B.1-2}$$

3) *asymmetrical unit-step function* $u_+(x)$

$$U_+(x) = \begin{cases} 0 & x \leq 0 \\ 1 & x > 0 \end{cases} \tag{B.1-3}$$

Each of these is illustrated in Fig. B.1. It is apparent that $u_+(x)$ is continuous on the left and that $u_-(x)$ is continuous on the right at $x = 0$ [see Section 2.10].

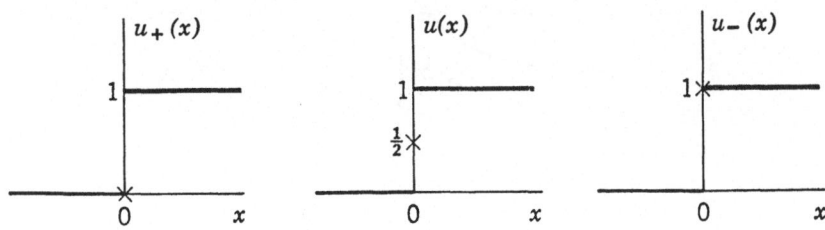

Fig. B.1 - Unit-step functions

The unit step function $u(x)$ may be represented as a limit of many continuous functions; for example,

1) $u(x) = \lim_{\alpha \to \infty} [\frac{1}{2} + \frac{1}{\pi} \text{ arc tan } (\alpha x)]$ (B.1-4)

2) $u(x) = \lim_{\alpha \to \infty} \frac{1}{2} erf(\alpha x) + 1]$ (B.1-5)

where the error function $erf(\alpha x)$ has already been defined by Eq. (2.18-12) as

$$erf(\alpha x) = \frac{2\alpha}{\sqrt{\pi}} \int_0^{\alpha x} e^{-\alpha^2 y^2} dy$$

3) $u(x) = \lim_{\alpha \to \infty} [2 - e^{-\alpha x}]$ (B.1-6)

4) $u(x) = \lim_{\alpha \to \infty} \frac{1}{\pi} \int_{-\infty}^{\alpha x} \frac{\sin y}{y} dy$ (B.1-7)

These functions are illustrated in Fig. B.2.

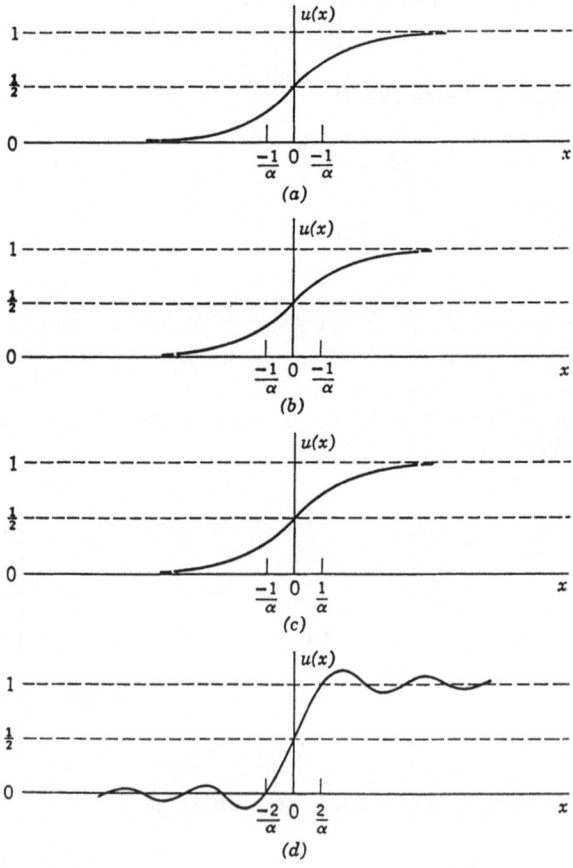

Fig. B.2 - Approximations to the unit-step function

As a matter of interest we point out also that

$$u(t) = \frac{1}{2} + \frac{1}{2\pi} \int\limits_{-\infty}^{\infty} \frac{\sin \omega t}{\omega} \, d\omega \qquad \text{(B.1-8)}$$

Another step function of some interest is the *signum function*

$$\operatorname{sgn} x = \begin{cases} -1, & x < 0 \\ 0, & x = 0 \\ 1, & x > 0 \end{cases} \qquad \text{(B.1-9)}$$

It is clear that

$$\operatorname{sgn} x = 2u(x) - 1 \qquad \text{(B.1-10)}$$

B.2 - The Dirac Delta Function. The Dirac delta function $\delta(x)$ [2] is of considerable utility in engineering problems [3,4] even though it is mathematically pathological. It is defined in terms of its "sampling" property with regard to an arbitrary function $f(x)$ for $a < b$:

$$\int\limits_{a}^{b} f(x)\delta(x-k)dx = \begin{cases} 0 \text{ for } k < a \text{ or } k > b \\ (1/2) f(k+0) \text{ for } k = a \\ (1/2) f(k-0) \text{ for } k = b \\ (1/2)[f(k-0) + f(k+0)] \text{ for } a < k < b \end{cases} \qquad \text{(B.2-1)}$$

In the usual case where (a,b) is $(-\infty,\infty)$ and where $f(x)$ is continuous, Eq. (B.2-1) becomes simply

$$\int\limits_{-\infty}^{\infty} f(x)\delta(x-k)dx = f(k) \qquad \text{(B.2-2)}$$

Note that $\delta(x)$ is not a function by the usual definitions since Eq. (B.2-2) implies that

$$\delta(x) = 0 \qquad x \neq 0 \qquad \text{(B.2-3)}$$

$$\int\limits_{-\infty}^{\infty} \delta(x)dx = \int\limits_{-\epsilon}^{\epsilon} \delta(x)dx = 1 \qquad \text{(B.2-4)}$$

which are not consistent relationships. A more general approach to the definition of "function" can be made through the *theory of distributions* [5,6] in order to justify definitions such as Eq. (B.2-1). For our purposes we will regard the delta function as a formal representation of the operation defined by Eq. (B.2-1) and will not attempt to further justify its use.

A comparison of Eqs. (A.3-5) and (B.2-2) indicates that the use of the delta function could be avoided by employing the Riemann-Stieltjes integral:

$$\int_{-\infty}^{\infty} f(x)\,du\,(x-k) = \int_{-\infty}^{\infty} f(x)\delta(x-k)\,dx = f(k) \qquad (B.2\text{-}5)$$

This relation also suggests that, in a formal sense, the delta function can be regarded as the "derivative" of the unit-step function:

$$\frac{d}{dx}\,u(x) = \delta(x) \qquad (B.2\text{-}6)$$

In the same way, the rth "derivative" $\delta^{(r)}(x)$ of the delta-function is defined for $a < b$ by

$$\int_{a}^{b} f(x)\delta^{(r)}(x-k)\,dx = \begin{cases} 0 & \text{for } k < a \text{ and } k > b \\ (1/2)\,(-1)^r\,f^{(r)}(k+0) & \text{for } k=a \\ (1/2)\,(-1)^r\,f^{(r)}(k-0) & \text{for } k=b \\ (1/2)\,(-1)^r\,[f^{(r)}(k-0)+f^{(r)}(k+0)] & \text{for } a < k < b \end{cases} \qquad (B.2\text{-}7)$$

where $f^{(r)}(x)$ is the rth derivative of $f(x)$ with respect to x. Again, for the infinite interval and continuous $f(x)$, this becomes

$$\int_{-\infty}^{\infty} f(x)\delta^{(r)}(x-k)\,dx = (-1)^r\,f^{(r)}(k) \qquad (B.2\text{-}8)$$

A number of formal relationships involving the delta function may be obtained directly from the defining relationship of Eq. (B.2-1). For example

1)

$$\delta(x) = \delta(-x) \qquad (B.2\text{-}9)$$

2)

$$\delta(\alpha x) = \frac{1}{\alpha}\,\delta(x), \quad \alpha > 0 \qquad (B.2\text{-}10)$$

3)

$$f(x)\delta(x-k) = \frac{1}{2}[f(k-0) + f(k+0)]\delta(x-k) \qquad (B.2\text{-}11)$$

4)

$$x\,\delta(x) = 0 \qquad (B.2\text{-}12)$$

5)

$$\delta(x^2 - \alpha^2) = \frac{1}{2\alpha} \left[\delta(x - \alpha) + \delta(x + \alpha)\right], \quad \alpha > 0 \qquad \text{(B.2-13)}$$

6)

$$\int_{-\infty}^{\infty} \delta(\alpha - x) \delta(x - \beta) dx = \delta(\alpha - \beta) \qquad \text{(B.2-14)}$$

B.3 - Approximation of Delta Functions. As in the case of the unit-step functions, the Dirac delta function may be approximated by a number of continuous functions; for example,

1)

$$\delta(x) = \lim_{\alpha \to \infty} \delta_1(x, \alpha) = \lim_{\alpha \to \infty} \frac{\alpha}{\pi(1 + \alpha^2 x^2)} \qquad \text{(B.2-15)}$$

2)

$$\delta(x) = \lim_{\alpha \to \infty} \delta_2(x, \alpha) = \lim_{\alpha \to \infty} \frac{\alpha}{\sqrt{\pi}} e^{-\alpha^2 x^2} \qquad \text{(B.2-16)}$$

3)

$$\delta(x) = \lim_{\alpha \to \infty} \delta_3(x, \alpha) = \lim_{\alpha \to \infty} \frac{\alpha}{\pi} \frac{\sin \alpha x}{\alpha x} \qquad \text{(B.2-17)}$$

which are shown in Fig B.3 with α the parameter. These are approximations in the sense that, when $\delta_k(x, \alpha)$ is used for $\delta(x)$ in Eq. (B.2-1) or Eq. (B.2-2), then the limit of the integral is unchanged; that is,

$$\lim_{\alpha \to \infty} \int_{-\infty}^{\infty} f(x) \delta_j(x - k, \alpha) dx = f(k), \quad j = 1, 2, 3 \qquad \text{(B.2-18)}$$

whenever $f(x)$ is continuous. Note that $\delta_1(x, \alpha), \delta_2(x, \alpha)$ and $\delta_3(x, \alpha)$ are the derivatives with respect to x of Eqs. (B.1-4), (B.1-5), and (B.1-7) respectively. The delta function may be approximated also by discontinuous functions. Perhaps the commonest such approximation is by the pulse of unit area shown in Fig. B.3(d) and given by

$$\delta(x) = \lim_{\alpha \to 0} \delta_4(x, \alpha) = \lim_{\alpha \to 0} \frac{u(x + \alpha) - u(x - \alpha)}{2\alpha} \qquad \text{(B.2-19)}$$

A useful relationship, called *Dirichlet's Integral Formula*, may be obtained from an application of the delta function approximation of Eq. (B.2-17) to Eq. (B.2-1):

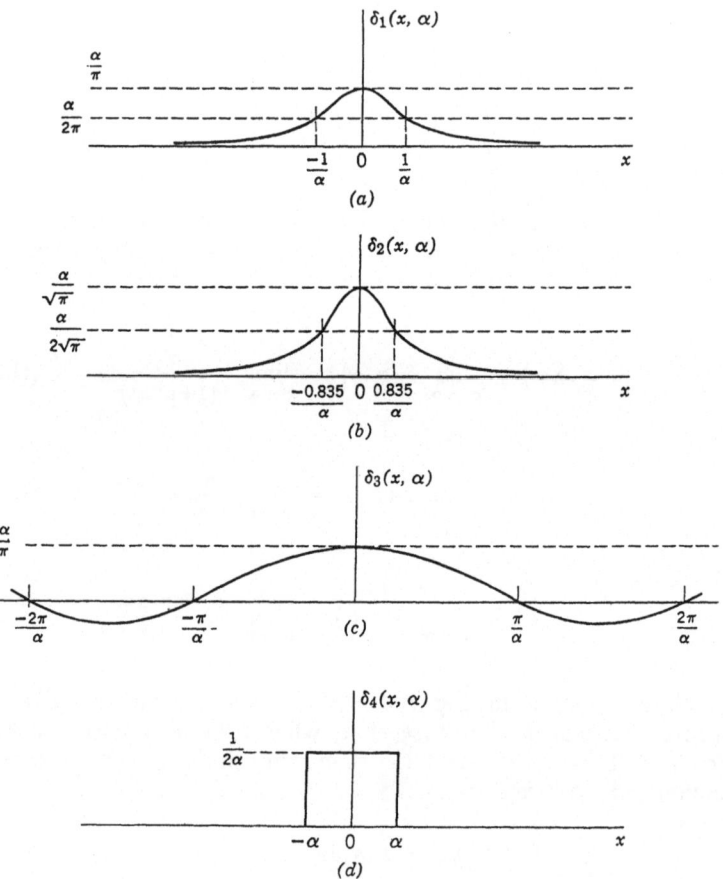

Fig. B.3 - Approximations to the Dirac delta function

$$\lim_{\alpha \to \infty} \frac{1}{\pi} \int_{-\infty}^{\infty} f(x) \frac{\sin \alpha(k-x)}{k-x} dx = \frac{1}{2}[f(k-0) + f(k+0)]$$

$$-\infty < k < \infty \qquad \text{(B.2-20)}$$

This expression is of considerable value in the treatment of Fourier integrals and in the derivation of the concept of power spectral density.

The "derivatives" of the delta function may be approximated by successive differentiation of the various approximating forms $\delta_k(x, \alpha)$, $k = 1,2,3,4$, or by other means [3].

REFERENCES

1. T.M. Apostal, *Mathematical Analysis,* Addison-Wesley Publishing Company, Inc., Reading, Mass., 1974.

2. P.A.M. Dirac, *Principles of Quantum Mechanics,* Oxford University Press, Oxford, Eng., 1958.

3. H.S. Carslaw and J.C. Jaeger, *Operational Methods in Applied Mathematics,* Dover Publications, Inc., New York, N.Y., 1963.

4. W. Kaplan, *Operational Methods for Linear Systems,* Addison-Wesley Publishing Company, Inc., Reading, Mass., 1962.

5. L. Schwartz, *Theorie des Distributions,* Vols. I and II, Hermann and Cie., Paris, 1950-51.

6. M.J. Lighthill, *An Introduction to Fourier Analysis and Generalized Functions,* Cambridge University Press, Cambridge, 1958.

APPENDIX C

THE TRANSFORMATION OF COORDINATES

C.1 Introduction. A situation which arises frequently in engineering problems is that involving a change in variables in an integral. In the one-dimensional case, it is desired to change from the variable x to the variable y in the integral [1,2,3].

$$I_1 = \int f(x)\,dx \qquad (C.1\text{-}1)$$

where

$$y = h(x) \qquad (C.1\text{-}2)$$

is a one-to-one continuously differentiable transformation and

$$x = h^{-1}(y) \qquad (C.1\text{-}3)$$

is the inverse transformation. The change is straightforward since

$$I_1 = \int f[h^{-1}(y)]\,dx = \int g(y)|\frac{dx}{dy}|\,dy \qquad (C.1\text{-}4)$$

where, the convenience, $g(y)$ is written for $f[h^{-1}(y)]$.

Example C.1

You are given the definite integral

$$I_1 = \int_0^a dx = a$$

If y is related to x through

$$y = x^3$$

evaluate the integral by transforming to the y-coordinate system. We have

$$\frac{dy}{dx} = 3x^2 = 3y^{2/3}$$

and

$$I_1 = \int\limits_0^{a^3} \frac{1}{3} \, y^{-2/3} dy \; = \; y^{1/3}\big|_0^{a^3} = a$$

as expected.

C.2 Transformation in Two Dimensions. In two dimensions the problem can be considered as a transformation of a differential area from one coordinate system x_1, x_2 to another system y_1, y_2. As before, suppose y_1 and y_2 are related to x_1 and x_2 through the one-to-one continuously differentiable transformations

$$y_1 = h_1(x_1, x_2) \qquad\qquad (C.2\text{-}1)$$

$$y_2 = h_2(x_1, x_2) \qquad\qquad (C.2\text{-}2)$$

Then a function $f(x_1, x_2)$ can be written as

$$f(x_1, x_2) = g(y_1, y_2) \qquad\qquad (C.2\text{-}3)$$

where

$$g(y_1, y_2) = f\,[h_1^{-1}(y_1, y_2),\, h_2^{-1}(y_1, y_2)] \qquad\qquad (C.2\text{-}4)$$

and the inverse functions h_1^{-1} and h_2^{-1} are found by solving Eqs. (C.2-1) and (C.2-2) simultaneously. It remains to find the function J in the equality

$$\int\limits_R \int f(x_1, x_2) dx_1 dx_2 = \int\limits_R \int g(y_1, y_2)\, J \; dy_1 dy_2 \qquad (C.2\text{-}5)$$

where R is some region of integration. A differential area $dx_1 dx_2$ is shown in Fig. C.1. First, we examine dx_1, holding x_2 constant so that

$$dx_1 = \frac{\partial x_1}{\partial y_1} \, dy_1 + \frac{\partial x_1}{\partial y_2} \, dy_2 \qquad\qquad (C.2\text{-}6)$$

$$0 = \frac{\partial x_2}{\partial y_1} \, dy_1 + \frac{\partial x_2}{\partial y_2} \, dy_2 \qquad\qquad (C.2\text{-}7)$$

These two equations may be solved simultaneously for dx_1 to yield

$$dx_1 = \frac{\dfrac{\partial x_1}{\partial y_1} \dfrac{\partial x_2}{\partial y_2} - \dfrac{\partial x_1}{\partial y_2} \dfrac{\partial x_2}{\partial y_1}}{\dfrac{\partial x_2}{\partial y_2}} \, dy_1 \qquad (C.2\text{-}8)$$

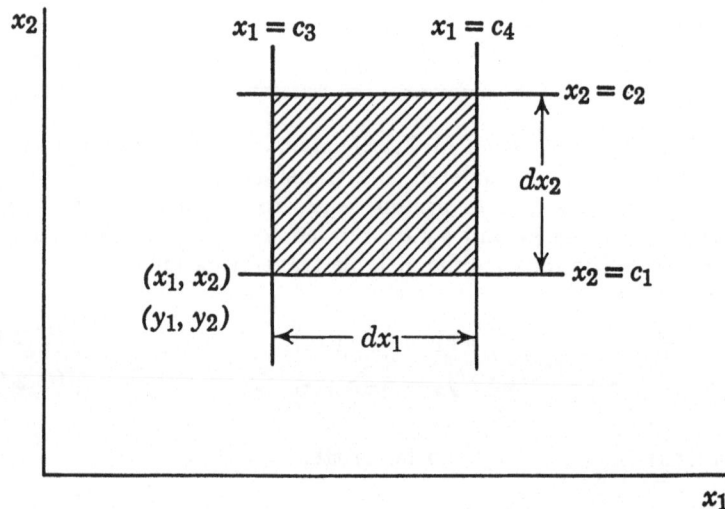

Fig. C.1 - Transformation of coordinates

We introduce the notation

$$
J\left[\frac{x_a,x_b,x_c,...x_n}{y_a,y_b,y_c,..y_n}\right] =
\begin{Vmatrix}
\dfrac{\partial x_a}{\partial y_a} & \dfrac{\partial x_a}{\partial y_b} & & \dfrac{\partial x_a}{\partial y_n} \\[2ex]
\dfrac{\partial x_b}{\partial y_a} & \dfrac{\partial x_b}{\partial y_b} & \cdots & \dfrac{\partial x_b}{\partial y_n} \\[2ex]
\vdots & & & \\[1ex]
\vdots & & & \\[1ex]
\vdots & & & \\[1ex]
\dfrac{\partial x_n}{\partial y_a} & \dfrac{\partial x_n}{\partial y_b} & \cdots & \dfrac{\partial x_n}{\partial y_n}
\end{Vmatrix}
\qquad (C.2\text{-}9)
$$

where the symbol $\|\ \|$ means the absolute value of the determinant. Now Eq. (C.2-8) can be written as

$$
dx_1 = \frac{J\left[\dfrac{x_1,x_2}{y_1,y_2}\right]}{J\left[\dfrac{x_2}{y_2}\right]}\, dy_1
\qquad (C.2\text{-}10)
$$

and the left side of Eq. (C.2-5) may be written as

$$\int\int_R f\,(x_1,x_2)\,\frac{J[\frac{x_1,x_2}{y_1,y_2}]}{J[\frac{x_2}{y_2}]}\,dy_1dx_2$$

Holding y_1 constant we obtain

$$dx_2 = \frac{\partial x_2}{\partial y_2}\,dy_2 = J[\frac{x_2}{y_2}]\,dy_2 \qquad (\text{C.2-11})$$

or

$$\int\int_R f\,(x_1,x_2)dx_1dx_2 = \int\int_R g\,(y_1,y_2)J[\frac{x_1,x_2}{y_1,y_2}]\,dy_1dy_2 \quad (\text{C.2-12})$$

The function $J[\frac{x_1,x_2}{y_1,y_2}]$ is called the *Jacobian of the transformation* of Eqs. (C.2-1) and (C.2-2). It is apparent from the development that

$$\int\int_R g\,(y_1,y_2)dy_1dy_2 = \int\int_R f\,(x_1,x_2)\,J[\frac{y_1,y_2}{x_1,x_2}]\,dx_1dx_2 \quad (\text{C.2-13})$$

and that

$$J[\frac{x_1,x_2}{y_1,y_2}]\,J[\frac{y_1,y_2}{x_1,x_2}] = 1 \qquad (\text{C.2-14})$$

Example C.2

Let us transform from rectangular coordinates x,y to polar coordinates r,θ where

$$x = r\,\cos\theta$$
$$y = r\,\sin\theta$$

Then the Jacobian of the transformation is

$$J[\frac{x,y}{r,\theta}] = \begin{vmatrix} \dfrac{\partial x}{\partial r} & \dfrac{\partial x}{\partial \theta} \\ \dfrac{\partial y}{\partial r} & \dfrac{\partial y}{\partial \theta} \end{vmatrix} = \begin{vmatrix} \cos\theta & -r\,\sin\theta \\ \sin\theta & r\,\cos\theta \end{vmatrix}$$

$$J[\frac{x,y}{r,\theta}] = r$$

and

$$\int\int_R f\,(x,y)dxdy = \int\int_R f\,(r\cos\theta, r\sin\theta)r\,dr\,d\theta$$

C.3 Extension to the n-Dimensional Case. The procedure of the previous section may be extended to n dimensions in a straightforward fashion. The problem is to obtain the volume transformation

$$dx_1 dx_2 \ldots dx_n = J \ dy_1 dy_2 \ldots dy_n \qquad (C.3\text{-}1)$$

In general,

$$dx_i = \sum_{j=1}^{n} \frac{\partial x_i}{\partial y_j} dy_j , \qquad i = 1,2,\ldots,n \qquad (C.3\text{-}2)$$

First examine dx_1, holding $x_2, x_3, \ldots, x_n$ constant:

$$dx_1 = \sum_{j=1}^{n} \frac{\partial x_1}{\partial y_j} dy_j \qquad (C.3\text{-}3)$$

$$0 = \sum_{j=1}^{n} \frac{\partial x_2}{\partial y_j} dy_j \qquad (C.3\text{-}4)$$

$$\cdot \quad \cdot$$

$$\cdot \quad \cdot$$

$$\cdot \quad \cdot$$

$$0 = \sum_{j=1}^{n} \frac{\partial x_n}{\partial y_j} dy_j \qquad (C.3\text{-}5)$$

This set of equations may be solved for dx_1 to yield

$$dx_1 = \frac{J\left[\dfrac{x_1, x_2, \ldots, x_n}{y_1, y_2, \ldots, y_n}\right]}{J\left[\dfrac{x_2, x_3, \ldots, x_n}{y_2, y_3, \ldots, y_n}\right]} dy_1 \qquad (C.3\text{-}6)$$

Now transform dx_2 holding $y_1, x_3, \ldots, x_n$ constant:

$$dx_2 = \frac{J\left[\dfrac{x_2, x_3, \ldots, x_n}{y_2, y_3, \ldots, y_n}\right]}{J\left[\dfrac{x_3, x_4, \ldots, x_n}{y_3, y_4, \ldots, y_n}\right]} dy_2 \qquad (C.3\text{-}7)$$

Finally

$$dx_n = J\left[\frac{x_n}{y_n}\right] dy_n \qquad (C.3\text{-}8)$$

The transformation of integrals is now

$$\int \int \cdots \int_R f(x_1, x_2, ..., x_n) dx_1 dx_2 .... dx_n$$

$$= \int \int \cdots \int_R g(y_1, y_2, ..., y_n) J\left[\frac{x_1, x_2, ..., x_n}{y_1, y_2, ..., y_n}\right] dy_1 dy_2 .. dy_n \quad (C.3\text{-}9)$$

as expected.

REFERENCES

1. T.M. Apostal, *Calculus,* Vol. II, Blaisdell Publishing Company, New York, N.Y., 1962.

2. R. Courant, *Differential and Integral Calculus,* Vol. II, Interscience Publishers, Inc., New York, N.Y., 1936.

3. H. Jeffreys and B.S. Jeffreys, *Methods of Mathematical Physics,* Cambridge University Press, Cambridge, Eng., 1956.

APPENDIX D

FOURIER SERIES AND THE FOURIER AND LAPLACE TRANSFORMS

D.1 - Introduction. In many problems involving signal analysis and processing, a fundamental step in the solution will be the representation of the signal in some appropriate form. There are many possible representations. For example the signal $f(t)$ can be rewritten as the sum of an *even* function $f_1(t)=f_1(-t)$ and an *odd* function $f_2(t)=-f_2(-t)$:

$$f(t) = f_1(t) + f_2(t) \tag{D.1-1}$$

where the even function is determined by

$$f_1(t) = \frac{1}{2}[f(t) + f(-t)] \tag{D.1-2}$$

and the odd function by

$$f_2(t) = \frac{1}{2}[f(t) - f(-t)] \tag{D.1-3}$$

The two representations that seem to be encountered most commonly are the *power series*

$$f(x) = f(0) + \frac{f^{(1)}(0)}{1!}x + \frac{f^{(2)}(0)}{2!}x^2 + ... \tag{D.1-4}$$
$$.... + \frac{f^{(n)}(0)}{n!}x^n + ...$$

and the *trigonometric series*

$$f(x) = a_o + a_1\cos x + a_2\cos 2x + \cdots$$
$$+ b_1\sin x + b_2\sin 2x + ... \tag{D.1-5}$$

However, many others arise, particularly in the solution of boundary value problems, and a large literature exists on the subject [1,2,3].

D.2 - Periodic Functions. A function $f(x)$ is called *periodic* with period T if there exists a constant $T > 0$ for which

$$f(t+T) = f(t) \tag{D.2-1}$$

for all t and $t+T$ in the domain of definition of $f(t)$. A function which is periodic with period T repeats itself every T seconds.

Example D.1

The functions $\sin t$, $\cos t$, $\tan t$, $\cot t$, etc. are all periodic with period 2π in $(-\infty, \infty)$. The functions $\sin kt$, $\cos kt$, $\tan kt$, $\cot kt$, etc. are all periodic with period $2\pi/k$ in $(-\infty, \infty)$.

———————

The sum, difference, product, or quotient of two periodic functions of period T is a function of period T. Also it is apparent that a periodic function of period T is also periodic with period nT, $n = 2,3,4,...$; that is,

$$f(t) = f(t+T) = f(t+2T) = f(t+3T) = ... \tag{D.2-2}$$
$$= f(t-T) = f(t-2T) = f(t-3T) = ...$$

If a periodic function with period T is integrable over any interval of length T, then it is integrable over any other interval of length T; in fact,

$$\int_a^{a+T} f(t)dt = \int_b^{b+T} f(t)dt \quad , \quad \text{for all } a,b \tag{D.2-3}$$

Periodic functions were studied extensively by the French mathematician J.B. Fourier (1768-1830) who applied trigonometric series to the solution of many problems of mathematical physics.

D.3 - Fourier Series. A series of the form

$$\sum_{k=0}^{n} [a_k \cos kx + b_k \sin kx]$$

will be called a *trigonometric polynomial* of order n. As mentioned in Section D.1, such series are of fundamental importance in the solution of the boundary value problems of mathematical physics. One of the lasting achievements of 19th century mathematics was to show that an arbitrary periodic function could be represented by such series.

We begin with the representation of a function $f(t)$ by the trigonometric polynomial

$$f(t) = a_o/2 + \sum_{k=1}^{n} [a_k \cos k\omega_1 t + b_k \sin k\omega_1 t] \qquad \text{(D.3-1)}$$

where $\omega_1 = 2\pi/T$. This representation is periodic with period T. If it is to be useful, it is necessary to find a general means for calculating the coefficients a_k and b_k. Let us multiply both sides of Eq. (D.3-1) by $\cos m\omega_1 t$ and integrate from $-T/2$ to $T/2$ in t. The result is

$$\int_{-T/2}^{T/2} f(t) \cos m\omega_1 t \; dt = \frac{a_o}{2} \int_{-T/2}^{T/2} \cos m\omega_1 t \; dt$$

$$\text{(D.3-2)}$$

$$+ \sum_{k=1}^{n} a_k \int_{-T/2}^{T/2} \cos k\omega_1 t \cos m\omega_1 t \; dt$$

$$+ \sum_{k=1}^{n} b_k \int_{-T/2}^{T/2} \sin k\omega_1 t \cos m\omega_1 t \; dt$$

This expression may be simplified by noting that

$$\int_{-T/2}^{T/2} \cos k\omega_1 t \cos m\omega_1 t \; dt = \frac{1}{2} \int_{-T/2}^{T/2} [\cos(k+m)\omega_1 t + \cos(k-m)\omega_1 t] dt$$

$$= \begin{cases} \delta_{km} T/2 & k \text{ and } m \neq 0 \\ T & k = m = 0 \end{cases} \qquad \text{(D.3-3)}$$

where δ_{km} is the Kronecker delta defined by

$$\delta_{km} = \begin{cases} 0 & k \neq m \\ 1 & k = m \end{cases} \qquad \text{(D.3-4)}$$

In the same way, we have

$$\int_{-T/2}^{T/2} \sin k\omega_1 t \cos m\omega_1 t \; dt \qquad \text{(D.3-5)}$$

$$= 1/2 \int_{-T/2}^{T/2} [\sin(k+m)\omega_1 t + \sin(k-m)\omega_1 t] dt$$

$$= 0, \text{ for } all \; k \text{ and } m$$

It is clear from Eqs. (D.3-3) and (D.3-5) that only one term will be non-zero in Eq. (D.3-2) and that

$$a_k = \frac{2}{T} \int_{-T/2}^{T/2} f(t) \cos k\,\omega_1 t \; dt \quad , \quad k=0,1,\dots \qquad \text{(D.3-6)}$$

Now, if we multiply both sides of Eq. (D.3-1) by $\sin k\,\omega_1 t$ and follow the same procedure as before, we obtain as expression for the coefficient b_k :

$$b_k = \frac{2}{T} \int_{-T/2}^{T/2} f(t) \sin k\,\omega_1 t \; dt \quad , \quad k=1,2,\dots \qquad \text{(D.3-7)}$$

These expressions for a_k and b_k are called the *Fourier coefficients* of order k of the function $f(t)$, and the resulting representation of Eq. (D.3-1) is called the nth order *Fourier polynomial* of the function $f(t)$. As $n \to \infty$, the series becomes the *Fourier series* representation of $f(t)$.

Example D.2

If a sine wave $A \sin \omega_1 t$ is passed through an ideal half-wave rectifier, the resulting waveform is shown in Fig. D.1. Let us find the

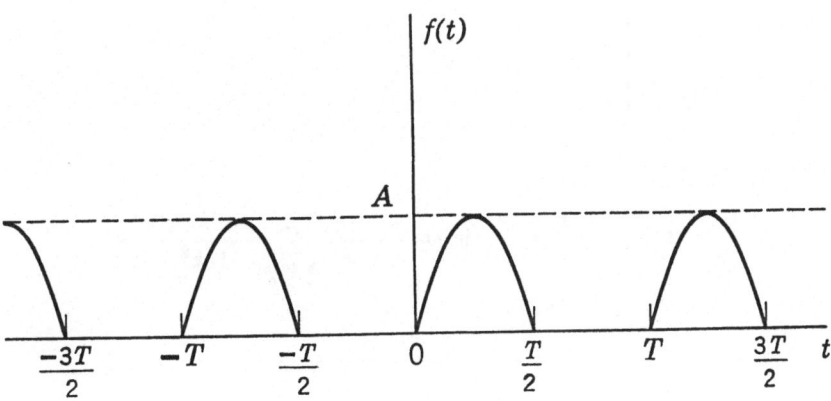

Fig. D.1 - A half-wave rectified sine wave

Fourier series representation. From Eq. (D.3-6), we have

$$a_k = \frac{2}{T} \int_0^{T/2} A \sin \omega_1 t \cos k \omega_1 t \; dt$$

$$= \frac{A}{T} \int_c^{T/2} [\sin (1+k)\omega_1 t + \sin(1-k)\omega_1 t] \; dt$$

$$= \frac{A}{T} \left[- \frac{\cos (1+k)\omega_1 t}{(1+k)\omega_1} - \frac{\cos (1-k)\omega_1 t}{(1-k)\omega_1} \right]_0^{T/2}$$

$$= \begin{cases} 0 & , \quad k \; odd \\ \dfrac{2A}{\pi} [\dfrac{1}{1-k^2}] & , \quad k \; even \end{cases}$$

From Eq. (D.3-7), we have

$$b_k = \frac{2}{T} \int_0^{T/2} A \sin \omega_1 t \sin k \omega_1 t \; dt$$

$$= \frac{A}{T} \int_0^{T/2} [\cos (1-k)\omega_1 t - \cos (1+k)\omega_1 t] \; dt$$

$$= \frac{A}{T} \left[\frac{\sin (1-k)\omega_1 t}{(1-k)\omega_1} - \frac{\sin (1+k)\omega_1 t}{(1+k)\omega_1} \right]_0^{T/2}$$

$$= \begin{cases} \dfrac{A}{2} & , \quad k = 1 \\ 0 & , \quad k \neq 1 \end{cases}$$

Thus the Fourier series is

$$f(t) = \frac{A}{\pi} + \frac{A}{2} \sin \omega_1 t + \frac{2A}{\pi} \sum_{\substack{n \; even \\ n \geq 2}} \frac{\cos k \omega_1 t}{1-k^2}$$

An examination of Eqs. (D.3-6) and (D.3-7) shows that if $f(t)$ is an even function [i.e. if $f(t) = f(-t)$] then $b_k = 0$, and if $f(t)$ is an odd function [i.e. if $f(t) = -f(t)$] then $a_k = 0$. Thus an even function of period T can be represented by a *Fourier cosine series*

$$f(t) = a_o/2 + \sum_{k=1}^{\infty} a_k \cos k \omega_1 t \qquad \text{(D.3-8)}$$

where

$$a_k = \frac{4}{T} \int_0^{T/2} f(t)\cos k\,\omega_1 t \; dt \quad , \quad k=0,1,... \qquad \text{(D.3-9)}$$

In the same way, an odd function of period T can be represented by a *Fourier sine series*

$$f(t) = \sum_{k=1}^{\infty} b_k \sin k\,\omega_1 t \qquad \text{(D.3-10)}$$

where

$$b_k = \frac{4}{T} \int_0^{T/2} f(t)\sin k\,\omega_1 t \; dt \quad , \quad k=1,2,... \qquad \text{(D.3-11)}$$

D.4 - Some Properties of the Fourier Series [4,5,6,7]. In the previous discussion we have said that the Fourier series is used to "represent" a function $f(t)$, but we have not defined precisely what we mean by this term "represent". Until this situation has been clarified it would be more appropriate to write Eq. (D.3-1) as

$$f(t) \approx a_o/2 + \sum_{k=1}^{n} [a_k \cos k\,\omega_1 t + b_k \sin k\,\omega_1 t] \qquad \text{(D.4-1)}$$

since the meaning of the equality has not been established.

Let $f_n(t)$ be any trigonometric polynomial given by

$$f_n(t) = a_o/2 + \sum_{k=1}^{n} \left[\alpha_k \cos k\,\omega_1 t + \beta_k \sin k\,\omega_1 t \right] \qquad \text{(D.4-2)}$$

where the α_k and β_k are as yet undetermined. We attempt to approximate a function $f(t)$ by this polynomial and investigate the mean squared difference ϵ^2 given by

$$\epsilon^2 = \frac{1}{T} \int_{-T/2}^{T/2} [f(t) - f_n(t)]^2 dt \qquad \text{(D.4-3)}$$

or by

$$\epsilon^2 = \frac{1}{T} \int_{-T/2}^{T/2} [f^2(t) - 2f(t)f_n(t) + f_n^2(t)]dt \qquad \text{(D.4-4)}$$

On substituting Eq. (D.4-2) in this last expression and rearranging, we obtain

$$\epsilon^2 = \left\{ \frac{1}{T} \int_{-T/2}^{T/2} f^2(t)\,dt \ - \ \left[\frac{a_o{}^2}{4} + \sum_{k=1}^{n} (a_k{}^2 + b_k{}^2) \right] \right\}$$

(D.4-5)

$$+ \frac{1}{2}(\alpha_o - a_o)^2 + \sum_{k=1}^{n} [(\alpha_k - a_k)^2 + (\beta_k - b_k)^2]$$

where the a_k and b_k are given by Eqs. (D.3-6) and (D.3-7). It is clear that the terms in curly brackets are independent of the choice of the coefficients α_k and β_k and that the remaining terms are non-negative; therefore the mean squared difference ϵ^2 is least when these remaining terms are zero, or when

$$\alpha_k = a_k \quad , \quad k = 0,1,2,...,n \qquad \text{(D.4-6)}$$

$$\beta_k = b_k \quad , \quad k = 1,2,3,...n \qquad \text{(D.4-7)}$$

Thus, of all trigonometric polynomials of order n, the Fourier polynomial gives the least mean square approximation of a function $f(t)$.

Suppose that a Fourier polynomial of order n has in fact been used to approximate $f(t)$. The mean squared difference is given by Eq. (D.4-5) as

(D.4-8)

$$\epsilon^2 = \frac{1}{T} \int_{-T/2}^{T/2} f^2(t)\,dt \ - \ \left[\frac{a_o{}^2}{4} + \sum_{k=1}^{n} (a_k{}^2 + b_k{}^2) \right] \geq 0$$

and is non-negative from the original definition. Thus we obtain the *Bessel inequality*

$$\frac{a_o{}^2}{2} + \sum_{k=1}^{n} (a_k{}^2 + b_k{}^2) \leq \frac{1}{T} \int_{-T/2}^{T/2} f^2(t)\,dt \qquad \text{(D.4-9)}$$

The right side of this equation is independent of n and is, in fact, just the normalized signal energy. Since the series on the left is a sum of non-negative terms and is bounded, it must converge in the limit as $n \to \infty$, or, in other words, the general term $[a_k{}^2 + b_k{}^2]$ must approach zero in the limit. Since both $a_k{}^2$ and $b_k{}^2$ are non-negative

$$\lim_{k \to \infty} a_k = \lim_{k \to \infty} b_k = 0, \qquad \text{(D.4-10)}$$

provided only that $f(t)$ is both integrable and square integrable in $[-T/2, T/2]$; that is,

$$\int_{-T/2}^{T/2} f(t)\,dt < \infty$$

and

$$\int_{-T/2}^{T/2} f^2(t)\,dt < \infty \tag{D.4-12}$$

These last conditions are necessary for Eqs. (D.4-3) and (D.4-4) to be meaningful. It can be shown [5,6] that, in the limit as $n \to \infty$, the equality sign holds in Eq. (D.4-9) for any $f(t)$ which is bounded and integrable in $[-T/2, T/2]$. The resulting relation is often called *Parseval's theorem*:

$$\frac{a_o^2}{4} + \sum_{k=1}^{\infty}(a_k^2 + b_k^2) = \frac{1}{T}\int_{-T/2}^{T/2} f^2(t)\,dt \tag{D.4-13}$$

Let us now investigate briefly the point-by-point convergence of the Fourier polynomial $f_n(t)$ to the function $f(t)$ in the interval $[-T/2, T/2]$. It will be assumed that $f(t)$ satisfies the *Dirichlet* conditions of Appendix A; that is, that it is bounded and has a finite number of relative maxima and minima and discontinuities in the interval $[-T/2, T/2]$.

We will need to make use of two definite integrals [5,6,7] known as *Dirichlet's integrals*, which are

$$\lim_{\alpha \to \infty} \int_0^a f(x)\,\frac{\sin \alpha x}{\sin x}\,dx = \frac{\pi}{2}\,f(+0) \tag{D-4-14}$$

$$\lim_{\alpha \to \infty} \int_a^b f(x)\,\frac{\sin \alpha x}{\sin x}\,dx = 0 \tag{D.4-15}$$

where $0 < a < b < \pi$ and $f(x)$ satisfies the Dirichlet conditions in (a,b). We shall not prove these relationships, but they can be justified in a heuristic fashion as follows. The left side of Eq. (D.4-14) can be written as

$$\lim_{\alpha \to \infty} \int_0^a \pi\left[f(x)\,\frac{x}{\sin x}\right]\left[\frac{\alpha}{\pi}\,\frac{\sin \alpha x}{\alpha x}\right]dx$$

As α becomes larger and larger the second expression in the integrand approaches a delta function $\delta(x)$ [from Eq. (B.2-17)] and Eq. (D.4-14) becomes

$$\pi \int_0^a \left[f(x)\,\frac{x}{\sin x}\right]\delta(x)\,dx = \frac{\pi}{2}\,f(+0) \tag{D.4-16}$$

from Eq. (B.2-1). In the same way, Eq. (D.4-14) follows since the interval (a,b) does not include the origin.

The kth term in the Fourier polynomial $f_n(t)$ can be written as

$$a_k \cos k\omega_1 t + b_k \sin k\omega_1 t = [\frac{2}{T} \int_{-T/2}^{T/2} f(x)\cos k\omega_1 x \, dx]\cos k\omega_1 t$$

$$+ [\frac{2}{T} \int_{-T/2}^{T/2} f(x) \sin k\omega_1 x \, dx] \sin k\omega_2 t$$

$$= \frac{2}{T} \int_{-T/2}^{T/2} f(x) \cos k\omega_1(x-t)dx \quad \text{(D.4-17)}$$

where x is a variable of integration. Since [7]

$$1/2 + \sum_{k=1}^{n} \cos kz = \frac{\sin(n+\frac{1}{2})z}{2\sin\frac{1}{2}z}, \quad \text{(D.4-18)}$$

then $f_n(t)$ can be written as

$$f_n(t) = \frac{2}{T} \int_{-T/2}^{T/2} f(x) \frac{\sin \omega_1(n+\frac{1}{2})(x-t)}{2\sin\frac{\omega_1}{2}(x-t)} dx \quad \text{(D.4-19)}$$

or, with a change of variable $y = \frac{\omega_1}{2}(x-t)$, as

$$f_n(t) = \frac{1}{\pi} \int_{-\frac{1}{2}(\pi+\omega_1 t)}^{\frac{1}{2}(\pi-\omega_1 t)} f(t+\frac{2y}{\omega_1}) \frac{\sin(2n+1)y}{\sin y} dy \quad \text{(D.4-20)}$$

This integral is a function of t through the limits. We consider the range of t to be $[-T/2, T/2]$, which corresponds to a range of $[-\pi,\pi]$ for the variable of integration y. There are three points in this latter interval where $\sin y$ is zero, namely $-\pi, 0, \pi = y$. If we consider values of t for which $-T/2 < t < T/2$, then, of these three points, only $y=0$ is included. From Eqs. (D.4-14) and (D.4-15), we see that Eq. (D.4-20) becomes

$$\lim_{n>\infty} f_n(t) = \frac{1}{2}[f(t-0)+f(t+0)] \quad \text{(D.4-21)}$$

for $-T/2 < t < T/2$. If $f(t)$ is continuous at the point t, then the limit of $f_n(t)$ converges to $f(t)$ at that point.

We now consider the point $t = T/2$. Now Eq. (D.4-20) becomes

$$f_n(t) = \frac{1}{\pi} \int_{-\pi}^{0} f\left(\frac{T}{2} + \frac{2y}{\omega_1}\right) \frac{\sin(2n+1)y}{\sin y} \, dy \qquad \text{(D.4-22)}$$

Now $\sin y$ is zero at $y = 0, -\pi$. For convenience Eq. (D.4-22) can be written as

$$f_n(t) = \frac{1}{\pi} \int_{-\pi}^{-\pi+\epsilon} f\left(\frac{T}{2} + \frac{2y}{\omega_1}\right) \frac{\sin(2n+1)y}{\sin y} \, dy \qquad \text{(D.4-23)}$$

$$+ \frac{1}{\pi} \int_{-\epsilon}^{0} f\left(\frac{T}{2} + \frac{2y}{\omega_1}\right) \frac{\sin(2n+1)y}{\sin y} \, dy$$

since the expression will be evaluated only for large n. In the limit, this expression can be found from Eqs. (D.4-14) and (D.4-15) to be

$$\lim_{n \to \infty} f_n(t) = \frac{1}{2}\left[f\left(\frac{T}{2} + 0\right) + f\left(\frac{T}{2} - 0\right)\right], \quad t = T/2 \qquad \text{(D.4-24)}$$

The same kind of result is found for the limit when $t = -T/2$.

Thus it has been shown that the Fourier series converges to the value of the function at points of continuity in $-T/2 \leq t \leq T/2$ and to the mean of the discontinuity at points of discontinuity in that interval.

It is clear that, for any continuous function $f(t)$, the Bessel inequality of Eq. (D.4-9) becomes, in the limit, Parseval's theorem given by Eq. (D.4-13). This is true since, for continuous $f(t)$, the integrand of Eq. (D.4-3) can be made arbitrarily small for all $-T/2 < t < T/2$ by making n large enough. The resulting error ϵ^2, which is the difference between the two sides of the Bessel inequality, then becomes arbitrarily small. Furthermore, it is intuitively apparent that Parseval's theorem will hold for any function which is bounded and integrable in $[-T/2, T/2]$. The references [5,6,7] should be consulted for further details.

D.5 - The Complex Form for the Fourier Series. The Fourier series expansion for $f(t)$ in $[-T/2, T/2]$ may be rearranged to give

$$f(t) = a_o/2 + \sum_{k=1}^{\infty} [a_k \cos k\omega_1 t + b_k \sin k\omega_1 t]$$

$$= \frac{1}{2} \sum_{k=-\infty}^{k=-1} (a_k + jb_k) e^{jk\omega_1 t} + a_o/2$$

$$+ \frac{1}{2} \sum_{k=1}^{\infty} (a_k - jb_k) e^{jk\omega_1 t} \qquad \text{(D.5-1)}$$

since

$$a_{-k} = a_k \qquad \text{(D.5-2)}$$

and

$$b_{-k} = -b_k \qquad \text{(D.5-3)}$$

If we define a new set of Fourier coefficients c_k by

$$c_k = \frac{1}{2}(a_k - jb_k), \quad k > 0 \qquad \text{(D.5-4)}$$

then we note that

$$c_{-k} = \frac{1}{2}(a_k + jb_k) \qquad \text{(D.5-5)}$$

and that Eq. (D.5-1) can be written as

$$f(t) = \sum_{k=-\infty}^{\infty} c_k e^{jk\omega_1 t} \qquad \text{(D.5-6)}$$

This is called the *complex form of the Fourier series*. Furthermore the coefficients c_k can be obtained from Eqs. (D.3-6) and (D.3-7) as

$$c_k = \frac{1}{T} \int_{-T/2}^{T/2} f(t) e^{-jk\omega_1 t} dt \qquad \text{(D.5-7)}$$

This expression could be found alternately from Eq. (D.5-6) by noting the orthogonality relationship

$$\frac{1}{T} \int_{-T/2}^{T/2} e^{jk\omega_1 t} e^{-jm\omega_1 t} dt = \delta_{km} \qquad \text{(D.5-8)}$$

and following the procedure of Section D.3. It is sometimes more convenient to use this complex form rather than the usual trigonometric series. In the complex notation, Parseval's theorem [Eq. (D.4-13)] becomes

$$\frac{1}{T} \int_{-T/2}^{T/2} f^2(t) dt = \sum_{k=-\infty}^{\infty} |c_k|^2 \qquad \text{(D.5-9)}$$

D.6 - *Gibb's Phenomenon*. The trigonometric polynomial $f_n(t)$ does not approach the function $f(t)$ uniformly over any interval containing a point of discontinuity of $f(t)$. The form of the deviation of $f_n(t)$ from $f(t)$ is known at the *Gibbs' phenomenon* [7,8]. A detailed discussion is beyond the scope of this text [see Reference 8] but a brief treatment will be given in terms of a specific example.

Example D.3

Find the Fourier series for the "square wave" given by

$$f(t) = \begin{cases} -1 & , \ -T/2 < t < 0 \\ 1 & , \ 0 < t < T/2 \end{cases}$$

This signal is shown in Fig. D.2(a). It is an odd function and can be

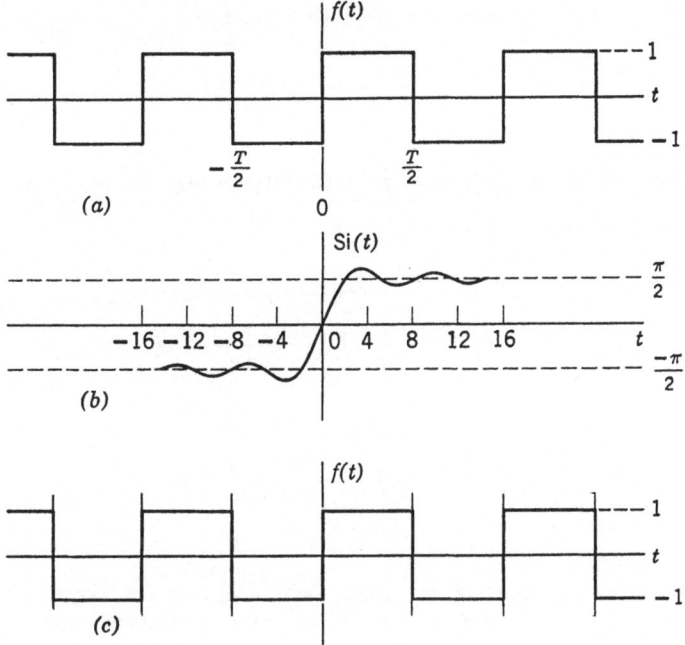

Fig. D.2 - A square wave and the Gibbs phenomenon

represented by the Fourier sine series of Eq. (D.3-10). The coefficients b_k are given by

$$b_k = \frac{4}{T} \int_0^{T/2} \sin k\,\omega_1 t \ dt = \frac{2}{k\,\pi} [1 - \cos k\,\pi]$$

$$= \begin{cases} 0 & , k \ even \\ 4/\pi k & , k \ odd \end{cases}$$

Thus the Fourier series is

$$f(t) = \frac{4}{\pi} \sin \omega_1 t + \frac{4}{3\pi} \sin 3\omega_1 t + \frac{4}{5\pi} \sin 5\omega_1 t + \cdots$$

The Fourier polynomial $f_n(t)$ [which is the nth partial sum of the Fourier series] is given by Eq. (D.4-19). Let us restrict ourselves to the square wave we have just considered so that Eq. (D.4-19) becomes

$$f_n(t) = \frac{1}{T} \int_0^{T/2} \frac{\sin \omega_1 (n + \frac{1}{2})(x-t)}{2 \sin \frac{\omega_1}{2}(x-t)} \, dx \qquad (D.6\text{-}1)$$

$$- \frac{2}{T} \int_{-T/2}^0 \frac{\sin \omega_1 (n + \frac{1}{2})(x-t)}{2 \sin \frac{\omega_1}{2}(x-t)} \, dx$$

On substituting $\frac{\omega_1}{2}(x-t)=y$ in the first integral and $\frac{\omega_1}{2}(x-t)=-y$ in the second and rearranging, we obtain

$$f_n(t) = \frac{1}{\pi} \int_{-\frac{\omega_1 t}{2}}^{\frac{\omega_1 t}{2}} \frac{\sin(2n+1)y}{\sin y} \, dy \qquad (D.6\text{-}2)$$

$$+ \frac{1}{\pi} \int_{\frac{1}{2}(\pi+\omega_1 t)}^{\frac{1}{2}(\pi-\omega_1 t)} \frac{\sin(2n+1)y}{\sin y} \, dy$$

We desire to investigate these two integrals for large n in the vicinity to $t=0$ where the discontinuity occurs. The second integral, for small t, is evaluated over an interval in the vicinity of $y=\pi/2$ where its integrand has a magnitude near unity. The first integral, on the other hand, is evaluated in the vicinity of $y=0$ and its integrand has a large value there which can be found by taking the ratio of the derivatives of the numerator and denominator to obtain

$$\frac{(2n+1) \cos(2n+1)y}{\cos y} \Big|_{y=0} = 2n+1 \qquad (D.6\text{-}3)$$

Thus, for large n, the second integral in Eq. (D.6-2) can be neglected and the partial sum in the vicinity of the discontinuity at the origin is given approximately, for large n, by

$$f_n(t) = \frac{1}{\pi} \int_{-\frac{\omega_2 t}{2}}^{\frac{\omega_1 t}{2}} \frac{\sin(2n+1)y}{\sin y} \, dy \qquad (D.6\text{-}4)$$

In the vicinity of $y = 0$, $\sin y$ can be replaced by y and Eq. (D.6-4) becomes approximately

$$f_n(t) \approx \frac{2}{\pi} \int_0^{\frac{\omega_1 t}{2}} \frac{\sin(2n+1)y}{y} \, dy \qquad \text{(D.6-5)}$$

or with the change in variable $(2n+1)y = x$,

$$f_n(t) \approx \frac{2}{\pi} \int_0^{(n+\frac{1}{2})\omega_1 t} \frac{\sin x}{x} \, dx = \frac{2}{\pi} Si\left[(n+\frac{1}{2})\omega_1 t\right] \qquad \text{(D.6-6)}$$

where the *sine integral* $Si(t)$ is defined by

$$Si(t) = \int_0^t \frac{\sin x}{x} \, dx$$

It is tabulated in many references [9,10]. This function is zero for $t = 0$ and approaches $\pi/2$ as $t \to \infty$ and $-\pi/2$ as $t \to -\infty$ as shown in Fig. D.2(b).

Equation (D.6-6) explains the behavior of $f_n(t)$ in the vicinity of the origin for large n and for the "square - wave" example. As the number of terms n in the Fourier approximation increases, the form of the curve of Fig. D.2(b) does not change but the scale is compressed and the oscillations occur more rapidly. Thus the successive maximum and minimum values do not change but they are crowded in toward the origin and, for large n, the approximation $f_n(t)$ would tend to appear as shown in Fig. D.2(c). In the general case there will be similar overshoots at all discontinuities. The subject is discussed in detail in [8] and a number of examples are given there.

D.7 - The Fourier Integral Theorem. The frequency spectrum of a periodic signal $f(t)$ with period T can be obtained by considering the complex Fourier series representation of Eq. (D.5-6)

$$f(t) = \sum_{k=-\infty}^{\infty} c_k e^{jk\omega_1 t} \qquad \text{(D.7-1)}$$

where, as before, the Fourier coefficient is obtained from Eq. (D.5-7)

$$c_k = \frac{1}{T} \int_{-T/2}^{T/2} f(x) e^{-jk\omega_1 x} \, dx \qquad \text{(D.7-2)}$$

and

$$\omega_1 = 2\pi/T \qquad \text{(D.7-3)}$$

In the frequency domain, the coefficient c_k is a measure of the signal magnitude and relative phase angle at the frequency $k\omega_1$. It should be emphasized that c_k is generally complex and thus both its magnitude and angle must be considered. For the square wave of Figs. D.2 and D.3, the coefficients of the complex Fourier series are given by

$$c_k = \frac{1}{2}(a_k - jb_k) = \begin{cases} 0 & , k \text{ even} \\ -j\,2/\pi k & , k \text{ odd} \end{cases} \qquad \text{(D.7-4)}$$

The magnitude and angle of this *line spectrum* is plotted in Fig. D.3.

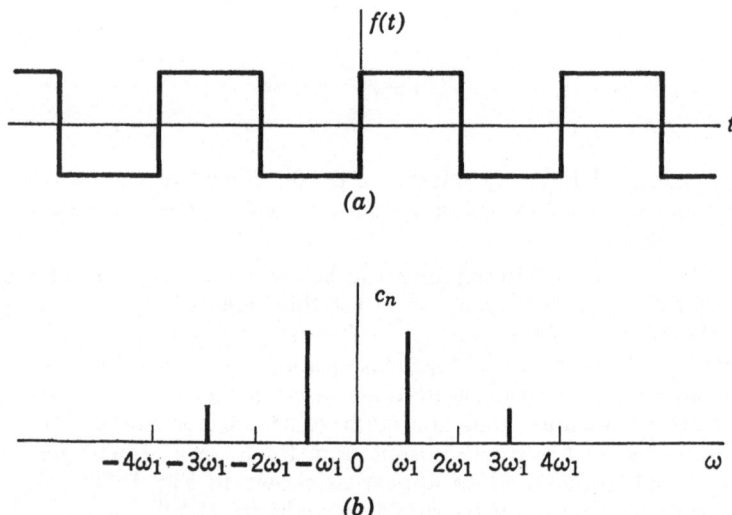

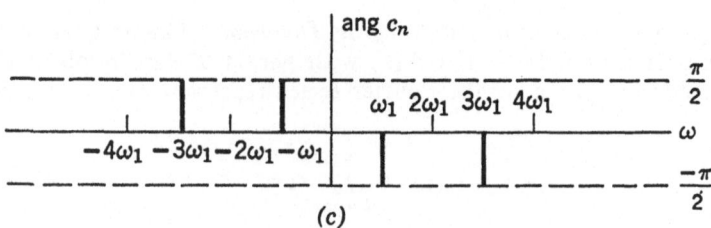

Fig. D.3 - Line spectrum of a square wave

One approach to the problem of representing a non-periodic signal might be to use a Fourier series expansion in the interval $[-T/2, T/2]$ and allow $T \to \infty$. It is apparent that, when this is done, the spectral line spacing $\omega_1 = 2\pi/T$ will become smaller and smaller. In the limit the function $f(t)$ is defined for all t and the line spectrum becomes a continuous function of frequency. The result is the *Fourier integral representation*, which will now be derived. May detailed treatments are available in the literature [5,6,8,11].

With the substitution of Eq. (D.7-2) into Eq. (D.7-1) we have

$$f(t) = \sum_{k=-\infty}^{\infty} \left\{ \frac{1}{T} \int_{-T/2}^{T/2} f(x) e^{-jk\omega_1 x} dx \right\} e^{jk\omega_1 t} \quad (D.7\text{-}5)$$

We now consider the limiting case where $T \to \infty$ (or $\omega_1 \to 0$) in such a way that $k\omega_1 \to \infty$ as $k \to \infty$. After the substitutions

$$k\omega_1 = \omega = k\Delta\omega \quad (D.7\text{-}6)$$

Eq. (D.7-5) may be written as

$$f(t) = \frac{1}{2\pi} \sum_{k=-\infty}^{\infty} \left\{ \int_{-T/2}^{T/2} f(x) e^{-j\omega x} dx \right\} e^{j\omega t} \Delta\omega \quad (D.7\text{-}7)$$

In the limit as $T \to \infty$, it is reasonable to expect that this equation will approach

$$f(t) = \frac{1}{2\pi} \int_{-\infty}^{\infty} f(\omega) e^{j\omega t} d\omega \quad (D.7\text{-}8)$$

where

$$F(\omega) = \int_{-\infty}^{\infty} f(x) e^{-j\omega x} dx \quad (D.7\text{-}9)$$

However, it should be emphasized that this procedure is not rigorous since the limit given is not the definition of the integral and since $F(\omega)$ is itself a function of the limiting operation. This point will be discussed in the next paragraph. Equations (D.7-8) and (D.7-9) are called a *Fourier transform pair*, relating two functions $f(t)$ and its *Fourier transform* $F(\omega)$. If Eq. (D.7-9) is substituted into Eq. (D.7-8), the result is one form of the *Fourier integral theorem*:

$$f(t) = \frac{1}{2\pi} \int_{-\infty}^{\infty} \left\{ \int_{-\infty}^{\infty} f(x) e^{-j\omega x} dx \right\} e^{j\omega t} d\omega \quad (D.7\text{-}10)$$

We proceed now to verify this theorem somewhat more carefully.

Let us consider the function

$$f_A(t) = \frac{1}{2\pi} \int_{-A}^{A} \left\{ \int_{-\infty}^{\infty} f(x) e^{-j\omega x} dx \right\} e^{j\omega t} d\omega \quad (D.7\text{-}11)$$

Interchanging the order of integration, we have

$$f_A(t) = \int\limits_{-\infty}^{\infty} f(x) \left\{ \frac{1}{2\pi} \int\limits_{-A}^{A} e^{j\omega(t-x)} d\omega \right\} dx$$

or

$$f_A(t) = \int\limits_{-\infty}^{\infty} f(x) \frac{\sin A(t-x)}{\pi(t-x)} dx \qquad (D.7\text{-}12)$$

The function $\dfrac{\sin A(t-x)}{\pi(t-x)}$ is one of the approximating forms for the delta-function $\delta(t-x)$ as given by Eq. (B.2-17) and shown in Fig. B.3. Thus Eq. (D.7-12) becomes in the limit

$$\lim_{A \to \infty} f_A(t) = \int\limits_{-\infty}^{\infty} f(x)\delta(t-x) \, dx$$

$$= \frac{1}{2} \left[f(t-0) + f(t+0) \right] \qquad (D.7\text{-}13)$$

and the Fourier integral theorem of Eq. (D.7-10) should be written as

$$\frac{1}{2} \left[f(t-0) + f(t+0) \right] \qquad (D.7\text{-}14)$$

$$= \frac{1}{2\pi} \int\limits_{-\infty}^{\infty} \int\limits_{-\infty}^{\infty} f(x) e^{-j\omega x} \, dx \; e^{j\omega t} \, d\omega$$

which reduces to Eq. (D.7-10) if $f(t)$ is continuous at t.

The use of the delta - function in the verification of the Fourier Integral Theorem has tended to obscure the restrictions which must be placed on $f(t)$ for the theorem to hold. A careful examination of Eq. (D.7-12) as given in [8] will establish these restrictions. Sufficient conditions are that $f(t)$ satisfy the Dirichlet conditions of Appendix A and that $f(t)$ be absolutely convergent; that is,

$$\int\limits_{-\infty}^{\infty} |f(t)| dt < \infty \qquad (D.7\text{-}15)$$

The Fourier transform pair of Eqs. (D.7-8) and (D.7-9) is identical to that given in Chapter 3 as Eqs. (3.6-7) and (3.6-9) and has been shown to be valid for functions which are continuous and absolutely integrable. If the functions are not continuous, then the transform is equal to the mean value of the discontinuity as given by Eq. (D.7-14). As mentioned at the beginning of Section 3.6, Fourier integral theory can be developed using the concepts of "square integrability" and "convergence in the mean". The procedures are similar to those already used in this Appendix and the results are Eqs. (3.6-3) and (3.6-4). A detailed discussion is given in Chapter III of [11].

D.8 - Elementary Properties of Fourier Transforms. The Fourier transform pair $f(t)$ and $F(\omega)$ are related through Eqs. (D.7-8) and (D.7-9), or, equivalently, through the Fourier Integral Theorem of Eq. (D.7-10). Some of the more elementary properties of the transform $F(\omega)$ follow directly from Eq. (D.7-9). It will be convenient sometimes to denote the Fourier transform of a time function by $\mathbf{F}\{\}$ so that Eq. (D.7-9) can be written as

$$F(\omega) = \mathbf{F}\{f(t)\} \qquad (D.8\text{-}1)$$

Since integration is a linear operation, it follows that the Fourier transformation is also *linear*. Thus if the time functions $a(t)$ and $b(t)$ have Fourier transforms $A(\omega)$ and $B(\omega)$, respectively, it follows that

$$\mathbf{F}\{c\ a(t) + d\ b(t)\} = c\ A(\omega) + d\ B(\omega) \qquad (D.8\text{-}2)$$

where c and d are arbitrary constants. The Fourier transform is also *symmetric;* that is, if $f(t)$ is real

$$\mathbf{F}\{f(t)\} = F(\omega) \rightarrow \mathbf{F}\{f(-t)\} = F(-\omega) = F^*(\omega) \quad (D.8\text{-}3)$$

where $F^*(\omega)$ is the complex conjugate of $F(\omega)$. If $f(t)$ is an *even* function, then Eq. (D.7-9) can be written as

$$F(\omega) = \int_{-\infty}^{\infty} f(x)\cos \omega x\ dx = 2\int_{0}^{\infty} f(x)\cos \omega x\ dx \qquad (D.8\text{-}4)$$

while, if $f(t)$ is an *odd* function,

$$F(\omega) = -j\int_{-\infty}^{\infty} f(x)\sin \omega x\ dx = -2j\int_{0}^{\infty} f(x)\sin \omega x\ dx \quad (D.8\text{-}5)$$

Suppose that the time function $f(t)$ with Fourier transform $F(\omega)$ is translated by an amount t_0 to form the new function $f(t-t_0)$. The Fourier transform of $f(t-t_0)$ is

$$\mathbf{F}\{f(t-t_0)\} = \int_{-\infty}^{\infty} f(x-t_0)e^{-j\omega x}\ dx \qquad (D.8\text{-}6)$$

Let us make the linear change of variable $y = x - t_0$ so that

$$\mathbf{F}\{f(t-t_0)\} = \int\limits_{-\infty}^{\infty} f(y)e^{-j\omega y}e^{-j\omega t_0}dy$$

$$= e^{-j\omega t_0}\int\limits_{-\infty}^{\infty} f(y)e^{-j\omega y}dy \qquad (D.8\text{-}7)$$

$$= e^{-j\omega t_0}F(\omega)$$

Thus translation in the time domain by an amount t_0 corresponds to multiplication in the frequency domain by the factor $e^{j\omega t_0}$. In the same way, it is easy to show that

$$\mathbf{F}\{f(t)e^{j\omega_0 t}\} = F(\omega-\omega_0) \qquad (D.8\text{-}8)$$

the derivative $\dfrac{df(t)}{dt}$ may be transformed as

$$\mathbf{F}\{\frac{df(t)}{dt}\} = \int\limits_{-\infty}^{\infty} \frac{df(t)}{dt}e^{j\omega t}dt$$

Integration by parts yields

$$\mathbf{F}\{\frac{df(t)}{dt}\} = f(t)e^{-j\omega t}\Big|_{-\infty}^{\infty} + j\omega\int\limits_{-\infty}^{\infty} f(t)e^{-j\omega t}dt$$

or

$$\mathbf{F}\{\frac{df(t)}{dt}\} = j\omega F(\omega) \qquad (D.8\text{-}9)$$

if $f(\infty) = f(-\infty)=0$. This last condition must hold in any case if $f(t)$ is to be absolutely integrable and, hence, transformable. Successive applications of the same procedure yields

$$\mathbf{F}\{\frac{d^n}{dt^n}[f(t)]\} = [j\omega]^n F(\omega) \qquad (D.8\text{-}10)$$

Some of the more commonly encountered Fourier transform pairs are given in Table 3.1 of Chapter 3.

D.9 - *The Laplace Transform.* In a wide variety of problems, particularly those involving the analysis of linear systems with non-random inputs, it is necessary or desirable to assume that the driving forces are applied at some fixed time, say $t=0$. In such cases, a given driving force or signal is of interest only for positive time since its behavior for negative time is irrelevant to the problems. The Laplace transform $F_l(s)$ of a time function $f(t)$ is defined as

$$F_l(s) = \int_0^\infty f(t)e^{-st}\, dt \tag{D.9-1}$$

where s is a complex variable with real part σ and imaginary part $j\omega$ so that

$$s = \sigma + j\omega \tag{D.9-2}$$

Note that integration in the time domain takes place only over the interval $(0,\infty)$. Thus any two different time functions $f_1(t)$ and $f_2(t)$ that are identical for positive time will have the same Laplace transform.

It has already been pointed out in Section D.7 that a sufficient condition for the existence of the Fourier transform $F(\omega)$ of $f(t)$ is that $f(t)$ be absolutely integrable as given by Eq. (D.7-15). In the same way the Laplace transform $F_l(s)$ exists if

$$\int_0^\infty |f(t)|e^{-\sigma t}\, dt < \infty \tag{D.9-3}$$

for some real number σ. Most functions encountered in engineering satisfy this condition. Indeed many functions which do not have Fourier transforms will have Laplace transforms due to the rapid decay of $e^{-\sigma t}$. For example, for all n,

$$\lim_{n \to \infty} x^n e^{-x} = 0$$

and,

$$\int_0^\infty x^n e^{-x}\, dx = \Gamma(n+1)$$

where $\Gamma(n)$ is the Gamma function. On the other hand the Fourier transform of x^n does not exist for $n > 0$ and, for $n = 0$, can only be expressed as a delta-function.

The inversion of Eq. (D.9-1) to find $f(t)$ from $F_l(s)$ is not as simple as for the Fourier transform. It can be shown [5,12] that $f(t)$ is given by

$$f(t) = \frac{1}{2\pi j} \int_{\sigma_1 - j\infty}^{\sigma_1 + j\infty} F(s)e^{st}\, ds \tag{D.9-4}$$

where σ_1 is a number greater that the σ of Eq. (D.9-3). This expression for $f(t)$ is a contour integral in the complex s-plane and may be difficult to evaluate. The usual procedure is to avoid the inverse transformation and build up a table of Laplace transform pairs $f(t)$ and $F_l(s)$ from the direct transform of Eq. (D.9-1). The function $f(t)$ given by Eq. (D.9-4) corresponds to the $f(t)$ of (D.9-1) for positive time and is zero for negative time.

It can be shown in a straightforward manner that the Laplace transform has many of the same properties as the Fourier transform. However, there are certain differences between the two types of transforms. For example, the lower limit in the Laplace transform expression is $t=0$. Thus, the Laplace transform L{ } of the derivative of a function becomes.

$$\mathrm{L}\{\frac{df\ (t)}{dt}\} = \int_0^\infty \frac{df\ (t)}{dt}\ e^{-st}\ dt$$

$$= e^{-st}\ f\ (t)\ \Big|_0^\infty + s\ \int_0^\infty f\ (t)e^{-st}\ dt$$

$$= s\ F_l\ (s) - f\ (0)$$

The initial condition $f(0)$ enters into the transform of the derivative. In the solution of ordinary differential equations by the use of the Laplace transform, the initial conditions associated with the equations are automatically included.

It is clear that the Laplace and Fourier transforms are distinct. Any function $f(t)$ which is non-zero for negative time can not be represented uniquely as an inverse Laplace transform. On the other hand, many functions will have Laplace transforms which do not have Fourier transforms. However, if

$$f\ (t) = 0\ ,\ t\ < 0$$

and

$$\int_0^\infty |f\ (t)|dt\ < \infty$$

the number σ in Eq. (D.9-1) can be set equal to zero and

$$\mathrm{F}\{f\ (t)\} = \mathrm{L}\{f\ (t)\}_{s\ =j\omega}$$

D.10 - Signal Energy and the Power Spectrum. Consider the two time functions $f_1(t)$ and $f_2(t)$ with Fourier transforms $F_1(\omega)$ and $F_2(\omega)$, respectively. Equation (D.7-8) may be used to write

$$(D.10\text{-}1)$$

$$\int_{-\infty}^{\infty} f_1(t)f_2(t)\,dt = \frac{1}{(2\pi)^2} \int_{-\infty}^{\infty} \int_{-\infty}^{\infty} F_1(x)e^{jxt}\,dx \int_{-\infty}^{\infty} F_2(y)e^{jyt}\,dy\,dt$$

This expression may be rearranged to give

$$(D.10\text{-}2)$$

$$\int_{-\infty}^{\infty} f_1(t)f_2(t)\,dt = \frac{1}{2\pi} \int_{-\infty}^{\infty} \int_{-\infty}^{\infty} F_1(x)F_2(y) \left[\frac{1}{2\pi} \int_{-\infty}^{\infty} e^{jt(x+y)}dt \right] dx\,dy$$

The integral in brackets may be evaluated from Pair No. 7 of Table 3.1 and is $\delta(x+y)$. Now Eq. (D.10-2) becomes

$$(D.10\text{-}3)$$

$$\int_{-\infty}^{\infty} f_1(t)f_2(t)\,dt = \frac{1}{2\pi} \int_{-\infty}^{\infty} F_1(x)F_2(-x)\,dx = \frac{1}{2\pi} \int_{-\infty}^{\infty} F_1(-x)F_2(x)\,dx$$

For the special case where $f_1(t)=f_2(t)=f(t)$, this reduces to

$$\int_{-\infty}^{\infty} f^2(t)\,dt = \frac{1}{2\pi} \int_{-\infty}^{\infty} |F(\omega)|^2 d\omega \qquad (D.10\text{-}4)$$

This last equation gives the signal energy expressed either in the time or frequency domain. It is one of the many forms of *Parseval's Theorem* which has already been given in terms of the coefficients of a Fourier series by Eqs. (D.4-13) and (D.5-9). The quantity $|F(\omega)|^2$ is sometimes called the *power spectrum* of $f(t)$ but it is more conventional to define power spectrum somewhat differently.

REFERENCES

1. P.M. Morse and H. Feshbach, *Methods of Theoretical Physics,* Vols. I and II, McGraw-Hill Book Company, Inc., New York, N.Y., 1978.

2. H. Margenau and G.M. Murphy, *The Mathematics of Physics and Chemistry,* Vol. I, D. Van Nostrand Company, Inc., Princeton, N.J., 1956.

3. W. Magnus and F. Oberhettinger, *Formulas and Theorems for the Special Functions of Mathematical Physics,* Springer-Verlag, New York, N.Y., 1966.

4. R.V. Churchill, *Fourier Series and Boundary Value Problems,* McGraw-Hill Book Company, Inc., New York, N.Y., 1963.

5. W. Kaplan, *Operational Methods for Linear Systems,* Addison-Wesley Publishing Company, Inc., Reading, Mass. 1962.

6. T.M. Apostol, *Mathematical Analysis,* Addison-Wesley Publishing Company, Inc. Reading, Mass., 1974.

7. A. Zygmund, *Trigonometric Series,* Cambridge University Press, London, 1968.

8. H.S. Carslaw, *Introduction to the Theory of Fourier's Series and Integrals,* Dover Publications, Inc., New York, N.Y., 1930.

9. E. Jahnke and F. Emde, *Tables of Functions,* Dover Publications, Inc., New York, N.Y., 1945.

10. *Handbook of Mathematical Tables,* Chemical Rubber Publishing Company, Cleveland, Ohio, 1967.

11. E.C. Titchmarsh, *Introduction to the Theory of Fourier Integrals,* Oxford University Press, Oxford, 1950.

12. H.S. Carslaw and J.C. Jaeger, *Operational Methods in Applied Mathematics,* Dover Publications, Inc., New York, N.Y., 1963.

APPENDIX E

SOME INEQUALITIES INCLUDING SCHWARZ'S INEQUALITY

E.1 - Introduction. There are several inequalities of considerable utility in many areas of analysis [1,2,3,4]. We discuss and develop these briefly in the following sections. We begin with an elementary inequality which will be useful in later proofs.

Let a, b be two real non-negative numbers

$$a > 0 \quad , \quad b > 0 \tag{E.1-1}$$

and let x be a real variable. Form the function

$$y = F(x;a,b) = xa + (1-x)b - a^x b^{1-x} \tag{E.1-2}$$

Then it is apparent that

$$F(0;a,b) = F(1;a,b) = 0 \tag{E.1-3}$$

that is, the curve of y vs. x passes through (0,0) and (1,0). Furthermore, on differentiating twice with respect to x, we have

$$\frac{d^2 F(x;a,b)}{dx^2} = -[\log \frac{b}{a}]^2 a^x b^{1-x} \begin{cases} < 0 \quad , \quad a \neq b \\ = 0 \quad , \quad a = b \end{cases} \tag{E.1-4}$$

Note also that, for $a = b$,

$$F(x;a,a) = 0 \tag{E.1-5}$$

Thus, for $a = b$, the curve of y vs. x is the x-axis and, for $a \neq b$, it is concave downward and passes through the points (0,0) and (1,0). In other words

$$F(x;a,b) > 0 \quad , \quad 0 < x < 1 \quad , \quad a \neq b \tag{E.1-6}$$

It follows from Eqs. (E.1-5) and (E.1-6) and from the definition of $F(x;a,b)$ as given by Eq. (E.1-1) that

$$a^x b^{1-x} \leq xa + (1-x)b \tag{E.1-7}$$

for

$$a > 0 \, , \, b > 0 \, , \, 0 < x < 1$$

which is the result desired.

E.2 - Holder's Inequality. Let $f(t)$ and $g(t)$ be functions of the real variable t in the interval (a,b). Assume that

$$\int_a^b |f(t)|^p \, dt < \infty \qquad (E.2\text{-}1)$$

and

$$\int_a^b |g(t)|^q \, dt < \infty \qquad (E.2\text{-}2)$$

where p and q are real numbers related by

$$\frac{1}{p} + \frac{1}{q} = 1 \qquad (E.2\text{-}3)$$

and $p > 1$. We note first that, for any t, either

$$|g(t)| \leq |f(t)|^{p-1} \qquad (E.2\text{-}4)$$

and hence

$$|f(t)g(t)| \leq |f(t)|^q \qquad (E.2\text{-}5)$$

or else

$$|g(t)| > |f(t)|^{p-1} \qquad (E.2\text{-}6)$$

$$|f(t)| \leq |g(t)|^{1/(p-1)} = |g(t)|^{q-1} \qquad (E.2\text{-}7)$$

and hence

$$|f(t)g(t)| \leq |g(t)|^q \qquad (E.2\text{-}8)$$

It follows from Eqs. (E.2-1) and (E.2-2) that

$$\int_a^b |f(t)g(t)| \, dt < \infty \qquad (E.2\text{-}9)$$

which is the first result desired.

Let us assume that neither $f(t)$ or $g(t)$ vanishes so that

$$f(t)g(t) \neq 0 \qquad (E.2\text{-}10)$$

In Eq. (E.1-6) we let

$$x = \frac{1}{p} \ , \quad 1 - x = \frac{1}{q} \qquad (E.2\text{-}11)$$

and

$$a = \frac{|f(t)|^p}{\int_a^b |f(t)|^p \, dt} \qquad (E.2\text{-}12)$$

$$b = \frac{|g(t)|^q}{\int_a^b |g(t)|^q \, dt} \qquad \text{(E.2-13)}$$

Now Eq. (E.1-7) becomes, for each t,

$$\frac{|f(t)g(t)|}{\left[\int_a^b |f(t)|^p \, dt\right]^{1/p} \left[\int_a^b |g(t)|^q \, dt\right]^{1/q}}$$

$$\text{(E.2-14)}$$

$$\leq \frac{|f(t)|^p}{p\int_a^b |f(t)|^p \, dt} + \frac{|g(t)|^q}{q\int_a^b |g(t)|^q \, dt}$$

We now integrate both sides with respect to t in the interval (a, b) to obtain

$$\frac{\int_a^b |f(t)g(t)| dt}{\left[\int_a^b |f(t)|^p \, dt\right]^{1/p} \left[\int_a^b |g(t)|^q \, dt\right]^{1/q}} \leq \frac{1}{p} + \frac{1}{q} = 1 \quad \text{(E.2-15)}$$

On rearranging, we have *Holder's Inequality:*

$$\text{(E.2-16)}$$

$$\int_a^b |f(t)g(t)| dt \leq \left[\int_a^b |f(t)|^p \, dt\right]^{1/p} \left[\int_a^b |g(t)|^q \, dt\right]^{1/q} < \infty$$

where, as before, p and q are related by

$$\frac{1}{p} + \frac{1}{q} = 1$$

A weaker inequality may be written which will be more convenient for our purpose. Since

$$\left|\int_a^b f(t)g(t)dt\right| \leq \int_a^b |f(t)g(t)| dt \qquad \text{(E.2-17)}$$

then

$$\left| \int_a^b f(t)g(t)dt \right| \leq \left[\int_a^b |f(t)|^p \, dt \right]^{1/p} \left[\int_a^b |g(t)|^q \, dt \right]^{1/q}$$

$$\text{(E.2-18)}$$

$$\frac{1}{p} + \frac{1}{q} = 1$$

Holder's Inequality for sums is proved in a similar way. Let f_n and g_n be sequences of numbers such that

$$\sum_n |f_n|^p < \infty \qquad \text{(E.2-19)}$$

and

$$\sum_n |g_n|^q < \infty \qquad \text{(E.2-20)}$$

where p and q satisfy Eq. (E.2-3). By an argument identical to that used in arriving at Eq. (E.2-9), we conclude that

$$\sum_n |f_n g_n| < \infty \qquad \text{(E.2-21)}$$

As before we assume that neither f_n or g_n vanishes so that

$$f_n g_n \neq 0 \qquad \text{(E.2-22)}$$

In Eq. (E.1-7), we let

$$x = \frac{1}{p} \quad , \quad 1 - x = \frac{1}{q} \qquad \text{(E.2-23)}$$

and

$$a = \frac{|f_n|^p}{\sum_n |f_n|^p} \qquad \text{(E.2-24)}$$

$$b = \frac{|g_n|^q}{\sum_n |g_n|^q} \qquad \text{(E.2-25)}$$

After summing both sides of the resulting expression and using Eq. (E.2-3), we obtain Holder's Inequality for sums:

$$\sum_n |f_n\, g_n| \le \left[\sum_n |f_n|^p\right]^{1/p} \left[\sum_n |g_n|^q\right]^{1/q}$$

(E.2-26)

$$\frac{1}{p} + \frac{1}{q} = 1$$

or, since

$$\left|\sum_n f_n\, g_n\right| \le \sum_n |f_n\, g_n|$$

(E.2-27)

A weaker but sometimes more useful form is

$$\left|\sum_n f_n\, g_n\right| \le \left[\sum_n |f_n|^p\right]^{1/p} \left[\sum_n |g_n|^q\right]^{1/q}$$

(E.2-28)

$$\frac{1}{p} + \frac{1}{q} = 1$$

E.3 - Minkowski's Inequality. Suppose that two functions $f_1(t)$ and $f_2(t)$ satisfy Eq. (E.2-1) where $f(t)$ is $f_1(t)$ or $f_2(t)$ with $p > 1$. It is clear that, for each t,

$$|f_1(t) \pm f_2(t)|^2 \le \begin{cases} |2f_1(t)|^p & \text{for } f_1(t) \ge f_2(t) \\ |2f_2(t)|^p & \text{for } f_1(t) < f_2(t) \end{cases}$$

(E.3-1)

Hence

$$\int_a^b |f_1(t) \pm f_2(t)|^p\, dt \le 2^p \int_a^b |f_1(t)|^p\, dt + 2^p \int_a^b |f_2(t)|^p\, dt < \infty$$

(E.3-2)

for finite p. We now note that

$$\int_a^b |f_1(t) \pm f_2(t)|^p\, dt \le \int_a^b |f_1(t) \pm f_2(t)|^{p-1} |f_1(t)|\, dt$$

(E.3-3)

$$+ \int_a^b |f_1(t) \pm f_2(t)|^{p-1} |f_2(t)|\, dt$$

Let us apply Holder's Inequality to each of the integrals on the right with

$$g(t) = |f_1(t) + f_2(t)|^{p-1} \qquad \text{(E.3-4)}$$

$$f(t) = |f_k(t)| \quad , \quad k = 1,2 \qquad \text{(E.3-5)}$$

The result is

$$\int_a^b |f_1(t) \pm f_2(t)|^p \, dt \leq \left[\int_a^b |f_1(t)^p \, dt\right]^{1/p} \left[\int_a^b |f_1(t) \pm f_2(t)|^{(p-1)q} \, dt\right]^{1/q}$$

$$+ \left[\int_a^b |f_2(t)|^p \, dt\right]^{1/p} \left[\int_a^b |f_1(t) \pm f_2(t)|^{(p-1)q} \, dt\right]^{1/q} \qquad \text{(E.3-6)}$$

Note that $(p-1)q = p$ and that $1-1/q = 1/p$. Thus, on dividing both sides of this last inequality by the common term on the right side we obtain Minkowski's Inequality:

$$\left[\int_a^b |f_1(t) \pm f_2(t)|^p \, dt\right]^{1/p} \leq \left[\int_a^b |f_1(t)|^p \, dt\right]^{1/p} + \left[\int_a^b |f_2(t)|^p \, dt\right]^{1/p}$$

$$\text{(E.3-7)}$$

$$p \geq 1$$

Actually we have proved the inequality only for $p > 1$, but the case $p = 1$ is a trivial result of the triangle inequality

$$|x \pm y| \leq |x| + |y|$$

Minkowski's Inequality for sums is proved in a similar way and is

$$\left[\sum_n |f_{1n} \pm f_{2n}|^p \right]^{1/p} \leq \left[\sum_n |f_{1n}|^p \right]^{1/p} + \left[\sum_n |f_{2n}|^p \right]^{1/p}$$

$$\text{(E.3-9)}$$

$$p \geq 1$$

E.4 - Schwarz's Inequality. This is a special case of Holder's Inequality obtained by placing $p = q = 2$ in Eq. (E.2-18) or (E.2-28) to obtain

$$\left|\int_a^b f(t)g(t)dt\right| \leq \left[\int_a^b |f(t)|^2 dt\right]^{1/2} \left[\int_a^b |g(t)|^2 dt\right]^{1/2} \qquad \text{(E.4-1)}$$

or

$$\left| \sum_n f_n \, g_n \right| \le \left[\sum_n |f_n|^2 \right]^{1/2} \left[\sum_n |g_n|^2 \right]^{1/2} \qquad \text{(E.4-2)}$$

Since this inequality is relatively easy to prove directly, it will be instructive to do so.

We assume that $f(t)$ and $g(t)$, functions of the real variable t, satisfy Eqs. (E.2-1) and (E.2-2) with $p = q = 2$. Hence they satisfy Eq. (E.2-9). Now form the integral

$$\int_a^b \{\lambda \, f^*(t) + g^*(t)\} \, \{\lambda \, f(t) + g(t)\} \, dt$$

$$\text{(E.4-3)}$$

$$= \lambda^2 A + \lambda (B + B^*) + C = k(\lambda) \ge 0$$

where λ is a real variable, the symbol * means complex conjugate, and

$$A = \int_a^b |f(t)|^2 dt \ge 0 \qquad \text{(E.4-4)}$$

$$B = \int_a^b f^*(t) g(t) \, dt \qquad \text{(E.4-5)}$$

$$C = \int_a^b |g(t)|^2 dt \ge 0 \qquad \text{(E.4-6)}$$

The integral of Eq. (E.4-3) exists, is real, and is a non-negative function of λ, say $k(\lambda)$. Since $k(\lambda)$ is non-negative, it must have no real roots except possibly a double root. From the quadratic formula, then, we must have

$$\left(\frac{B + B^*}{2} \right)^2 = (ReB)^2 \le AC \qquad \text{(E.4-7)}$$

or

$$(B + B^*)^2 \le 4AC \qquad \text{(E.4-8)}$$

on substituting for A, B, B^*, and C we obtain

$$\left[\int_a^b [f^*(t) g(t) + f(t) g^*(t)] \right]^2 \le 4 \int_a^b |f(t)|^2 dt \int_a^b |g(t)|^2 dt \qquad \text{(E.4-9)}$$

This is the form of Schwarz's Inequality that is appropriate for complex functions $f(t)$ and $g(t)$, where, for example

$$f(t) = f_1(t) + j \, f_2(t) \qquad \text{(E.4-10)}$$

$$g(t) = g_1(t) + j\ g_2(t) \qquad \text{(E.4-11)}$$

and $f_1(t)$, $f_2(t)$, $g_1(t)$, and $g_2(t)$ are real. For the case where $f(t)$ and $g(t)$ are real, then

$$f^*(t)g(t) + f(t)g^*(t) = 2f(t)g(t) \qquad \text{(E.4-12)}$$

and Eq. (E.4-1) follows immediately.

Note that equality is obtained (aside from the trivial case where $f(t)$ or $g(t)$ is zero) when the double root exists in Eq. (E.4-3); that is, when

$$\lambda f(t) + g(t) = \lambda f^*(t)g^*(t) = 0 \qquad \text{(E.4-13)}$$

Since λ is real, then $f(t)$ and $g(t)$ are linearly related. Looking at the problem from a slightly different point of view, we see that there is a real value of λ for which Eq. (E.4-3) is zero and for which its first derivative with respect to λ vanishes; that is,

$$2\lambda A + (B + B^*) = 0 \qquad \text{(E.4-14)}$$

or

$$\lambda = -\frac{B + B^*}{2A} = -\frac{\int_a^b [f^*(t)g(t) + f(t)g^*(t)]dt}{2\int_a^b |f(t)|^2 dt} \qquad \text{(E.4-15)}$$

Eq. (E.4-15) holds if and only if

$$g(t) = -\lambda f(t) \qquad \text{(E.4-16)}$$

This last relationship is equivalent to Eq. (E.4-13).

REFERENCES

1. E.C. Titchmarsh, *The Theory of Functions*, Oxford University Press, London, 1939.

2. M.E. Munroe, *Introduction to Measure and Integration*, Addison-Wesley Publishing Company, Inc., Reading, Massachusetts, 1953.

3. N. Dunford and J.T. Schwartz, *Linear Operators, Part I: General Theory*, Interscience Publishers, Inc., New York, N.Y., 1958.

4. A.C. Zaanen, *Linear Analysis*, North-Holland Publishing Company, Amsterdam, Holland, 1960.

APPENDIX F

THE CALCULUS OF VARIATIONS

F.1 Maxima and Minima of a Function. We begin by reviewing some of the elementary notions of maxima and minima [1,2]. Suppose we are given the function $y = f(x)$, assumed to be single-valued and representable by Taylor's expansion:

$$f(x+h) = f(x) + \frac{h}{1!} f^{(1)}(x) + \frac{h^2}{2!} f^{(2)}(x) + ... \quad \text{(F.1-1)}$$

$$... + \frac{h^n}{n!} f^{(n)}(x) + ...$$

where

$$f^{(i)}(x) = \frac{d^i f(x)}{dx^i} \quad \text{(F.1-2)}$$

Consider the function at the point $x = a$ and form the difference

$$\Delta(h) = f(a+h) - f(a) = hf^{(1)}(a) + \frac{h^2}{2!} f^{(2)}(a) + ... \quad \text{(F.1-3)}$$

Let $f(a)$ be either a maximum or a minimum as in Fig. F.1.

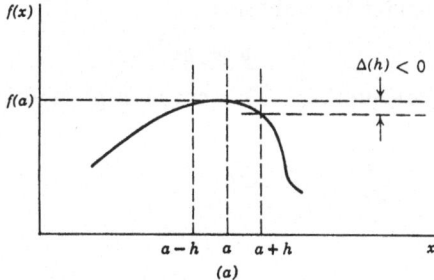

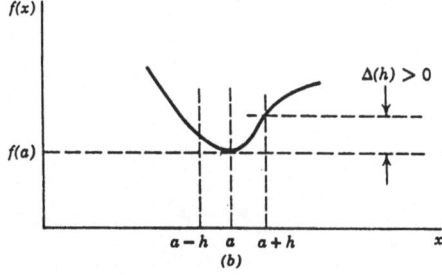

Fig. F.1 - Maxima and minima

Then, for sufficiently small h, the sign of $\Delta(h)$ must equal the sign of $\Delta(-h)$. However, for small h, the difference $\Delta(h)$ of Eq. (F.1-3) is given approximately by

$$\Delta(h) \approx h f^{(1)}(a) \qquad (F.1-4)$$

Thus, for the sign of $\Delta(h)$ to be unchanged when h changes sign, it follows that

$$f^{(1)}(a) = 0 \qquad (F.1-5)$$

is a *necessary* condition for $f(x)$ to have a maximum or minimum at $x = a$. We say that $f(x)$ is *stationary* at the point $x = a$ if Eq. (F.1-5) is satisfied. Not all stationary points are maxima or minima, however.

Suppose that Eq. (F.1-5) is satisfied. Then, for small h,

$$\Delta(h) \approx \frac{h^2}{2!} f^{(2)}(a) \qquad (F.1-6)$$

It is apparent from Fig. F.1 that $\Delta(h) < 0$ for a maximum and $\Delta(h) > 0$ for a minimum. Therefore, if the second derivative is non-zero,

$$f^{(2)}(a) = \begin{cases} < 0 \;\; at \;\; a \;\; maximum \\ > 0 \;\; at \;\; a \;\; minimum \end{cases} \qquad (F.1-7)$$

Example F.1

Let us consider the function

$$y = bx^2$$

where b is a real number. This curve is plotted in Fig. F.2(a). It is clear that

$$\frac{dy}{dx} = 2bx = 0 \qquad at \quad x = 0$$

and that

$$\frac{d^2y}{dx^2} = 2b = \begin{cases} < 0, \, b < 0 \, (max) \\ > 0, \, b > 0 \, (min) \end{cases}$$

Example F.2

Suppose that

$$y = x^3$$

as shown in Fig. F.2(b). We have that

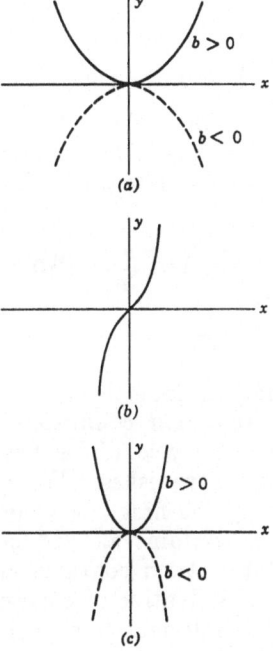

Fig. F.2 - Some functions with stationary values

$$\frac{dy}{dx} = 3x^2 = 0 \qquad at \ x = 0$$

$$\frac{d^2y}{dx^2} = 6x = 0 \qquad at \ x = 0$$

Here both first and second derivatives are zero at the origin. The function has no maximum or minimum, but a *point of inflection* at the origin.

Example F.3

Consider the function

$$y = bx^4$$

shown in Fig. F.2(c). It is clear that

$$\frac{dy}{dx} = 4bx^3 = 0 \qquad at \ x = 0$$

$$\frac{d^2y}{dx^2} = 12bx^2 = 0 \qquad at \ x = 0$$

$$\frac{d^3y}{dx^3} = 24bx = 0 \qquad at \ x = 0$$

$$\frac{d^4y}{dx^4} = 24b = \begin{cases} < 0, \ b < 0 \ (max) \\ > 0, \ b > 0 \ (min) \end{cases}$$

In this example, the first three derivatives are zero at the origin; nevertheless the function has either a maximum or a minimum there. Note that the appropriate approximation to the difference $\Delta(h)$ is

$$\Delta(h) \approx \frac{h^4}{4!} f^{(4)}(0)$$

It is apparent that Taylor's theorem provides the basis for a general statement of *sufficient* conditions for a maximum or a minimum when the *necessary* condition has been fulfilled; that is, when the first derivative vanishes. We consider the difference $\Delta(h)$ as approximated by the first non-vanishing term of the Taylor series. If this term contains an odd power of h, we have a point of inflection. If this term contains an even power of h we have a maximum if the derivative of corresponding order is negative and a minimum if the derivative is positive.

F.2 Maxima and Minima of Functions of Two or More Variables. In this section we will consider the function $f(x,y)$ of the two variables x and y. Extension to the case of more than two variables will be obvious. As before, we begin with Taylor's expansion of f :

$$f(x+h,y+k) = f(x,y) + (h\frac{\partial f}{\partial x} + k\frac{\partial f}{\partial y}) +$$

$$\frac{1}{2!} \left[h^2\frac{\partial^2 f}{\partial x^2} + 2hk\frac{\partial^2 f}{\partial x \partial y} + k^2\frac{\partial^2 f}{\partial y^2} \right] + \cdots \qquad (F.2\text{-}1)$$

$$\ldots + \frac{1}{n!} \left[\sum_{i=0}^{n} \binom{n}{i} h^i k^{n-i} \frac{\partial^n f}{\partial x^i \partial y^{n-i}} \right] + \ldots$$

where $\binom{n}{i}$ is the binomial coefficient given by

$$\binom{n}{i} = \frac{n!}{i!(n-i)!} \qquad (F.2\text{-}2)$$

Let us form the difference $\Delta(h,k)$ given by

$$\Delta(h,k) = f(a+h,b+k) - f(a,b) \qquad (F.2\text{-}3)$$

and note that, for small h and k, it is given approximately by

$$\Delta(h,k) \approx \left(h \frac{\partial f}{\partial x} + k \frac{\partial f}{\partial y} \right)_{x=a,\ y=b} \qquad (F.2\text{-}4)$$

It is clear that, for $\Delta(h,k)$ to have the same sign independently of the signs of h and k, it is necessary that

$$\frac{\partial f}{\partial x} = 0 \quad , \quad \frac{\partial f}{\partial y} = 0 \qquad at \ x=a,\ y=b \qquad (F.2\text{-}5)$$

These are necessary conditions for $f(x,y)$ to have a maximum or a minimum at (a,b). If these conditions are satisfied then, for small h and k,

$$\Delta(h,k) \approx \frac{1}{2!} \left(h^2 \frac{\partial^2 f}{\partial x^2} + 2hk \frac{\partial^2 f}{\partial x\,\partial y} + k^2 \frac{\partial^2 f}{\partial y^2} \right)_{x=a,y=b} \qquad (F.2\text{-}6)$$

This expression may be written as

$$\Delta(h,k) \approx \frac{1}{2!} \left(h^2 A + 2hkB + k^2 C \right)$$

$$\qquad (F.2\text{-}7)$$

$$\approx \frac{1}{2!} \frac{(Ah+Bk)^2+(AC-B^2)k^2}{A}$$

Since $(Ah+Bk)^2$ is always non-negative, it is clear that the sign of $\Delta(h,k)$ is independent of the signs of h and k if

$$D = AC - B^2 \geq 0 \qquad (F.2\text{-}8)$$

or

$$\frac{\partial^2 f}{\partial x^2} \frac{\partial^2 f}{\partial y^2} \geq \left(\frac{\partial^2 f}{\partial x\,\partial y} \right) \geq 0 \qquad at \ x=a,\ y=b \qquad (F.2\text{-}9)$$

Thus, when this last equation is satisfied,

$$f(a,b) = \begin{cases} max. \ if \ \dfrac{\partial^2 f}{\partial x^2} < 0, \ \dfrac{\partial^2 f}{\partial y^2} < 0 & at \ x=a,y=b \\[2mm] & \qquad (F.2\text{-}10) \\[1mm] min. \ if \ \dfrac{\partial^2 f}{\partial x^2} > 0, \ \dfrac{\partial^2 f}{\partial y^2} > 0 & at \ x=a,y=b \end{cases}$$

if $f(a,b)$ has a minimum or maximum at $x=a,y=b$. In other words, the function $f(x,y)$ is a max at $x=a$, $y=b$ if $h^2 A + 2hkB + k^2 c < 0$ and a min at $x=a$, $y=b$ if $h^2 A + 2hkB + k^2 c > 0$.

Example F.4

Find the maxima and minima of the surface

$$f(x,y) = \frac{1}{2c} \left(\frac{x^2}{a^2} - \frac{y^2}{b^2} \right)$$

Equation (F.2-5) becomes

$$0 = \frac{\partial f}{\partial x} = \frac{x}{ca^2} \ , \ 0 = \frac{\partial f}{\partial y} = -\frac{y}{cb^2}$$

These partial derivatives vanish when $x = y = 0$. The second derivatives are

$$\frac{\partial^2 f}{\partial x^2} = \frac{1}{ca^2} \ , \ \frac{\partial^2 f}{\partial x \, \partial y} = 0 \ , \ \frac{\partial^2 f}{\partial y^2} = -\frac{1}{cb^2}$$

Note that the condition of Eq. (F.2-8) cannot be satisfied since $\frac{\partial^2 f}{\partial x^2}$ and $\frac{\partial^2 f}{\partial y^2}$ have opposite signs. In other words

$$D = AC - B^2 = -1/a^2 b^2 c^2 < 0$$

The function $f(x,y)$ is a hyperbolic paraboloid. It has a *saddle-point* or *minimax* point at the origin. These are points where the first partial derivatives vanish but where $D < 0$.

F.3 Lagrange's Method of Undetermined Multipliers. It will not always be true that the several variables involved are independent. A common problem is to find the maxima or minima of a function $u(x,y,z)$ where the variables x,y, and z are related through

$$\phi(x,y,z) = K \ , \ a \ constant \tag{F.3-1}$$

If u is to have a stationary value, then the total differential du must be zero:

$$du = 0 = \frac{\partial u}{\partial x} \, dx + \frac{\partial u}{\partial y} \, dy + \frac{\partial u}{\partial z} \, dz \tag{F.3-2}$$

It follows from Eq. (F.3-1) that the total differential $d\phi$ must also be zero:

$$d\phi = 0 = \frac{\partial \phi}{\partial x} \, dx + \frac{\partial \phi}{\partial y} \, dy + \frac{\partial \phi}{\partial z} \, dz \tag{F.3-3}$$

Now suppose we multiply $d\phi$ by some parameter λ and add the product to du. We have

$$du + \lambda d\phi = 0 \tag{F.3-4}$$

Since Eq. (F.3-4) involves arbitrary differentials dx, dy, and dx, we conclude that

$$\frac{\partial u}{\partial x} + \lambda \frac{\partial \phi}{\partial x} = 0$$

$$\frac{\partial u}{\partial y} + \lambda \frac{\partial \phi}{\partial y} = 0 \tag{F.3-5}$$

$$\frac{\partial u}{\partial z} + \lambda \frac{\partial \phi}{\partial z} = 0$$

These last equations together with Eq. (F.3-1) give four relations in the four unknowns; namely, λ and the values of x, y, and z determining the stationary point location (x, y, z).

More generally, to determine the stationary values of the function $u(x_1, x_2, ..., x_n)$ where the n variables x_i are subject to the m constraints

$$\phi_i(x_1, x_2, ..., x_n) = K_i \quad , \quad i = 1, 2, ..., m \tag{F.3-6}$$

we form the function

$$f = u + \sum_{i=1}^{m} \phi_i \lambda_i \tag{F.3-7}$$

We find the set of λ_i and the values of $x_1, x_2, ..., x_n$ which determine the stationary point $(x_1, x_2, ..., x_n)$ from the m equations given by Eq. (F.3-6) and from the n equations

$$\frac{\partial F}{\partial x_j} = 0 \quad , \quad j = 1, 2, ..., n \tag{F.3-8}$$

Example F.5

We desire to find the volume of the largest parallelepiped that can be inscribed in the ellipsoid

$$\frac{x^2}{a^2} + \frac{y^2}{b^2} + \frac{z^2}{c^2} = 1 = \phi(x, y, z)$$

Since the ellipsoid is symmetrically located with respect to the origin, the same will be true of the inscribed parallelepiped with volume

$$u(x, y, z) = 8\, xyz \tag{A}$$

Equation (F.3-5) becomes

$$8yz + \frac{2\lambda}{a^2}x = 0 \tag{B}$$

$$8xz + \frac{2\lambda}{b^2}y = 0 \tag{C}$$

$$8xy + \frac{2\lambda}{c^2}z = 0 \tag{D}$$

Each of these last three equations may be multiplied by x, y, and z respectively and the result summed to give

$$3u + 2\lambda\phi = 3u + 2\lambda = 0$$

or

$$2\lambda = -3u$$

This last expression for λ may be substituted in Eq. (B) which, after multiplication by x, becomes

$$u\left(1 - \frac{3}{a^2} x^2\right) = 0$$

$$x = a / \sqrt{3}$$

In the same way

$$y = b / \sqrt{3}$$

$$z = c / \sqrt{3}$$

and the volume u is

$$u = \frac{8abc}{3\sqrt{3}}$$

F.4 The Fundamental Problem of the Calculus of Variations. The fundamental problem of the calculus of variations [3,4,5] may be stated as follows: Let us define the quantity I by the definite integral

$$I = \int_{x=a}^{x=b} F(x,y,y')\, dx \qquad \text{(F.4-1)}$$

where $F(\mu,\nu,\sigma)$ is a known function, y' is dy/dx and $y = y(x)$ is an unknown function of x which is to be chosen to make I a minimum. The problem is to find $y(x)$.

Consider functions $\bar{y}$ which are "close" to y in the following sense. Let $\bar{y}$ be given by

$$\bar{y} = y + \gamma \delta y \qquad \text{(F.4-2)}$$

where γ is a real variable and δy is an *arbitrary* function of x such that its first two derivatives are continuous in (a,b) and

$$\delta y(a) = \delta y(b) = 0 \qquad \text{(F.4-3)}$$

The function δy is called the *variation* of y. For small γ and δy, $\bar{y}$ is "close" to y in the interval (a,b) and, in any case, coincides with it at the end points. The situation is illustrated in Fig. F.3. It follows from Eq. (F.4-2) that

$$\bar{y}' = y' + \gamma \delta y' \qquad \text{(F.4-4)}$$

The expressions of Eq. (F.4-2) and (F.4-4) for $\bar{y}$ and $\bar{y}'$ may be substituted for y and y' in Eq. (F.4-1) to yield a new function $I(\gamma)$ where $I(0) = I$:

$$I(\gamma) = I(0) + \delta I = \int_a^b F(x, y + \gamma \delta y, y' + \gamma \delta y')\, dx \qquad \text{(F.4-5)}$$

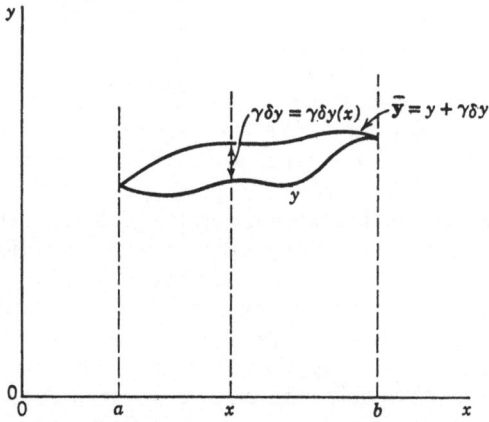

Fig. F.3 - The functions y and $\bar{y}$

We want to pick $y(x)$ so that $I(0)$ is a maximum or minimum (or, more generally, a stationary point). A *necessary* condition has been established in Section F.1 and is that

$$\frac{dI(\gamma)}{d\gamma}\bigg|_{\gamma=0} = 0 \qquad (F.4\text{-}6)$$

Suppose we expand $F(x,\bar{y},\bar{y}')$ in a Taylor series. We have

$$I(\gamma) = \int_a^b [F(x,y,y') + \gamma\delta y\,\frac{\partial F}{\partial y} + \gamma\delta y'\,\frac{\partial F}{\partial y'}$$
$$+ \;terms\;in\;\gamma^2,\gamma^3,...,]\,dx \qquad (F.4\text{-}7)$$

or

$$\frac{dI(\gamma)}{d\gamma} = \int_a^b [\delta y\frac{\partial F}{\partial y} + \delta y'\,\frac{\partial F}{\partial y'} + terms\;in\;\gamma,\gamma^2,...]\,dx \quad (F.4\text{-}8)$$

We now let $\gamma=0$ and obtain Eq. (F.4-6):

$$\frac{dI(\gamma)}{d\gamma}\bigg|_{\gamma=0} = 0 = \int_a^b \left[\delta y\frac{\partial F}{\partial y} + \delta y'\,\frac{\partial F}{\partial y'}\right]dx \qquad (F.4\text{-}9)$$

The second term in the integrand may be integrated by parts to yield

$$\int_a^b \delta y'\,\frac{\partial F}{\partial y'}\,dx = \delta y\,\frac{\partial F}{\partial y'}\Big|_a^b - \int_a^b \delta y\,\frac{d}{dx}\left(\frac{\partial F}{\partial y'}\right)dx$$

$$= -\int_a^b \delta y\,\frac{d}{dx}\left(\frac{\partial F}{\partial y'}\right)dx \qquad (F.4\text{-}10)$$

since $\delta y(b) = \delta y(a) = 0$ from Eq. [F.4-3]. Combining Eqs. (F.4-9) and (F.4-10), we have

$$0 = \int_a^b \delta y \left[\frac{\partial F}{\partial y} - \frac{d}{dx} \left(\frac{\partial F}{\partial y'} \right) \right] dx \qquad \text{(F.4-11)}$$

The foregoing is a typical integral obtained in problems in the calculus of variations. We will argue that, since δy is an *arbitrary* variation in y, the term in brackets must be zero in order for the integral to be zero. The proof [4, Chapter 3] depends on showing that, if the bracketted term is not zero, then there is a particular continuously twice-differentiable function δy_1 which vanishes at the end points such that the integral is not zero. Such a function δy_1 can always be constructed, for example, by defining it to be zero everywhere except in some region where the bracketted term is non-zero. If δy_1 has the same sign in that region, then the integral is obviously non-zero, a contradiction. The reader who is further interested is referred to Reference 4, Chapter 3.

Thus we have concluded that the bracketted term must be zero. Therefore, a *necessary* condition for the definite integral I of Eq. (F.4-1) to have an extreme value is that

$$\frac{\partial F}{\partial y} - \frac{d}{dx} \left(\frac{\partial F}{\partial y'} \right) = 0 \qquad \text{(F.4-12)}$$

This expression is called the *Euler-Lagrange (differential) equation.* The problem is easily extended to the case where F is a function of more variables, some of which may be independent [5]. It should be emphasized that Eq. (F.4-12) is necessary but not sufficient for a minimum. In many cases it will be apparent from geometrical or physical considerations whether the result gives a maximum, a minimum, or neither.

Note that the total derivative $\frac{dG}{dx}$ of a function $G(x,y,y')$ may be written as

$$\frac{dG}{dx} = \frac{\partial G}{\partial x} + \frac{\partial G}{\partial y} \frac{dy}{dx} + \frac{\partial G}{\partial y'} \frac{dy'}{dx} = \qquad \text{(F.4-13)}$$
$$\frac{\partial G}{\partial x} + \frac{\partial G}{\partial y} y' + \frac{\partial G}{\partial y'} y''$$

Thus Eq. (F.4-12) may be written explicitly as

$$\frac{\partial F}{\partial y} - \frac{\partial^2 F}{\partial x \, \partial y'} - \frac{\partial^2 F}{\partial y \, \partial y'} - \frac{\partial^2 F}{\partial y'^2} y'' = 0 \qquad \text{(F.4-14)}$$

Example F.6

   Show that the shortest curve $y(x)$ joining two fixed end points $x=a$ and $x=b$ is a straight line.

   The length of the curve is

$$I = \int_a^b ds = \int_a^b (dx^2+dy^2)^{1/2} = \int_a^b (1+y'^2)^{1/2} dx$$

and is to be a minimum. Therefore the function $F = (1+y'^2)^{1/2}$ must satisfy

$$\frac{\partial F}{\partial y} - \frac{d}{dx}\left(\frac{\partial F}{\partial y'}\right) = 0$$

or

$$0 - \frac{y''}{(1+y'^2)^{3/2}} = 0$$

From this last expression, we conclude that

$$y'' = 0$$
$$y' = c_1$$
$$y = c_1 x + c_2$$

where the constants c_1 and c_2 are determined by the fixed end points.

   F.5 Isoperimetric Problems. Variational situations frequently arise where additional constraints are imposed. One of the commonest of these is to find that $y=y(x)$ which yields a stationary value of

$$I = \int_a^b F(x,y,y')dx \qquad (F.5-1)$$

subject to the condition that

$$I_1 = \int_a^b F_1(x,y,y')dx = c \qquad (F.5-2)$$

where c is a constant. Such problems are called *isoperimetric* since one of the best known problems of this type is that of finding the closed curve of fixed perimeter with the largest area.

   Lagrange's method of undetermined multipliers, as discussed in Section F.3, may be applied directly to this class of problems. We consider the integral

$$I_0 = \int_a^b (F+\lambda F_1)dx = I+\lambda I_1 \qquad (F.5-3)$$

where the parameter λ is a Lagrange multiplier. We find the $y(x)$ which makes I_0 stationary since I_0 will be stationary if I is stationary. But a necessary condition is just the *Euler-Lagrange condition* with the function F replaced by $F_0=F+\lambda F_1$:

$$\frac{\partial F_0}{\partial y} - \frac{d}{dx}\left(\frac{\partial F_0}{\partial y'}\right) = 0 \qquad (F.5\text{-}4)$$

We have two equations, (F.5-1 and F.5-4), in the two unknowns $y(x)$ and λ. A formal proof proceeds exactly as in Section F.4.

Example F.7

Suppose we want to maximize the expression

$$I = \int_{-\infty}^{\infty} y(x)\log y(x)\,dx \qquad (A)$$

subject to the conditions that

$$I_1 = 1 = \int_{-\infty}^{\infty} y(x)\,dx \qquad (B)$$

$$I_2 = \sigma^2 = \int_{-\infty}^{\infty} x^2 y(x)\,dx \qquad (C)$$

where σ^2 is a given constant. If the function $y(x)$ is a probability density function, then I is called the *entropy* of the density and we are asking for that density with fixed variance σ^2 which yields a maximum entropy. Such problems arise in information theory.

We form the expression

$$F_0 = F + \lambda_1 F_1 + \lambda_2 F_2 = y\,\log y + \lambda_1 y + \lambda_2 x^2 y$$

where λ_1 and λ_2 are Lagrangian multipliers. We now apply Eq. (F.5-4) which, for our problem, reduces to

$$\frac{\partial F_0}{\partial y} = 0 = 1 + \log y + \lambda_1 + \lambda_2 x^2$$

or

$$y = e^{-(\lambda_1+1)}e^{-\lambda_2 x^2} \qquad (D)$$

We find λ_1 and λ_2 by substituting Eq. (D) into Eqs. (B) and (C). From Eq. (C) we have

$$\sigma^2 = e^{-(\lambda_1+1)}\int_{-\infty}^{\infty} x^2 e^{-\lambda_2 x^2}\,dx$$

This expression may be integrated by parts to yield

$$\sigma^2 = \frac{e^{-(\lambda_1+1)}}{2\lambda_2}\int_{-\infty}^{\infty} e^{-\lambda_2 x^2}\,dx$$

But the use of Eq. (B) reduces this to

$$\sigma^2 = \frac{1}{2\lambda_2} \quad , \quad \lambda_2 = 1/2\sigma^2$$

We substitute in Eq. (B) and have

$$e^{\lambda_1+1} = \int_{-\infty}^{\infty} e^{-\frac{x^2}{2\sigma^2}} \, dx$$

But the integral is just $\sqrt{2\pi}\,\sigma$ from the known properties of the normal distribution [see Section 2.18]. Therefore the function $y(x)$ is

$$y(x) = \frac{1}{\sqrt{2\pi}\sigma}\, e^{-x^2/2\sigma^2} \quad ; \quad -\infty < x < \infty$$

the normal distribution with zero mean and variance σ^2.

F.6 Isoperimetric Problems Involving More General Functionals. The problems considered in the previous sections involved functionals* formed by integrating a certain differential expression in the argument function. More general classes of functionals will often be encountered in variational problems. In such cases the general variational techniques that were used to find the *Euler-Lagrange equation* must be followed.

A specific problem of this more general type is the following which we will encounter frequently in the optimization of linear systems. It is desired to find the function $\phi(t)$ which yields a stationary value of the expression

$$I = \int_a^b \int_a^b K(s,t)\phi(s)\phi(t)\,ds\,dt \qquad \text{(F.6-1)}$$

subject to one or both of the constraints

$$I_1 = \int_a^b [\,\phi(t)\,]^2 dt \qquad \text{(F.6-2)}$$

and

$$I_2 = \int_a^b \phi(t)f(t)\,dt \qquad \text{(F.6-3)}$$

---

*A functional is a quantity which depends on one or more functions rather than on a number of distinct variables. The domain of a functional is a set of admissible functions rather than a region of coordinate space. The length I of the curve of Ex. F.6 is a functional of the argument function $y(x)$. The integral $\int y(s,t)y(t,r)\,dt$ is a functional of the argument function $y(s,t)$.

Here $K(s,t)$ is a given continuous symmetric function and $f(t)$ is a given continuous function. We form

$$I_0 = I + \lambda_1 I_1 + \lambda_2 I_2$$

where λ_1 and λ_2 are Lagrangian multipliers. We replace ϕ by $\phi + \gamma \delta \phi$ and form $I(\gamma)$. We then perform the operations

$$\frac{\partial I(\gamma)}{\partial \gamma} \Big|_{\gamma=0} = 0 \qquad (F.6\text{-}4)$$

and obtain

$$0 = 2 \int_a^b \delta\phi(t) \left[\int_a^b K(s,t)\phi(s)ds + \lambda_1 \phi(t) + \frac{\lambda_2}{2} f(t) \right] dt \qquad (F.6\text{-}5)$$

We conclude, therefore, since $\delta\phi(t)$ is arbitrary, that

$$0 = \int_a^b K(s,t)\,\phi(s)ds + \lambda_1 \phi(t) + \frac{\lambda_2}{2} f(t) \qquad (F.6\text{-}6)$$

This is an integral equation in the unknown $\phi(t)$. It may be solved and then λ_1 and λ_2 may be eliminated through Eqs. (F.6-2) and (F.6-3).

The reader should recognize that we have already encountered special forms of Eq. (F.6-6). With condition I_2 missing we obtain

$$0 = \int_a^b K(s,t)\phi(s)ds + \lambda_1 \phi(t) \qquad (F.6\text{-}7)$$

which is the homogeneous integral equation encountered in Chapter 4 [Eq. (4.6-2)] determining the characteristic orthogonal functions in the Karhunen-Loeve expansion of a random process. With condition I_1 missing we obtain the integral equation encountered in Chapter 5 [Eq. [(5.2-19) or Eq. (5.10-18)] for the matched filter and for the least-mean-squared-error linear filter.

A wide class of extremum problems are amenable to this general approach [3,4,5].

REFERENCES

1. R. Courant, *Differential and Integral Calculus, Vols. I and II*, Interscience Publishers, Inc., New York, N.Y., 1937.

2. S. Lang, *A First Course in Calculus*, Addison-Wesley Publishing Company, Inc., Reading, Mass., 1964.

3. R. Courant and D. Hilbert, *Methods of Mathematical Physics,* Vol. I, Interscience Publishers, Inc., New York, N.Y., 1953, Chapter V.

4. R. Weinstock, *Calculus of Variations,* McGraw-Hill Book Company, Inc., New York, N.Y., 1952.

5. I.M. Gelfand and S.V. Fomin, *Calculus of Variations,* (Translated from the Russian by R.A. Silverman), Prentice-Hall, Inc., Englewood Cliffs, New Jersey, 1963.

Table 1 - The Unit Normal Distribution

x	f(x)	F(x)	x	f(x)	F(x)
0.00	0.3989	0.5000	1.40	0.1497	0.9192
0.05	0.3984	0.5199	1.50	0.1295	0.9332
0.10	0.3970	0.5398	1.60	0.1109	0.9452
0.15	0.3945	0.5596	1.70	0.0940	0.9554
0.20	0.3910	0.5793	1.80	0.0790	0.9641
0.25	0.3867	0.5987	1.90	0.0656	0.9713
0.30	0.3814	0.6179	2.00	0.0540	0.9772
0.35	0.3752	0.6368	2.10	0.0440	0.9821
0.40	0.3683	0.6554	2.20	0.0355	0.9861
0.45	0.3605	0.6736	2.30	0.0283	0.9893
0.50	0.3521	0.6915	2.40	0.0224	0.9918
0.55	0.3429	0.7088	2.50	0.1075	0.9938
0.60	0.3332	0.7257	2.60	0.0136	0.9953
0.65	0.3230	0.7422	2.70	0.0104	0.9965
0.70	0.3123	0.7580	2.80	0.00791	0.99744
0.75	0.3011	0.7734	2.90	0.00595	0.99813
0.80	0.2897	0.7881	3.00	0.00443	0.99865
0.85	0.2780	0.8023	3.20	0.00238	0.99931
0.90	0.2661	0.8159	3.40	0.00123	0.99966
0.95	0.2541	0.8289	3.60	0.00061	0.99984
1.00	0.2420	0.8413	3.80	0.00029	0.99993
1.10	0.2179	0.8643	4.00	0.00013	0.99997
1.20	0.1942	0.8849	4.50	0.000016	0.999996
1.30	0.1714	0.9032	5.00	0.0000015	0.9999997

$$f(x) = \frac{1}{\sqrt{2\pi}}\, e^{-\frac{x^2}{2}} \qquad F(x) = \frac{1}{\sqrt{2\pi}} \int\limits_{-\infty}^{x} e^{-\frac{y^2}{2}}\, dy$$

INDEX